SHAKESPEARE

AND THE

DAWN OF MODERN SCIENCE

SHAKESPEARE

AND THE

DAWN OF MODERN SCIENCE

PETER D. USHER

CAMBRIA PRESS

Amherst, New York

Printed in the United States of America

Library of Congress Cataloging-in-Publication Data

Usher, Peter D.
Shakespeare and the dawn of modern science / Peter D. Usher.
p. cm.
Includes bibliographical references and index.
ISBN 978-1-60497-733-2 (alk. paper)
1. Shakespeare, William, 1564–1616—Criticism and interpretation. 2. Literature and science. 3. Astronomy in literature. 4. Shakespeare, William, 1564–1616—Symbolism. I. Title.

PR3047.U84 2010
822.3'3—dc22

2010040039

To Barbara

Table of Contents

List of Figures	**ix**
List of Tables	**xiii**
Foreword	**xv**
Preface	**xix**
Acknowledgments	**xxvii**
Chapter 1: The New Astronomy	**1**
Chapter 2: *Love's Labour's Lost*	**43**
Chapter 3: *Hamlet*	**69**
Chapter 4: *Cymbeline*	**171**
Chapter 5: *The Merchant of Venice*	**231**
Chapter 6: *The Winter's Tale*	**255**
Chapter 7: The Resolution Revolution	**307**
Notes	**331**
Works Cited	**359**
Illustration Credits	**375**
Index	**377**

List of Figures

Figure 1.1. The bounded geocentric model according to Peter Apian, from *Cosmographia* (1539). 5

Figure 1.2. Retrograde motion of Saturn relative to background stars. 6

Figure 1.3. (a) (Upper) If circle AB represents the Earth's orbit, an observer moving from A to B will detect a larger angle AOB when object O is closer. By convention, one-half of angle AOB is the parallax angle. (b) (Lower) Two objects "O" lying on the Firmament (the supposed eighth sphere of the stars) appear farther apart when the Earth is closer (angle ObO) than when it is more distant (angle OaO). 7

Figure 1.4. The bounded heliocentric model of Nicholas Copernicus, from *De Revolutionibus* (1543). 13

Figure 1.5. Planetary alignments for heliocentric orbits. The position of the Earth at E is shown relative to a Superior Planet (like Mars) in the outer orbit and an Inferior Planet (like Venus) in the inner orbit. Relative to the Sun-Earth direction, a Superior Planet can be at Opposition (O) or Conjunction (C); and an Inferior Planet can be at Inferior Conjunction (IC), Superior Conjunction (SC), Maximum Eastern Elongation (MEE), or Maximum Western Elongation (MWE). 14

Figure 1.6. The bounded geo-heliocentric model of Tycho Brahe, from *Liber Secundus* (1588). 19

Figure 1.7. The unbounded heliocentric model of Thomas Digges, from *A Perfit Description* (1576). 21

Figure 1.8. Relative sizes of the Earth, Sun, Moon and planets, from *Prognostication Everlasting* (1576). 27

Figure 1.9. Gainer's experimental telescope built according to Digges' design with materials and tools available in the sixteenth century (f/8, 4.5-inch aperture, 1-inch plano-convex eyepiece, and magnification 36). 31

Figure 1.10. Detail from Galileo's *Sidereus Nuncius* (1610) illustrating Jupiter and its four moons. 34

Figure 1.11. Cartoons illustrating Saturn's image: (a) (Uppermost) as it appeared to Galileo in 1610, (b) (middle) how he interpreted what he saw, and (c) (nethermost) how the planet would have appeared at the time if his spyglass had had better optics. 35

Figure 1.12. Saturn on January 14, 2007, imaged by a Toscano 8-inch (200 mm) telescope. The image is a sum of 504 separate images each of 1/25 second exposure and shows the Cassini gap separating the inner brighter B-ring and the outer fainter

A-ring. These rings lie respectively from 92,200 to 117,500 and from 121,000 to 136,200 kilometers (57,300 to 73,000 and 75,200 to 84,600 miles) from the center of Saturn yet are only a few hundred meters (yards) thick. The equatorial radius of Saturn is 60,000 kilometers (37,300 miles). For comparison, the Earth's equatorial radius is 6,400 kilometers (4,000 miles). 36

Figure 1.13. Saturn and its rings depicted by NASA's Voyager 2 spacecraft on July 21, 1981, when the spacecraft was 34 million kilometers (21 million miles) from the planet. 37

Figure 1.14. Illustration of successive phases of the "dark star" Venus seen from Earth as both orbit the Sun (see figure 1.5). (Top left) Venus is just past its farthest point from the Earth (Superior Conjunction), so its image is relatively small and gibbous. (Top middle) The image becomes larger and less gibbous as it approaches the Earth and reaches maximum elongation. (Top right) The image becomes crescent. (Bottom row) Venus appears larger and its phase more crescent as it nears its closest point to the Earth (Inferior Conjunction). 38

Figure 2.1. The Moon as seen through Gainer's reconstruction of the Digges telescope (exposure 1/20 sec, ISO 100, 5-megapixel camera, good seeing). 63

Figure 3.1. Image of the Sun showing sunspots. 81

Figure 3.2. Lunar Opposition and Conjunction with the Sun, illustrating the meaning of *Op-heli-a*. 87

Figure 3.3. First de Gheyn engraving of Tycho Brahe for his 40th birthday (1586), from Huizinga. 123

Figure 3.4. Second de Gheyn engraving of Tycho Brahe (1586). 123

Figure 3.5. On the left, Jupiter as it appeared to a ground-based observer in 2002. For comparison, the image on the right shows its appearance in 2000 as recorded by NASA's Cassini spacecraft. In both images, the oval blemish in the southern hemisphere is the Great Red Spot. In the Cassini image, the shadow of the Galilean moon Europa appears to the west of the Great Red Spot. Jupiter's equatorial radius is 71,400 kilometers (44,400 miles). For comparison, the Earth's equatorial radius is 6,400 kilometers (4,000 miles). 142

Figure 7.1. Reproduction of a sketch of Shakspere's coat of arms based on the original drawn in 1596 by the Garter King of Arms (see e.g., Michell 72–73). The punning crest features a falcon shaking a spear, and above are the words *Non, Sanz Droict* ("No, without right"), a variation of the intended *Non Sanz Droict* ("Not without right"). 318

List of Tables

Table 3.1.	Character identifications in *Hamlet*.	71
Table 3.2.	Model personifications in *Hamlet*.	72
Table 3.3.	Chronology of Act 1 of *Hamlet*: astronomical events and dates at Elsinore in November 1572.	101
Table 3.4.	Imputed roles in the players' play.	137
Table 3.5.	Tabular properties of Yorick and Yaughan.	158
Table 4.1.	Bird hierarchies in *Cymbeline*.	185
Table 4.2.	Correspondence between Galileo's chief results in *Sidereus Nuncius* and Giacomo's descriptions in *Cymbeline*.	203
Table 4.3.	Inferred correspondence between diamond ring exchanges and Saturn ring changes in *Cymbeline*.	229
Table 6.1.	Chronology of Hermione's pregnancy (± ½ day) in *The Winter's Tale*.	260
Table 6.2.	Ages and epochs on stage.	267
Table 6.3.	Determination of the epoch on stage of acts 1 to 3 of *The Winter's Tale*.	285

Table 6.4. Role of integer pairs [10, 23] and [19, 22] for basis year 1578. 289

Table 6.5. Principal dates in Act 4 of *The Winter's Tale*. 295

Table 7.1. A sample of published years of death of Leonard Digges the Elder. 312

FOREWORD

The centers of learning in the late sixteenth and early seventeenth centuries are usually associated with the great universities in Italy, France, and Germany, whereas London has been considered, more or less, a backwater of intellectual activity. It was, however, due to the influence of the well-educated and enlightened Queen Elizabeth that new openness to and encouragement of new ways of thinking began to thrive throughout England. This was particularly true in the fields of mathematics and natural philosophy.

It was in England, a generation before Galileo in Italy, that John Dee was applying empiricism as a means to investigate nature. This approach was anathema at many of the major philosophical centers. Although better known as a mystic and necromancer, Dee was also an exceptional mathematician, an experimenter in optics and ballistic trajectories, and Europe's foremost authority on mathematical methods of celestial navigation.

Leonard Digges and his son Thomas were influenced by Dee's empirical approach, and they contributed their own significant advances in mathematics, computational astronomy, and optics. Their surviving publications indicate that they constructed an elementary form of a reflecting telescope and used it for astronomical observations. Thomas Harriot was an outstanding mathematician of whom little is known because he was reluctant to publish his work. It is known, however, that he made telescopic observations of the Moon prior to those of Galileo.

Shakespeare, a participant in this new wave of intellectual curiosity, would have known of the major philosophical debates of this time and would have been influenced by some of those who addressed them. Among the foremost of these issues were the differences between heliocentric and geocentric model Universes and between the application of empiricism vs. pedantic repetition of Aristotelian ideas.

Shakespeare's plays were directed to two groups: patrons in the yard and patrons in the gallery. The patrons in the yard were, for the most part, poorly educated or illiterate, and allegories and metaphors related to contemporary philosophical questions would be lost on them. Patrons in the gallery, however, included the educated aristocracy, many of whom had traveled widely and had probably been involved in serious discussions of contemporary issues and new philosophies. It is not improbable that Shakespeare would address the prevalent awareness of these ideas and weave them into new metaphors and allegories in order to enrich his scripts. The patrons in the yard would be entertained by the drama, comedy, history and pathos, whereas those in the gallery would also have food for thought and discussion. There was something for everyone in Shakespeare's plays.

Peter Usher is an emeritus professor and researcher in astronomy and an avid long-time reader and investigator of Shakespeare's plays. He combines rigorous scholarly research and methods of scientific investigation to present a refreshing approach to interpreting Shakespeare's work. He asks provocative questions and offers perceptive observations of metaphorical and allegorical references in the plays.

Why is it, for example, that the would-be assassins of Hamlet are named after two of Tycho Brahe's ancestors, Rosencrantz and Guildenstern? Why does Hamlet state, "I am but mad north-north-west?

When the wind is southerly, I know a hawk from a handsaw" (Edwards 2.2.347–2.2.348)? What is the significance of these directions? In *Love's Labour's Lost*, what is the significance of the contrast between the inquiring empiricism of Berowne and the pedantic style of Holofernes?

Peter Usher contends that Shakespeare's writing must be interpreted within the emerging scientific zeitgeist of the Elizabethan period and that at least some of his works are an allegorical presentation of the conflict between old and newly emerging ideas. He presents his position convincingly. Regardless of any bias readers may have in interpreting Shakespeare's plays, they will find much to hold their interest.

—Michael K. Gainer
Emeritus Professor of Physics
St. Vincent College
Latrobe, Pennsylvania

Preface

> [Most people] are satisfied with appearances, as though they were realities, and are often more influenced by the things that seem than by those that are.
>
> —Niccolò Machiavelli

In the mid-sixteenth century, Nicholas Copernicus broke from the traditional view that the Earth rested immobile at the center of the Universe, asserting instead that the Sun was the immobile center. As the Earth lost its privileged position, so did humankind, prompting a reevaluation of all branches of learning in which humans considered themselves the center of attention.

Shakespeare's writing career began about a half century after this revolution in worldview. The bard was knowledgeable in many different areas of learning and was oftentimes ahead of his contemporaries, yet his Canon appears to lack a coherent account of contemporary cosmological thinking. It is simply not credible that a poet of this stature could remain ignorant of the cultural impact that the New Astronomy was having during his lifetime—or that he would refrain from using the literary devices at his command to address the topic if he was not ignorant of its significance.

With this difficulty in mind, in 1996 I suggested that Shakespeare's *Hamlet* is a cosmic allegory describing the competition between the four chief cosmic models that were vying for acceptance at the turn of the seventeenth century. Critics deem a work allegorical when they sense that an author has a meaning in mind that is broader than the work appears to have. Allegory caters to the ideals of attainable knowledge and political unity in a moral and fundamentally theological context by using elaborate symbolism and levels of meaning deeper than literal, making it a suitable means for relating appearance with reality. This technique of expressing something in such a way as to convey nonliteral meaning is especially useful when it comes to dealing with sensitive issues like new worldviews, false cosmologies, and the overthrow of corrupt regimes.

Understanding allegory with an astronomical component requires attention to science and a variety of other disciplines, among which are interpretation theory, history, and philosophy of science, as well as literature itself. Whereas the parts of a work enable understanding of the whole work, the whole, in turn, mediates understanding of the parts. This cycle of understanding is not a vicious circle, however. Genuine advances do occur as novel ideas and theories are either discarded or modified and as those that survive attain a measure of validation through analytic discourse and empirical test.

Every inquirer brings limitations and preconceptions to the quest and must work to ensure objectivity in interpretation. In previous work, I have drawn parallels between the hermeneutic-dialectical approach to understanding literature and the hypothetico-deductive inquiry into the Book of Nature, both of which strive for maximally reasonable interpretations. Text and context are the ultimate authorities in literary interpretation, and it is in accordance with these canons that I conduct this research.

Chapter 1 outlines the evolution of scientific cosmological thinking from the Pythagoreans of the sixth century BC to Galileo Galilei in the early seventeenth century. It places into historical context Michael

Gainer's recent observation that mathematician Thomas Digges' account of a two-element optical magnifier likely describes the perspective glass invented by his father, Leonard Digges. This purports to be the first astronomical telescope. Gainer reproduced the device according to available specifications and showed that it was capable of resolving detail on the Sun, Moon, and planets. Independent evidence suggests that Shakespeare knew the Digges family and used their publications in his writing, and it is plausible that the poet would also attend to telescopic observations of celestial objects.

Analysis in the ensuing five chapters reveals the scientific content of five Shakespearean plays, and the last chapter discusses the results.

Chapter 2 examines *Love's Labour's Lost*, which is a searing indictment of pedantry and religious intolerance. As its plot develops, Shakespeare addresses Navarre's concern with the Princess' visit and Berowne's concern with epistemology. Mythology and astronomical phenomena provide a chronology of events which combine to solve riddles posed by Dull and Holofernes. The Spanish Armada figures prominently and not just in the naming of Armado. Astronomical considerations explain the metaphysical significance of the slain deer, and the attributes of Rosaline indicate that in 1594, Shakespeare was aware of properties of the so-called Ancient Planets, information that he could not have known without telescopic aid.

Chapter 3 argues that Shakespeare's most famous play, *Hamlet*, is an allegory for the competition of the four chief cosmological models extant at the turn of the seventeenth century. It builds on work that I synthesized and presented in *Hamlet's Universe* prior to 2007 and it sheds new light on aspects of the play that had not then penetrated my consciousness. It explains the supernatural role of the Ghost, the meaning of Hamlet's love letter, why he is 30 years old, his reference to a "nutshell," and why Hamlet is "mad north-northwest." In addition, this chapter clarifies the epistemological foundation of the player's play and explains the significance of the most famous skull in literature. This novel interpretation establishes the relationship of the entire play to the scientific advances of the sixteenth century. Using multiple and previously unexplained conceits, Shakespeare describes properties of the Sun, Moon, planets, and stars that he could not have known without telescopic aid. Yet the date of the writing of *Hamlet* in about 1601 predates the normally accepted date of the first known astronomical telescopic observations by at least 9 years.

Galileo Galilei's short book *Sidereus Nuncius* (The Starry Messenger) of March 1610 is the world's first known publication devoted to astronomical telescopic observations and as such has had a considerable impact on world culture. Chapter 4 establishes that this book and some of Galileo's other astronomical writings are hitherto unrecognized sources of *Cymbeline*. The mooted date of the writing of *Cymbeline* in late 1610 agrees with estimates derived on purely literary considerations. Posthumus' vision of the descending deity Jupiter occurs at the behest of the ghosts of his deceased parents and two brothers. These four circling specters represent Galileo's telescopic detection of the four brightest Jovian moons. The mooted allegory explains alleged defects in the play, including the perplexing predictions of the Roman soothsayer. *Cymbeline* is a paean to the glories of the night sky revealed through telescopy, but chapter 4 reveals that the results that Shakespeare describes are not solely those of Galileo. This reading supports interpretations in the previous two chapters and lends credence to the overall contention of the present work.

Chapter 5 discusses the literal storyline of *The Merchant of Venice*, in which a penurious Venetian, Bassanio, seeks to finance his romantic pursuit of fair Portia, who lives in Belmont. His best friend Antonio is temporarily short on funds, so Bassanio borrows the money from Shylock, who terms the agreement a "merry bond." However, Antonio stands as surety, and in the event of a default, he must forfeit a pound of his flesh. In Belmont, Bassanio meets the condition laid down by Portia's father, namely that a candidate may win the hand of Portia only if he correctly chooses one of three caskets made of gold, silver and lead. A secondary romance between Bassanio's friend Graziano and Portia's companion Nerissa springs from this success. The subtext concerns principally these two chief romances.

Chapter 6 analyzes *The Winter's Tale*. Like Plutarch's *Lives of the Noble Grecians and Romans*, this play chronicles events in the lives of eminent personages. However, where Plutarch's work represents historical and mythological figures of his time, Shakespeare's play concerns those who contributed to scientific awakening in the later Renaissance. The play is remarkable for a high incidence of integers and an unusual leap of sixteen years within the timeline of the story. Analysis of words and numbers in the script requires auditors to hone their arithmetical skills in order to identify eminent contemporary natural philosophers and their major works. The tragic death of the young Prince Mamillius finds a ready explanation, as does the most famous stage direction in all of drama, "Exit, pursued by a bear." *The Winter's Tale*, like plays analyzed in previous chapters, has a surreal quality which helps explain the survival of Perdita and the revivification of Hermione.

Chapters 2–6 argue that Shakespeare describes details of celestial objects that no one else knew at the time. Leonard Digges' perspective glass is a reasonable means for resolving such detail. Chapter 7 documents events in Leonard's life, including his role in Wyatt's rebellion of 1554, his conviction for treason, the good sense that saved him from the gallows, and his eventual reinstatement in 1558 under Elizabeth I. Chapter 7 further notes the coincidence that neither Leonard Digges nor William Shakespeare left any artifactual evidence of their work. The rule "watch out especially for anything odd" (Hotson, *Hilliard* 182) impels investigation through hypothesis and test, and chapter 7 adheres to this rule, addressing an array of oddities associated with Leonard and the Stratford actor William Shakspere (1564–1616) who is widely regarded as the poet Shakespeare. Some have wondered whether Shakspere wrote Shakespeare, and they have proposed alternate candidates for the authorship of the Canon—but the unique facet of this work is that the cosmic allegorical context provides an entirely new perspective on these propositions.

The analysis of four more plays in addition to *Hamlet* adds considerably to the web of evidence previously assembled and published. On a simple model, connections between the parts that establish a theme increase quadratically, so the examination of four more plays substantially bolsters the case for a self-consistent and historically accurate network of astronomical descriptions and allusions. Antecedent presentations and publications have, on occasion, raised more hackles than eyebrows, but objectivity requires interpreters to follow evidence wherever it might lead. To that extent, this text is heuristic and exploratory. It must be questioned whether historical and literary coincidences are self-consistent and whether their incidence is sufficiently high to make a plausible case. Whether selection effects and other cognitive distortions are minimal must also be considered. However, in the end it is up to the reader to decide on how fine the interpretive mesh must be to constitute an acceptable level of plausibility. Healthy skepticism will obviate what Jonathan Swift observed, that "learned commentators view / In Homer more than Homer knew," and if the ideas put forth here are wrong, I shall be the first to engage in what Alexander Pope considered the "last and greatest art, the art to blot."

The interdisciplinarity of this endeavor needs no apology, as it is common to promote understanding in various fields by examining the works of contemporary writers. Shakespeare is for everyone, and "almost every writer about Shakespeare has at least a few grains to add to the general stock of knowledge" (Levi xviii). In understanding what is arguably Shakespeare's most famous play, *Hamlet*, both professional and "nonprofessional scholars" have made important contributions (Edwards 36), and as far as the chief theme of this book is concerned, in the great Apollonian democracy of the nine Muses, Urania, the goddess of Astronomy, is the equal of the other eight who bear responsibility for the arts of song, dance, poetry, tragedy, and comedy.

Throughout this book, I introduce propositions and evidence in order to incorporate them into the dialectic process. Of course, novel insights and theories remain provisional, but in order not to litter the text with expressions of that provisionality, and because I believe that the interpretations offered "hang together," I refrain from overusing words like "possibly," "perhaps," "may" and "might." The text is closely argued, so some familiarity with the plays would be helpful. However, some readers may prefer to soldier on in order to appreciate the overall flow of argument and the structure of the book. This in turn would promote retrospective understanding of the chapters.

I use the term "New Philosophy" to mean the development of a critical and objective method of understanding phenomena of the physical Universe and the accumulation of facts about nature in the post-Copernican era (Coffin 65). "New Astronomy" has the orthodox meaning of the forms of motion of celestial objects in the post-Copernican era. "New Physics" refers to post-Copernican physical concepts. I use "New Organon" in the Baconian sense to mean the idea, novel in the sixteenth and early seventeenth centuries, of acquiring knowledge empirically. This method of study by observation or experimentation contrasts with Aristotle's syllogistic *Organon*. I let "Old" apply to the aforementioned disciplines in the pre-Copernican era.

Astronomical data are from the *Horizons* ephemeris program of the Jet Propulsion Laboratories of the National Aeronautics and Space Administration, supplemented by *Voyager 4* Carina Software, and are of course proleptic (i.e., data associated with dates in the Old Style (Julian) calendar are calculated according to the corresponding dates in the New Style (Gregorian) calendar). Generally, "rotation" means spin as in the 24-hour period of the Earth, and "revolution" refers to orbits as in the annual journey of the Earth around the Sun. I capitalize object names and technical terms and regard "Cosmos," "Universe," and "World" as synonymous. "Worldview" refers to any concept of the Universe that is necessarily part of an overall worldview. The endnotes contain explanatory material, but when necessary, digressions occur in the main text. Standard abbreviations are *OED* (Oxford English Dictionary) and *DNB* (Oxford Dictionary of National Biography). Context decides whether male personal pronouns refer to both genders. On the first occurrence of the name of a historical figure, I list years of birth and death because these often prove useful in discussions. For reference, these are also listed in the Index. When capitalized, "Continent" refers to mainland Europe, "Canon" refers to the body of works by William Shakespeare, F1 and F2 stand for the First and Second Folio editions, and Q2 refers to the Second Quarto edition of a play.

ACKNOWLEDGMENTS

I thank Louis and Ronnie Lind for many kindnesses and Harvey and Esther Nathanson for interest and support. They and other members of the Oakland Philosophical Society provided a milieu important to the completion of this work. I am grateful to Stephanie Hughes, the Shakespeare Oxford Society, and its members for being receptive to new ideas. Throughout, Michael Gainer has been a constant source of help and encouragement, and I am much indebted to him for his experimental test of present theory and for contributing a Foreword to this monograph. Peter Andersen has kindly shared his thoughts and translated his research. Sylvia Sachs, Lester Berkovitz, and the Osher Academy of Lifelong Learning of the Carnegie Mellon University have graciously given me opportunities to present portions of this material in a classroom setting. I am grateful also to Madalon Amenta for her unflagging interest in this research and to Carl Croushore for his help. Others who have kindly lent assistance through the course of these studies include Gary Goldstein, Nina Green, Steven May, Allan Mills, and Steve Sohmer. Much appreciation is due to anonymous reviewers who have supported my research at critical junctures. Above all, I thank my wife Barbara whose love is my greatest source of strength and inspiration.

SHAKESPEARE

AND THE

DAWN OF MODERN SCIENCE

Chapter 1

The New Astronomy

> Yet sometimes it chanceth, that the opinion most generally received, is not most true.
>
> —Robert Recorde

The rise of science after the Middle Ages began with the rise of astronomy in the first half of the sixteenth century, when Nicholas Copernicus (1473–1543) proposed a Sun-centered model of the planetary system. This challenged the Earth-centered cosmology that had held sway since the dawn of history.

Some argue that William Shakespeare completely ignored the scientific revolutions that were taking place during his lifetime, and others suggest that he did not—but argument has had to rely on passages from his works lifted out of context. Recently I suggested "that *Hamlet* is an allegory for the competition between the cosmological models of the contemporaries Thomas Digges (*c.*1546–1595) of England and Tycho Brahe (1546–1601) of Denmark" (Usher, "New Reading" 1305).

It is the goal of this work to assess Shakespeare's alleged backwardness in the cosmic sciences within the broader context of five plays. This chapter contains an outline of the state of astronomical knowledge from about 1589 (at the start of Shakespeare's writing career) through subsequent developments to about 1610–1612—a period that includes the advent of astronomical telescopy. Before presenting the major Worldviews extant in 1589, it is helpful first to consider the ancient Greek cosmology of Pythagoras (*c*.582–*c*.507 BC) and his followers who flourished from the sixth to the fourth centuries BC.

PYTHAGOREAN DESIGN

Pythagoras lived the early part of his life on the Greek island of Samos off the coast of modern Turkey. As a young man, he fled the tyrant who ruled there, and after traveling widely, he settled in Croton, Italy. There, he founded a school of philosophy based in part on the tenets that belief in God is the best foundation for governance and that only through philosophy could humans develop a relation to the divine.

Like Socrates (469–399 BC), Pythagoras believed that lovers of wisdom are superior to those who seek fame or pleasure, and that mathematics is the key to acquiring that wisdom. Pythagoreans saw beauty in mathematics and came to value the concept of numbers as the essence of reality. They believed in a quasi-dimensionality of space, according to which the values 1, 2, 3, and 4 (representing point, line, area, and solid) could be added to reach a sum of 10, which Pythagoreans regarded as the perfect number. This notion was all the more credible to them because these four numbers appear in the ratios—4:3, 3:2, and 2:1—describing the chief musical harmonies.

Pythagoras believed that the Earth was round and that the planets produced music only he could hear because of the exemplary life that he led. Philolaus (*c*.480–*c*.385 BC) believed that the Earth revolved, along with the Moon, Sun, Mercury, Mars, Jupiter, and Saturn, as well as with a Counter-Earth, which revolved around an imaginary "Central Fire."

Earthlings could see the Sun but not the Central Fire and Counter-Earth, as these were always below their horizon. In the fifth century BC, Ecphantus and Hicetas are reputed to have advocated a rotating Earth in order to account for the apparent daily rotation of the sky, and later, Pythagoreans decided to define the Earth's revolutionary period as one year (Dicks 73). Archytas (428–347 BC) questioned whether the apparent sphere of stars confined physical space, and Plutarch (AD *c.*46–*c.*120) credited Heraclides Pontus (*c.*388–*c.*315 BC) and the Pythagoreans with the idea of multiple worlds. "Heraclydes and the Pythagoreans hold, that every Star is a world by it selfe, conteining an earth, an aire, and a skie, in an infinit celestiall nature" (Heninger 125). The realm of the stars brought the total number of cosmic entities to the desired magical value of 10. All told, by the middle of the fourth century BC, Pythagoras and his followers had introduced concepts that resurfaced two thousand years later as the essential ingredients of the so-called New Astronomy. These concepts are:

1. the Earth rotates on its axis,
2. the Earth revolves around the Sun,
3. the Universe is potentially infinite.

Coeval with the development of the Pythagorean Worldview, Anaxagoras (*c.*500–428 BC) and the atomists Leucippus (*c.*500–428 BC), Democritus (469–*c.*370 BC), and Epicurus (341–270 BC) proclaimed that the World is populated by an infinite number of atoms—an idea further promoted by the Roman poet Lucretius (*c.*99–*c.*55 BC). Surprisingly, these atomists could imagine an infinite Universe populated by monads but could not envision the rotundity of the Earth. Anaxagoras and Democritus thought that the Earth was flat; Democritus believed it was disc-shaped and hollow in the middle, and Leucippus thought it was shaped like a drum (Dicks 80–82). The belief in a flat Earth persisted even into the nineteenth and twentieth centuries, demonstrating how slow the pace of progress in cosmological thinking has been over the ages.

BOUNDED GEOCENTRICISM

Pythagoras notwithstanding, the fundamental question of the shape of the Earth remained unsettled at the time of Socrates, but his student Plato (428–347 BC) decided the issue in favor of sphericity. Initially, Pythagorean philosophy heavily influenced Plato, but over time the influence waned. Plato named Pythagoras and the Pythagoreans only once each and disregarded most other aspects of that school's cosmology (Dicks 63). He retained the bounded form of the Philolaic model, but he dispensed with its decadal design because he had placed the Earth at the center of the system, which meant he had to discard the Central Fire and Counter-Earth.

To Plato, the sky seemed like the surface of a rotating sphere surrounding a stationary, spherical Earth. This is the two-sphere construction of the Universe comprising a spherical Earth at the center of a spherical shell of stars. Between these lay the Ancient Planets, whose motion relative to the more distant stars was erratic—a mystery Plato left for his successors to explain. Plato had scant regard for empiricism, and in the history of experimental science "he must be counted a disaster" (Dampier 25). However, his formulations were essentially poetical and intended more as a simulacrum for philosophical and spiritual guidance than as a depiction of physical reality. As such, his World was justifiably homocentric or human centered, but human curiosity and superstition compelled succeeding generations to try to transform the symbolism into a mechanism for practical use. Thus physical geocentricism was born, an ideology which Plato's student Aristotle (384–322 BC) adopted and refined.

ARISTOTLE

Aristotle made frequent mention of the teachings of the Pythagoreans, but he did not find them credible. He knew of the Pythagorean notion of a moving Earth but rejected it outright in favor of a physical analogue of Plato's spiritual form. He accepted without question that a sphere of stars separated the natural space from the space reserved for deities; and even though Socrates had agreed that musical harmony is to the ear "what astronomy is to the eye," Aristotle could not accept the notion of music of the spheres, dismissing it as "absurd and extravagant" (Dicks 71).

The empirical observation that individual stars move along perfectly circular tracks at perfectly uniform rates led Aristotle to regard the circle and sphere as the core figures of a divine geometry, as he describes in his treatise, *On the Heavens*. As such, the figures became part of theological doctrine. Yet at the other end of the spectrum, it was evident that the Earth, though round, was not uniformly smooth, nor were its inhabitants free from sin. Thus, Aristotle developed a theo-cosmological model with a gradient of perfection beginning near zero at the corrupt Earth and increasing outward to the heavens. Between the extremes lay the seven Ancient Planets or Wanderers, whose order had been established by the ancient Babylonians (see figure 1.1).

G: Earth, Moon, Mercury, Venus, Sun, Mars, Jupiter, Saturn, stars.

FIGURE 1.1. The bounded geocentric model according to Peter Apian, from *Cosmographia* (1539).

Here, "stars" refer to the "fixed stars," which are fixed relative to one another and which form the backdrop against which the "wandering stars," or Ancient Planets, move.

The Ancient Planets are called "Wanderers" because they seem to wander relative to the stars in the sky. No Ancient Planet moves at a steady rate or follows a circular path. Most egregious of all is that relative to the stars, five of them—Mercury, Venus, Mars, Jupiter, and Saturn—routinely reverse their direction of travel. The Sun and Moon always move from west to east relative to the stars—a fact easily observed in the case of the Moon over a span of an hour or more—but the five renegade wanderers sometimes move in the opposite direction. Accounting for this "retrograde motion" remained the premier problem of astronomy into the seventeenth century (figure 1.2).

Aristotle argued against the notion of a moving Earth on several grounds. If the Earth moved in either rotation or revolution, an observer looking at two stars lying in a particular direction would see them sometimes closer together and sometimes further apart (figure 1.3a). The effect is like that which enables creatures with two or more eyes to see their immediate surroundings in three dimensions. One-half the difference between two viewing directions is called the parallax angle.[1] In the case of the Earth, the parallax angle would be larger and easier to

FIGURE 1.2. Retrograde motion of Saturn relative to background stars.

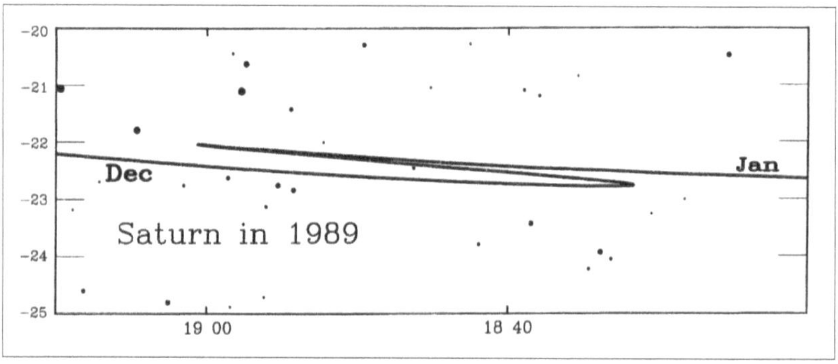

detect if the Earth were to revolve about the Sun, rather than if it were merely to rotate on its axis, because the Earth's orbit is necessarily larger than the Earth itself.

No one observed the parallax effect either daily or annually, implying either that the stars were extremely distant or that the Earth was stationary. Aristotle dismissed the first possibility because there would have to be a huge volume of space separating the Ancient Planets from Earth if the stars were so far away. He believed that this contradicted the doctrine of Final Cause, which held that creation was purposeful and everything in the Universe served a function. To him, "place" was a location where something should reside, so he argued that an intelligent Creator would not construct a Universe that was largely empty.

FIGURE 1.3. (a) (Upper) If circle AB represents the Earth's orbit, an observer moving from A to B will detect a larger angle AOB when object O is closer. By convention, one-half of angle AOB is the parallax angle. (b) (Lower) Two objects "O" lying on the Firmament (the supposed eighth sphere of the stars) appear farther apart when the Earth is closer (angle ObO) than when it is more distant (angle OaO).

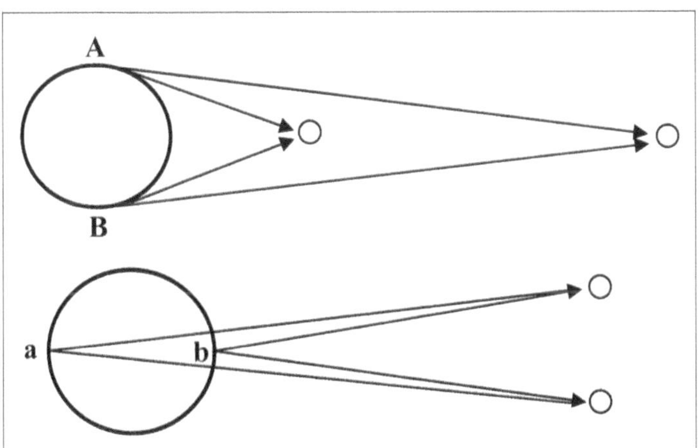

Moreover, it was plain to see that the sky turns about the Earth as do all planets on average. Because nothing can move of its own accord, Aristotle conceived of a Prime Mover (described in *Metaphysics*) whose job was to initiate motion in the physical Universe. Aristotle delegated the task of moving the several spheres to lesser deities whose combined effects account for the motion of each of the Ancient Planets. The decreasing gradient of celestial speed matches the decreasing gradient of celestial perfection from Heaven to Earth. Aristotle was supremely confident that he had read the Creator's mind correctly and concluded that the Earth must lie immobile at the center of the Universe.

Aristotle may have written other relevant material that is now lost, but available information indicates that he did not entertain the three rudimentary possibilities that later formed the basis of the New Astronomy. Aristotelians could not believe that they were viewing the celestial drama from a rotating and revolving platform called Earth. They did not separate observed celestial motions into their components because they attributed the motions of objects entirely to the objects themselves, not realizing that they themselves were in motion. Aristotelians thought that what they saw was reality and they made no allowance for the fact that they were located at the center of their own perception. In essence, the tale of the chief cosmological models that developed after Aristotle is one of a steady progression toward the restoration and refinement of Pythagorean principles.

THE GEOCENTRIC MODEL

The circles and spheres of divine geometry were useful also because they were much easier to handle mathematically than other figures. Claudius Ptolemy (AD 90–168) made full use of them in his thirteen-volume *Almagest* ("The Greatest"), which is a comprehensive account of astronomical data and theory extant at the time. Ptolemy followed the prescriptions of Aristotle and later geometricians, notably Hipparchus (*c.*190–*c.*120 BC), and applied various geometrical devices *ad hoc* so that when compounded and with judicious choices of parametric values, they provided results that mimicked the observed motion of Ancient Planets.

Ptolemy succeeded in predicting and accounting for positions of the Ancient Planets (*G*) with tolerable accuracy and without violating the doctrine of circular and spherical perfection. He accounted for planetary motions by compounding the effects of small circular motions (epicycles) centered on a larger circle (the deferent), abetted by uniform angular displacements around points off center from the center of the deferent (eccentrics and equants) (see Mitton 138). Because nature herself gave no rules on how to select and manipulate this geometry, the Ptolemaic model was not an explanation of physical phenomena but more of an algorithm geared to "saving phenomena" (i.e., to finding ways to reproduce or account for them). Ptolemy's celestial geometry was *ad hoc* because it relied on the aesthetic and mathematical appeal of circles and spheres as well as the theological belief in the central importance of humankind, but it lacked a self-consistent pattern of physical causes upon which to base understanding of the very phenomena that it was geared to "save." Nevertheless, the moderate success of the geocentric algorithm and the absence of an alternative led to its widespread adoption. In Toledo in 1080, scholars recalculated the ephemerides (i.e., tables of apparent directions in the sky for celestial objects at selected times), and the process repeated in 1252 under the supervision of Alfonso X (1221–1284), King of Castile and Leon. These became known as the Alfonsine Tables.

One product of geocentric theory was the capacity to determine the physical dimensions of the Universe as it was then conceived. In the first millennium AD, the size of the Earth was quite well known, and with the help of eclipses of the Sun and Moon, Ptolemy and his predecessors devised a way to estimate distances to the planets. Because the bounded geocentric model placed the sphere of the stars just beyond the distance of the outermost planet, Saturn, the belief was that the physical Universe ended at a distance of about 20,000 Earth radii from Earth (van Helden, *Measuring* 27). Ptolemaic values persisted into the sixteenth century with only minor modifications, but with the exception of the distance of the Moon, all derived distances were grossly underestimated. In modern units, the ancient value for the radius of the entire Universe was about 80 million kilometers (50 million miles)—about half the actual size of the Earth's orbit around the Sun and smaller than the sizes of some stars.

As for the Old Physics, Aristotle perpetuated the Pythagorean belief that all material consists of four primary and contrasting qualities—hot and cold, wet and dry—that in various proportions produce the four basic elements of Earth, Water, Air, and Fire (Dampier 33). Earlier reductionists held similar or even simpler views, like Heraclitus (*c.*540 – 480 BC), who believed that Fire was the basic element. Aristotle supposed that each of the four elements predominated in concentric regions, ordered with some overlap from the Earth outward into sublunary space, and that a quintessential fifth element, Aether, held sway in the superlunary realm. He allowed an exception to this rule by ascribing light from stars and the Ancient Planets to the interaction of Fire and Air, which friction ignited as the Ancient Planets moved around the Earth. However, Democritus, Plato, and later philosophers knew that the Moon borrows its light from the Sun and does not shine by Fire. They attributed this exception to the Moon's proximity to Earth, but in so doing they rendered the category of the seven Ancient Planets fundamentally heterogeneous from the start because the Moon then warranted a subcategory all its own.

Despite Aristotle's apparent expertise in the subject of classification, he paid scant attention to systematic differences in the properties of the seven Ancient Planets, which in retrospect, might have weakened his faith in geocentricism. For example, of the five Ancient Planets that undergo retrograde motion, Mercury and Venus never stray from the direction of the Sun by more than set amounts (about 22° and 45°), whereas the remaining three can lie at any angle to the Sun. These same three happen to be brightest when in the throes of retrograde motion, and the decreasing angular sizes of their retrograde loops correlate with the increasing times that they take to complete a circuit of the sphere of stars. However, the hold that homocentricism had on the human imagination was so strong that the ancient philosophers scarcely heeded this body of systematic contrary evidence. The apparent anomalies only received proper attention centuries later when Copernicus evaluated the observational data available to him in an impartial and objective manner and, mingling mathematics and aesthetics, initiated the age of the New Philosophy. As stated in the Preface, I use the term "New Philosophy" to mean the development of a critical and objective method of understanding phenomena of the physical Universe and the accumulation of facts about nature in the post-Copernican era (Coffin 65), and I let "Old" apply to these disciplines in the pre-Copernican era.

Aristotle made positive contributions in all areas he studied, but by modern consensus, his influence in physics and astronomy was inimical to the understanding of the physical World (Dampier 25–36). He believed that systematic data acquisition and experimentation were *infra dignitatem*, and although he considered empirical facts in the terrestrial sciences, his efforts in the cosmic sciences were ill conceived because they lacked procedures for dealing with intangibles. He was uncertain about how to deal simultaneously with observable phenomena both on Earth and in the sky, perhaps because he thought the latter ranged outward toward the divine abode of the gods and was immune to vulgar empirical inquiry. In particular, Aristotle and those who promoted the doctrine of hierarchical perfection were not concerned in the least that their ideas sprang from the presumably inerrant thinking of flawed terrestrials who resided at the supposed center of universal imperfection.

Scholars transmitted Aristotle's works via Spain and Sicily to Western Europe. The works arrived piecemeal, undergoing translations along the way, during which adherents developed ideas of their own based on material available at the time. Various schools arose whose accumulated body of thought is now known as Aristotelianism. Followers of the peripatetic philosopher accepted his cosmology uncritically, even though Aristotle himself had doubted its full validity. Some who sheltered their beliefs behind Aristotelianism were even disinclined to test his ideas lest the ideas be proven wrong. By the eleventh century, the schoolmen (or "scholastics" as they were called) had embarked on wide-ranging discussions of cosmological concepts, and by the thirteenth century, Church authorities had incorporated the basic Aristotelian World model into

Christian dogma, thanks to the intellectual prowess of Thomas Aquinas (1225–1274) and earlier scholars. There it sat essentially unchallenged until the sixteenth century, when philosophers began to attend rigorously to the need for systematic quantification of phenomena through mathematics and to the provisionality of hypotheses to explain them.

The Heliocentric Universe

Nicholas Copernicus hailed from Ermland, which was a sovereign province under the protection of the Polish crown. After attending the University of Krakow, he studied law and medicine in Italy. At the University of Bologna, he came under the influence of Domenico Maria da Novara (1454–1504), who was a professor of Mathematics and Astronomy and an ardent defector from the camp of Ptolemaic astronomy. Under his tutelage, Copernicus escaped Plato's geocentric cave to see the Sun in a new light.

Upon his return to Poland, Copernicus served as a church administrator and continued to study the motions of the Sun, Moon, and planets. In 1543 he published his seminal *De Revolutionibus Orbium Coelestium* ("On the Revolutions of the Celestial Orbs"), comprised of six books. Book 1 explains the essence of the heliocentric hypothesis in terms sufficiently simple for a literate layperson to get the gist of the argument. He made the Sun the center of the planetary orbits (as in sequence *H*), and only the Moon retained its old geocentric status (figure 1.4).

H: Sun, Mercury, Venus, Earth, Mars, Jupiter, Saturn, stars.

Copernicus gave the Earth two motions—a daily rotation on its axis and an annual revolution about the Sun. He accepted the empirical facts that the rotation of the Earth, the revolution of the Moon around the Earth, and the revolution of planets around the Sun were clockwise when viewed from the direction of South to North. On this basis, to a first approximation, he explained all major apparent anomalies in their motions.

Aristotelian-Ptolemaic theory demanded that the observer be stationary at the center of everything, which necessarily imputed observed motions

FIGURE 1.4. The bounded heliocentric model of Nicholas Copernicus, from *De Revolutionibus* (1543).

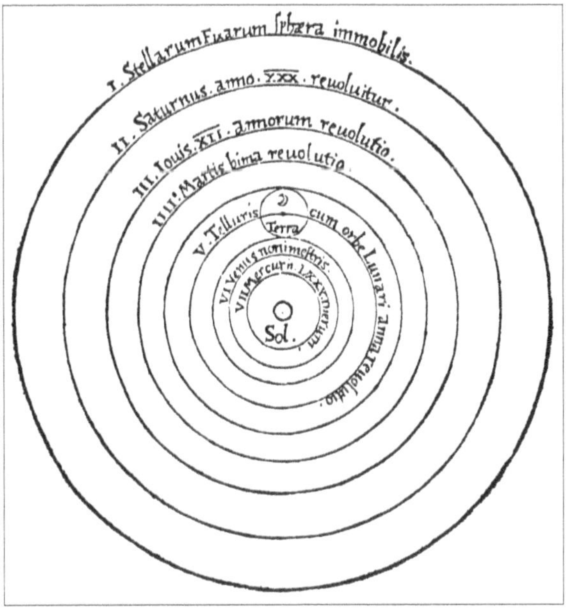

in the sky to the objects themselves. To Copernicus, on the other hand, retrograde motion was a matter of relative motion—an appearance resulting from the orbital motion of a planet relative to an observer on Earth who is also in motion.[2] In effect, the five epicycles that were required to explain retrograde motion in the Ptolemaic scheme fall victim to the single proposition that the Earth orbits the Sun. The aesthetic appeal of the Copernican solution lies in its economy of suppositions, which accords with the precepts of Giles of Rome (*c.* 1247–1316) and William of Occam (*c.* 1284–1349), whose instrument of logic ("Occam's razor") states that hypotheses are not to be multiplied without necessity—that is, that the preferred theory is the simplest one that explains the facts.

Concerning the fact that the planets shine, Copernicus criticized his geocentric predecessors who "do not admit that…planets have a certain opacity, like that of the Moon" (Copernicus 20), inferring that all the planets reflect and scatter sunlight just like the Moon. In addition, in an effect most noticeable for Mars, the outermost planets shine brightest by the reflected light of the Sun when they appear on or near the meridian at midnight. They are then opposite the Sun, in the configuration known as Opposition (see figure 1.5). Heliocentric theory attributes these planets' greater apparent brightness at Opposition to their greater proximity to Earth and the shorter distance that sunlight must travel to illuminate

FIGURE 1.5. Planetary alignments for heliocentric orbits. The position of the Earth at E is shown relative to a Superior Planet (like Mars) in the outer orbit and an Inferior Planet (like Venus) in the inner orbit. Relative to the Sun-Earth direction, a Superior Planet can be at Opposition (O) or Conjunction (C); and an Inferior Planet can be at Inferior Conjunction (IC), Superior Conjunction (SC), Maximum Eastern Elongation (MEE), or Maximum Western Elongation (MWE).

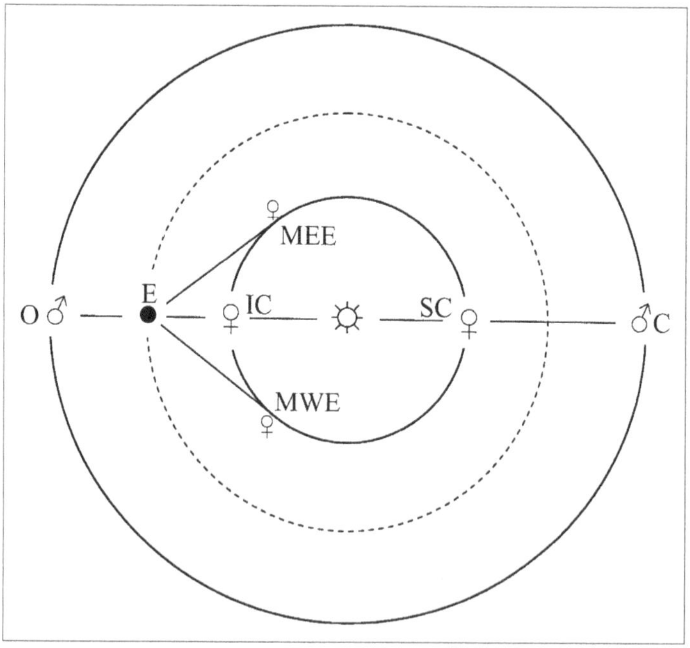

them compared to the same configuration in the geocentric model. The heliocentric hypothesis also explains other phenomena that the Old Astronomy could not without a slew of *ad hoc* hypotheses. For example, the apparent directional affinity of Mercury and Venus for the Sun (see figure 1.5) and the diminishing angular sizes of retrograde loops of the planets Mars, Jupiter, and Saturn are readily explained, the latter owing to the fact that apparent changes in the directions of ever more distant objects are progressively less affected by the motion of the Earth.

Copernicus realized that the stars could lie at different distances, but like Ptolemy he opted for a bounding sphere of stars and left it "to the philosophers" to determine the finitude of the Universe (Copernicus 17). However, by adhering to the concept of a revolving Earth, Copernicus had to make the sphere of stars very large because otherwise he would have observed a parallactic shift in their directions (figure 1.3b). If the stars were relatively close by, their angular separations would appear to increase and decrease as the Earth approached and receded from them, whereas the farther the stars are, the less pronounced the effect.

The lack of observed parallax meant that Copernicus either had to abandon the theory of heliocentricism or conclude that stars were much farther away than his predecessors had assumed. To Copernicus, heliocentricism was both mathematically and aesthetically pleasing because it answered questions about the planets efficiently, so the option of placing the stars far away was preferable to abandoning heliocentricism. Copernicus called the void between what was then the outermost planet, Saturn, and the now indeterminately large sphere of stars an *immensum*. This option faced strenuous opposition from philosophers and religious scholars who believed in the doctrine of First Cause (which stated that the Universe contained all possible existents) because to these sages, the empty space of the *immensum* served no purpose.

In about 1529, Copernicus circulated a paper titled *Commentariolis* ("Little Commentary") stating the essence of his new theory. In 1533, after the paper had made its way to Rome, Church leaders encouraged Copernicus to publish his theory in full, but Copernicus was loath to comply while he still had years to live. In 1531 actors had ridiculed him on stage, and in 1539 the Wittenberg theologian Martin Luther (1483–1546) called Copernicus a fool who "would overturn the whole science of astronomy," and Luther's disciple Philipp Schwarzerd (known as Melanchthon) (1497–1560) hinted that a wise government should not tolerate such blasphemy (Berry 97; Crombie 2: 176).

Copernicus took care to credit prior geokineticists Heraclides Pontus, Ecphantus, and especially the Pythagorean Philolaus, who he says Plato had journeyed to Italy to interview (Copernicus 13). Only in the draft version of *De Revolutionibus* does Copernicus mention the "ancient Copernicus," Aristarchus of Samos (*c.*310–250 BC), whose work had broader scope but was less accessible to him (Hall 74n; Koyré, *Astronomical* 100n21). Copernicus' magnum opus eventually appeared, thanks to the urging of the mathematician Georg Joachim (1514–1576), known as Rheticus. In the preface and dedication to *De Revolutionibus*, Copernicus explained that he had withheld publication for nearly 36 years "on account of the fear that I felt" (Copernicus 5). A well-meaning Lutheran minister named Andreas Osiander (1498–1552) took responsibility for printing the work and sought to protect Copernicus from charges of heresy by adding a preface explaining that the Copernican model—rather than representing reality—was useful for purposes of calculation only. This was yet another instance of proven advances catering to the sensibilities of the closed mind.

Copernicus foresaw the sort of objections that readers unfamiliar with mathematics would file.

> But if perchance there are certain "idle talkers" who take it upon themselves to pronounce judgment, although wholly ignorant of mathematics, and if by shamelessly distorting the sense of some passage in Holy Writ to suit their purpose, they dare to reprehend and to attack my work; they worry me so little that I shall even scorn their judgments as foolhardy. (7)

"Mathematics," Copernicus explained, "is written for mathematicians." He scorns Lactantius (*c.*260–340) who, he says, "speaks in an utterly childish fashion concerning the shape of the Earth" and "laughs at those who have affirmed that [it] has the form of a globe." In return, Copernicus affirms that scholars "need not be surprised if people like [Lactantius] laugh at us." Copernicus knew that his mode of inquiry was foreign to the world, and in hindsight his methodology is recognized as the start of the process of replacing the old organon of the Old Philosophy with the New Organon of the New Philosophy.

The Hybrid Model

In 1572 a Danish nobleman named Tycho Brahe observed a New Star that had burst forth upon the Firmament. The phenomenon was a source of consternation to sundry potentates, and in Denmark, King Frederick II (1534–1588) summoned Tycho (as he is most often called) to apply his astrological expertise to foretell its implications for national security (Christianson 18), as sudden celestial apparitions were routinely regarded as omens of ill fortune. Tycho's work on the New Star so impressed Frederick that in 1576, he ceded the small island of Hven (also known as Ven or Hveen) to Tycho so that he might build an astronomical observatory there. The island lies between modern Denmark and Sweden in the Oresund Sound, about 15 kilometers (10 miles) south-southeast of Helsingor. In 1580 Tycho named his completed abode Uraniborg (the Castle of the Heavens), and soon thereafter excavations began for Stjerneborg (the Castle of the Stars). From Hven on the north-northwest horizon, Tycho could see Helsingor where Frederick was building Kronborg Castle, soon to be immortalized in Shakespeare's *Hamlet* as Elsinore Castle.

Tycho objected to the Copernican model on several grounds, not the least being the seeming ponderousness and inertia of the Earth, which he believed argued against its having any motion. On top of that, Tycho was a dedicated observer who made accurate positional measurements of the stars, but he could not detect a stellar parallax for any of them and thus could find no direct evidence to suggest that the Earth moved. His inability to detect a stellar parallax angle to the limit of his observing accuracy presented the same dilemma to him as it did to Ptolemy and Copernicus, namely that either the Earth was stationary, or if it did move, the stars were at unacceptably large distances (figures 1.3a, 1.3b). Specifically, if heliocentricism were the case and Tycho adopted Copernicus' value of 1,142 Earth radii for the Earth's orbital radius, then the stars would have to lie more than 700 times farther than the average distance of the orbit of Saturn, which was then believed to be 10,550 Earth radii. Because the volume of a sphere increases as the cube of its radius, the volume of space surrounding the Solar System would have to exceed that occupied by Tycho's model by a factor of 700 cubed—over 300 million times (Thoren 279). In accordance with prevailing ideology, Tycho thought that such a huge amount of "wasted" space was absurd, and concluded therefore that the Universe was geocentric. The irony in this calculation is that, unwittingly, Tycho had made a great advancement because his accurate position measurements placed a lower limit on the size of the Copernican *immensum*.

Another argument that persuaded him against heliocentricism was the apparent angular sizes of stars. These vary with atmospheric conditions and appear smaller under conditions of diminished glare (as one might obtain during twilight), but no one took account of these facts. Tycho made the standard erroneous assumption that the appearance of a star was a measure of its real size, meaning that stars lying at the great distances dictated by the absence of detectable parallax had to be enormous. Rather than challenge his assumptions, Tycho concluded that this result was absurd too. Ergo, the Earth lies inert at the center of creation.

During his travels in 1575, Tycho acquired a book that pictured the Ptolemaic model, but with Mercury and Venus in orbit about the Sun. By 1583–1584, he had started to think in terms of this hybrid geo-heliocentric model whose origins date to Martianus Capella (*fl.*410–439), Heraclides Pontus, and the ancient Egyptians (Dreyer 167; Thoren 239–241). Accordingly, in 1583 he devised a model which appeared five years later in modified form in his *De Mundi aetherei recentioribus Phaenomenis Liber Secundus* ("Second Book concerning recent appearances in the Celestial World," or *Liber Secundus* for short). The model combined what he believed were the best features of geocentricism and heliocentricism. He opted for a stationary Earth for reasons stated and because he wished to reduce everything to its imagined stability. He put the two well-resolved Ancient Planets—the Sun and Moon—into geocentric orbits and allowed the remaining five wandering planets to orbit the Sun (figure 1.6). The fact that his arrangement led to planetary orbits that intersected was no impediment—he broke with Ptolemaic thinking and dispensed with crystalline spheres, letting planets move freely through space where he hoped (with judicious choice of orbital parameters) they need not collide. He justified smashing the concept of crystalline spheres using the observed properties of the Comet of 1577, which he found lay at least six times farther away than the Moon and crossed the spheres that were supposed to carry the planets in prior models.

Because Tycho believed that every cranny of the Universe served a purpose, he devised a plenum with as little wasted space as possible. His model was about 70% smaller than Ptolemy's, making it—comparatively

FIGURE 1.6. The bounded geo-heliocentric model of Tycho Brahe, from *Liber Secundus* (1588).

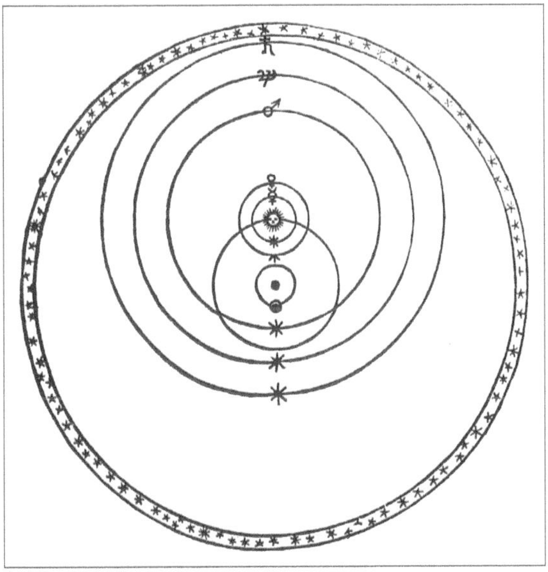

speaking—a *minutum*, which contrasted sharply with the Copernican *immensum*. In addition, it occurred to Tycho that the different apparent brightnesses of the stars could result from their being at different distances, and he was able to rationalize a stellar distribution of thickness of about 1,000 Earth radii. Of course, as the stars rotated in unison about the Earth, they had to maintain the same positions relative to one another somehow, and at the same time, the outermost layer of this shell needed material that was opaque in order to keep out information that might pour into physical space from the surrounding Empyrean.

Toward the end of the sixteenth century, Tycho fell into disfavor with the new Danish King Christian IV (1577–1648) and exiled himself. He ended up in Prague, capital of Bohemia, where he sought the aid of the mathematician Johannes Kepler (1571–1630). Tycho wanted Kepler to apply the data he had collected in order to put his hybrid World model on a more secure mathematical foundation, but Tycho died in 1601 before anything came of it, whereupon Kepler abandoned that project and instead used Tycho's data to improve the Copernican model. Kepler discovered that at least one planet (Mars) moved in an ellipse. This and a second related discovery appeared in *Astronomia nova...Commentariis de Motibus Stellae Martis* ("The New Astronomy...Commentary on the Motion of Mars") in 1609, in which he enunciated the two empirical laws of planetary motion now known as the Law of Ellipses and the Law of Areas. The former states that all planets move in ellipses with the Sun located at one of the foci of the ellipse, and the latter states that in so doing, an imaginary line joining a planet to the Sun sweeps out equal areas in equal times. A third discovery came ten years later in *Harmonices Mundi* ("The Harmony of the World"), and all three helped verify Isaac Newton's (1642–1727) theory of gravitation .

THE INFINITE UNIVERSE

In the fifteenth and sixteenth centuries, the infinitude of theological space was a common topic for debate, but few ventured to revive the doctrine that physical space was unlimited. Pythagorean concepts had reemerged in the thinking of Aristarchus, Seleucus the Babylonian (*fl.* 150 BC), Seneca (4 BC–AD 65), Aryabhatta (476–550), Brahmagupta (*c.* 598–*c.* 665), Hildegard of Bingen (1098–1179), Bernardus Silvestris (*fl.* 1147), and Copernicus, but they did not flourish fully until 1576 when Thomas Digges published his short essay, *A Perfit Description of the Caelestiall Orbes according to the most aunciente doctrine of the Pythagoreans, latelye revived by Copernicus and by Geometricall Demonstrations approved.* He appended it to an almanac titled *A Prognostication Everlasting* that his father Leonard Digges (*c.* 1521–1571?) had founded. Unlike conventional almanacs, this dealt with recurrent cyclic phenomena and therefore did not need continual updating.

As the title of Thomas' essay indicates, he credits Copernicus for reviving the Pythagorean theory of a moving Earth, but to this he adds the further notion of a physically infinite Universe (figure 1.7). This

FIGURE 1.7. The unbounded heliocentric model of Thomas Digges, from *A Perfit Description* (1576).

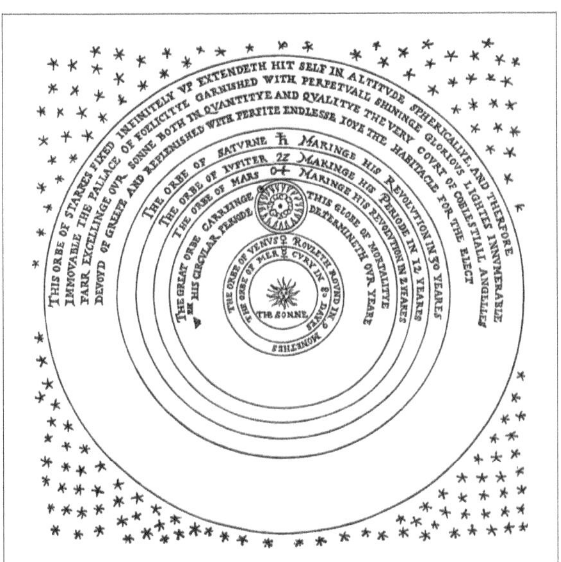

was the "real break with the past" (Heninger 130) as it completed the threefold constitution of the New Astronomy. Therewith, Digges' position became potentially precarious because without a starry Firmament to separate the Empyrean from the physical World, scientific inquiry would extend to all of space, including the abode of the gods.

Digges couched his model in reverent terms:

> ...this orbe of starres fixed infinitely up extendeth it self in altitude sphericallye and therefore immovable the pallace of foelicitye garnished with perpetuall shininge glorius lightes innumerable far excellinge our sonne both in quantitye and qualitye the very court of coelestiall angelles devoyd of greefe and replenished with perfite endlesse joye the habitacle for the elect. (Johnson 166; see figure 1.7)

From 1565, in chiefly poetic use, the word "orbe" in this excerpt can mean either the actual form of a celestial object or its apparent form (*OED*). Digges' phraseology mixes the two, confounding interpretation. The first few words define the celestial object in question ("this orbe of starres"), comprising the "perpetuall shininge glorius lightes." He posited that its actual form is of infinite extent ("fixed infinitely up," "lightes innumerable"), but its apparent form is spherical ("extendeth it self in altitude sphericallye"). It appears immobile ("and therefore immovable") because if an infinite distribution were to spin, its inconceivably distant extremities would have an impossibly large speed. The ensemble itself is "therefore immovable." Thus, the stars as a whole set the standard of rest, and their apparent circular motion is merely a reflection of the rotation of us, the observers on Earth.

Digges emphasized the role of the observer, "Herein can wee never sufficiently admire thys wonderfull & incomprehensible huge frame of goddes woorke proponed to our senses" (as cited in Johnson 165). Digges is using his sense of sight to admire God's handiwork and therefore his statement concerns appearances. He elaborated further on the apparent properties of the celestial lights that comprise the "Orbe."

> Of whiche lightes Celestiall it is to bee thoughte that we onely behoulde sutch as are in the inferioure partes of the same Orbe, and as they are hygher, so seeme they of lesse and lesser quantity, even tyll our sighte beinge not able farder to reach or conceyve, the greatest part rest by reason of their wonderfull distance unto us. (as cited in Johnson 165)

In other words, observers on Earth only behold that part of the "Orbe of starres fixed infinitely up" that lies "in the inferioure partes of the same Orbe," or closer to the observers. This is because the farther away the stars are ("as they are hygher"), so they seem fainter and fainter ("of lesse and lesser quantity") until observers see none at all ("our sighte beinge not able farder to reach or conceyve"). As a result, the rest of the stars are invisible owing to their distance ("not able farder to reach… the greatest part rest by reason of their wonderfull distance unto us"). Similarly, in 1600—a year before the nominal date for the writing of *Hamlet*—William Gilbert (1540–1603) published *De Magnete* ("On the Loadstone"), in which he foresaw the possibility of a multitude of stars that are "obscured because of [their] distance" (Gilbert 319). The devout Thomas Digges further defined the nature of the seen and unseen Universe, "And this may wel be thought of vs to be the gloriouse court of ye great god…"

Thus to Digges, the spherical shape of the distribution of stars is apparent and not real because the Diggesian observer (who finds the huge frame of God's work as admirable as it is incomprehensible) is necessarily at the center of that which he admires and finds incomprehensible.[3] Some modern scholars have failed to appreciate this and think that Digges embraced the absurdity of placing the Solar System at the center of an infinite spherical distribution.[4]

As a consummate mathematician, Thomas Digges would have encountered the notion of immeasurably large quantities and would have found these compatible with the parallel theological concept. He would have known from the 1440 work by Nicholas of Cusa (1401–1464) on "learned ignorance" that theological space has a center everywhere and a circumference nowhere. It would be within the bounds of reason if Digges thought that a leap to infinite physical space was thereby justified. In fact, he asserts that the stellar distribution extends infinitely up to the very court of the celestial angels and the "habitacle for the elect." He included God as an essential part of the quest to understand the natural World (Koyré, *Closed* 38–39) and criticized those who rely strictly on their senses without proper regard for the role of reason, the latter being a gift from God "to lighten the darkness of our understanding." Digges knew that humans have cognitive abilities superior to those in animals (whose primary activities are sleeping and eating) and advised us to put God's gift of brains to use in order to read the Book of Nature.

Copernicus and his predecessors placed a sharp division between natural and supernatural space, whereas Digges conflated the two. In 1953 Koyré posited that Digges introduces a radical new semiempirical form that shifts ignorance from the mysteries of the Empyrean to those at the outskirts of infinite space (*Closed* 38–39). Digges reconciled what he saw telescopically and posited theoretically with what he believed theologically. His real world is like Shakespeare's, which extends beyond human sensibility and ultimately embraces something supernatural (Beauregard 54). Digges reached an accommodation between natural and supernatural space and thereby designed a new frame of creation that preserves both the divine abode and an inconceivably, incalculably large physical space.

Like Copernicus, Digges subscribed to the Pythagorean and Platonic view that the Universe adheres to rules that are capable of aesthetically pleasing mathematical expression. His model combines the spiritual essence of Plato's Universal Form with a model of physical reality, and as such follows the ontological trajectory of Plato's Divided Line. Digges proceeded from Image (appearances) to the new Form (a model of reality) by way of Objects and their mathematical representation, while at the same time adhering to the spirituality of his personal Puritanical worldview. Through correlative reasoning, he ascends Plato's ladder of knowledge from Imagination to Hypotheses and thence to Theories concerning higher Forms of the intelligible world, all the while acceding to Plato's insistence on the divinity of Heaven (Dicks 131).

The Perspective Glass

The question arises, however, as to how Thomas Digges came by his new vision of the heavens. In 1571, with the publication of *A Geometrical Practical Treatise named Pantometria,* he and his father discussed basic designs for a two-element optical magnifier, or "perspective glass."

> But marveylous are the conclusions that may be performed by glasses concave and convex of circulare and parabolicall formes, using for multiplication of beames sometime the aide of Glasses transparent, whiche by fraction should unite or dissipate the images or figures presented by the reflection of other. (as cited in Johnson 175)

"Glasses transparent" refer to lenses, and "glasses…parabolicall" and "concave and convex of circulare formes" refer to mirrors. Digges continued by describing the result of combining these two kinds of "glasses."

> By these kinde of Glasses or rather frames of them, placed in due Angles, yee may not onely set out the proportion of an whole region, yea represent before your eye the lively image of every Towne, Village, &c. and that in as little or great a space or place as ye will prescribe, but also augment and dilate any parcell thereof, so that…yee may…discerne any trifle, or reade any letter…as plainely as if you were corporally present, although it be distante from you as farre as eye can discrie. (as cited in Johnson 175–176)

Digges explained why he refrained from divulging further details. "But of these conclusions I minde not here more to intreate, having at large in a separate volume by it selfe opened the miraculous effects of perspective glasses" (as cited in Johnson 176). He referred readers desiring additional information to a separate work; however, this work "never was made public" (*DNB*) and "has apparently been lost" (Johnson 176). In ensuing chapters, this matter will be discussed further.

In the Preface to *Pantometria*, Thomas announced the death of his father, Leonard, and praised his accomplishments in optics,

> ...my father by his continual painfull [painstaking] practises, assisted with demonstrations *Mathematicall*, was able, and sundrie times hath by proportional Glasses duely situate in convenient angles, not onely discovered things farre off, read letters, numbered peeces of money with the very coyne and superscription thereof, cast by some of his freends...in open fieldes...but also seven myles of[f] declared wat hath beene doon at that instante in private places. (as cited in van Helden, *Transactions* 30)

Thomas did not identify the "things farre off" nor the "many other matters farre more straunge and rare." If the Digges telescope could read the denomination of a coin at a distance of seven miles, then it would have to have outstanding ability to discern detail (i.e., a high resolving capacity or "resolution"). The claim implies a high resolution, which has led writers to dismiss it as fanciful. However, the fact remains that Digges wholeheartedly supported the Copernican model, which substantiates the possibility that—with optical aid—he resolved some of the Ancient Planets into disks, implying that they were spherical (figure 1.8) and not just points of light. This would justify Copernicus' proposition that they were planets like the Earth.

The Preface to *Pantometria* also refers to the "burning glasses" of Archimedes (287–212 BC) by which the Sun's rays are focused, creating sufficient heat to ignite material. Thomas wrote that his father "hath...sundrie times by the Sunne beames fired powder, and dischargde Ordinaunce half a myle and more distante" and refers to "many other matters farre more straunge and rare which I omitte as impertinente to this place" (as cited in van Helden *Transactions* 30). The operation would require well-figured long focal length mirrors, which should not be discounted, as evidence will accumulate in later chapters for the existence

FIGURE 1.8. Relative sizes of the Earth, Sun, Moon, and planets, from *Prognostication Everlasting* (1576).

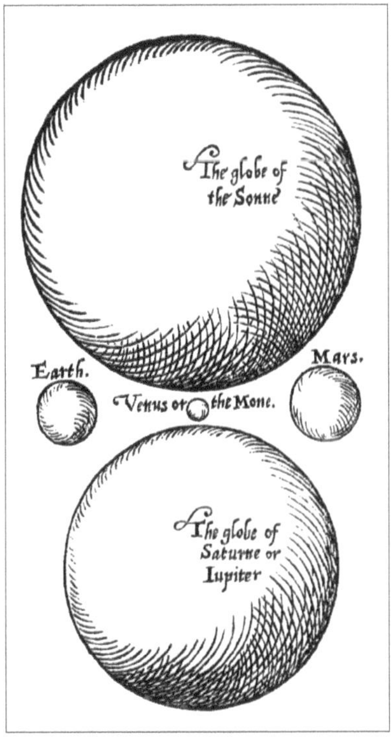

of mirrors with lesser (but still long) focal lengths. Thomas was concerned "for the security and defence of our natural country," adding that strangers would marvel at the descriptions (*DNB*).

Thomas Digges' scientific leap forward in 1576 overcame the centricity of perceptions and distinguished what was real from what was apparent. It shattered the outermost bounding sphere that the Ptolemaic, Copernican, and Tychonic World models had in common, and a year later, Tycho Brahe's observations of the Comet of 1577 enabled him to smash the celestial spheres that supposedly supported the planets. With the Copernican redefinition of the Sun as the center of the planetary system and Digges' depiction of an isotropic, homogeneous, and infinite distribution of stars, the New Astronomy was born. These birth pangs have received short shrift even in modern astronomy textbooks (Best, Maene, and Usher), thereby depriving humankind of that great intellectual leap from the narrow perspective of a bounded Universe to one of indefinite extent. Before considering Digges' contributions further, it is helpful to examine the origins of optical research in England.

Roger Bacon

In his *Arithmeticall Miltare Treatise, named Stratioticus* of 1579, Thomas Digges explained how his father's interest "grew by the aide he had by one old written booke of the same Bakons Experiments, that by straunge adventure, or rather Destinie, came to his hands" (as cited in van Helden, *Transactions* 30). From descriptions of experiments conducted by Roger Bacon (*c.*1214–1294), the conclusion seems inescapable that the friar was an original thinker and a meticulous experimentalist. Yet to this day, his reputation has suffered from the stigma of transgressions against received wisdom, including advocacy of experimentation, dabbling in the "natural magic" of scientific inquiry, and advocating that Aristotle's books be burnt (Berry 85–86).

Bacon disappeared from the public eye in the 1280s, probably because of incarceration (Clegg 163), but his *Opus Majus* ("Major Work") from about 1267 has survived. In it, Bacon wrote of "glasses so cast, that things at hand may appear at a distance, and things at a distance…at hand…and…starres shine in what place you please" (as cited in van Helden, *Transactions* 28). Evidence suggests that Bacon had assembled a functioning two-element telescope, a conclusion that Clegg and Johnson also reach.

On July 26, 1682, the minutes of the Royal Society recorded that Leonard Digges "had a method of discovering all objects pretty far distant" and "that this was by the help of a book…of Roger Bacon" (Birch 4: 156). The minutes stated that this was remarkable, as it occurred "more than thirty years before Metius or Galileo made their discovery of those glasses." And the minutes stated further that "Fracastorius had mentioned...his being able to discover things at a distance by the help of two spectacle glasses set at a distance one before the other; which was about the year 1530" (Birch 4: 156–157).

The Digges-Bourne Telescope

In 1578 a shipwright named William Bourne (1535–1585) shed some light on the topic of an Elizabethan telescope. In *Inuentions or Deuices. Very necessary for all generalls and captaines, or leaders of men, as wel as by sea as by land*, he reported that a lens-mirror combination magnifies distant objects. Specifically, he described a second optical component of a mirror lens having an off-axis concave mirror facing a light-gathering (objective) lens. He wrote of convex objective lenses up to 16 inches in diameter and a ¼-inch thick in the middle, and although he described what one sees through them, he omitted details. In effect, all he said was that "you shall see a small thing [at] a great distance...this is very necessary...as the viewing of an army...which I doo omit" (Bourne, as cited in van Helden, *Transactions* 30). Bourne had military uses in mind, as did the Diggeses and John Dee (1527–1608), who advocated placing a perspective glasses on every one of Her Majesty's ships.

In about 1585, at the urging of William Cecil, Lord Burghley (1520–1598), Bourne composed another work entitled *A treatise on the properties and qualities of glasses for optical purposes, according to the making, polishing, and grinding of them*. In this, he told of convex lenses made of fine white Venetian glass ground smooth by a concave iron tool. The lenses are thin, have long focal lengths, and are set in frames to help prevent fracture. Bourne made no direct mention of the quality of magnified images produced by a single concave lens, which for many people would appear fuzzy. He stated, however, that an image "*especially...* will be much amplifyed and *furdered*, by the receavinge of the beame that cometh thorow the glasse, somewhatt concave or hollowe inwards and well polysshed" (Bourne, as cited in van Helden, *Transactions* 33, emphasis added). The verb "to furder" means to further, assist, promote, or favor an action (*OED*), so it appears that a second element *especially favors* the amplified image, which may mean that it improves its quality. Bourne said that concave mirrors are "very necessary for perspective," and he mentioned the sizes of the lens and mirrors. A second element, the mirror-lens eyepiece, is therefore an indispensable element of the design. In the ninth and final chapter, he described a device with a convex objective lens and concave mirror for an eyepiece—the so-called Bourne-Digges telescope (Rienitz 110).

Bourne affirmed that Digges built a perspective glass "withowte any dowbte of the matter" (as cited in van Helden, *Transactions* 34), but details are sparse, and he deferred to Dee and Thomas Digges, asserting that they knew more than he did. Bourne described a concave glass that is well polished and well foiled for reflection from the back surface, yet his treatise divulges little that is new, and its occasional inconsistencies suggest that he was unfamiliar with some aspects of what he was describing.

GAINER'S EXPERIMENT

Digges' design describes a transparent glass (i.e., lens) that collimates rays formed by reflection from parabolic or circular mirrors, rendering an image visible to the eye. In 2008, Reeves noted that Albert van Helden and Sven Dupré had drawn attention to the significance of this description, and in the same year, Michael Gainer also realized the significance of the passage and put perception into practice by constructing a device according to that prescription (figure 1.9).

Gainer posited that the Digges design matches that of telescopes named for the astronomer and musician William Herschel (1738–1822) and concluded that building such a proto-Herschelian device in the sixteenth century was technically feasible. Certainly, Thomas Digges and others at the time "were not interested in mere theory," but had an appreciation for experimental excellence (Johnson 173). Given the skills and resources at his family's disposal, it is entirely feasible that Leonard built a perspective glass before or during the reign of Queen Elizabeth I

FIGURE 1.9. Gainer's experimental telescope built according to Digges' design with materials and tools available in the sixteenth century (f/8, 4.5-inch aperture, 1-inch plano-convex eyepiece, and magnification 36).

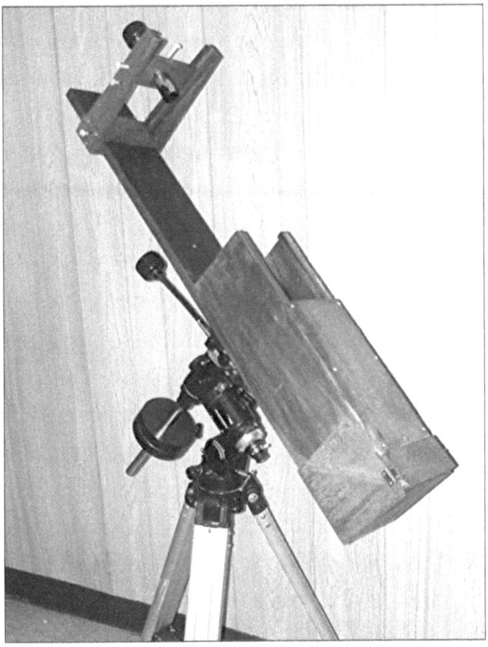

(1553–1603). The question is whether one or both of the Diggeses used such a device to study the heavens. Later chapters provide evidence in favor of that view, but the difficulty is that no one has discovered any direct physical evidence of a Digges telescope, or any associated records of its use, and consequently the existence of such a device remains mired in uncertainty (Ronan; Turner; van Helden, *Transactions* 14–15).

DIGGES' UNPUBLISHED WORKS

In the dedication of *Stratioticus*, Thomas wrote that he "spent many… yeares in reducing the Sciences Mathematicall, from Demonstratiue

Contemplations, to Experimentall Actions" (as cited in Johnson 170), in other words, putting theory into practice. He listed several books that he had begun writing and intended to publish, but never did. The list includes works on navigation, architecture, artillery, pyrotechnics, and fortification; but one title, *Commentaries vpon the Reuolutions of Copernicus*, is especially germane to the topic at hand. None of the promised works survived, if indeed they ever existed.

In the preface to *Stratioticus*, Thomas referred to Pythagorean "exclusiveness," by which he meant knowledge transmitted by word of mouth to an initiated few. Thomas Digges tried to justify his reluctance to publish by referring idiomatically to the precedent set by his father as *per manus tradere*—the practice of handing down knowledge from father to son, committing it to memory and propagating it orally to "a fewe selecte friendes" (Digges and Digges a.ii).

Thomas explained that he ceased publishing because "infernal furies" tormented him with "law-brabbles" and "infernal suits at law" (as cited in Hotson, *Appoint* 115). For instance, he lodged a flurry of lawsuits, including one to recover funds embezzled by a relative, but distractions other than lawsuits may have played a role as well. Johnson enumerated some, and one must wonder at their nature and extent. Thomas had enemies aplenty (Hotson, *Appoint* 115–117, 125), including his "archenemy" Roger Manwood (1525–1592), the Kentish Chief Justice and Lord Chief Baron of the Queen's Exchequer who claimed to have invented the perspective glass (Reeves 63–65).[5] Any chance of the pledged works being published perished with Thomas' death, but Digges may have still used the perspective glass in a systematic study of the heavens before the time of Galileo Galilei, whose telescopic data purport to be the first in history.

GALILEO GALILEI

From 1574 to 1579, Galileo Galilei attended a monastery school in Florence, and in 1581, he matriculated to the University of Pisa. Like his father Vincenzio Galilei (1520–1591), he had a critical attitude to ideas and an aversion to dogma, and though he planned to follow his father's wishes and study medicine, he found philosophy more to his liking. In 1585 Galileo (who goes often by his first name) dropped out of school because he failed to secure a scholarship. He tutored students intermittently, and he secured a position as Professor of Mathematics in 1589 at the University of Pisa, where he engendered the ire of his superiors by contradicting Aristotle.

In 1592 Galileo accepted a position as a mathematics professor at the University of Padua. During the summer of 1605, he tutored Cosimo II de' Medici (1590–1620)—scion of the Grand Duke of Tuscany—who upon becoming Grand Duke in 1609 appointed Galileo his chief mathematician and secured a research professorship for him at Pisa. Like Thomas Digges, Galileo was not a university graduate, yet both he and Thomas were creative thinkers who enjoyed poetry and the classics. They typified Renaissance humanists who were curious about physical phenomena and whose penchant for experimentation and observation inclined them to come to grips with empirical reality.

An event that drastically affected the course of history occurred in 1609 when Galileo learned that a Dutchman Hans Lippersey (1570–1619) had built a telescope using two lenses. Galileo built one for himself using spectacle lenses procured from optical shops and began to observe celestial objects. His astronomical discoveries occurred in the three-year interval from January 7, 1610 to December 1, 1612 (Drake, *Discoveries* 144–145; van Helden, *Sidereus* 13). He began by stating in a letter that the Moon's surface was not smooth as most people believed, and that planets appeared resolved into little disks. This letter of January 7, 1610 is currently accepted as the first document to report astronomical data gathered through a telescope. It served as a warm-up for his monograph *Sidereus Nuncius* ("The Starry Message" or "The Starry Messenger"), whose dedicatory letter is dated March 12, 1610. (Hereafter, the reference "*Sidereus*" is to van Helden's translation of *Sidereus Nuncius*.) On the same evening as the date on that letter, Galileo reported that he had observed three unusually bright images arrayed on either side of Jupiter, lying more or less along an imaginary line passing through the planet. These objects attracted attention because they were unusually bright, like "little full moons," and not at all "fulgurous and trembling" (Drake, *Work* 146–147) like stars (figure 1.10). By January 15, 1610, he had concluded that Jupiter had four moons (known today as "Galilean moons") bearing names that Kepler assigned in 1613: Io, Europa, Ganymede, and Callisto.

Sidereus Nuncius took the world by storm. The British Ambassador to Venice, Sir Henry Wotton (1568–1639), purchased a copy and predicted that Galileo was going to be "either exceedingly famous or exceedingly ridiculous" (Hall 320). Wotton was a gatherer of foreign intelligence and a friend of the metaphysical poet John Donne (1572–1631), who would soon have occasion to refer to Galileo's results.[6] The ambassador also promised to procure a Galilean telescope. Wotton was a Kentishman, yet in light of the present thesis, it seems he was unaware of the advances made in the previous century by his fellow countians, the Diggeses.

The 1610 discovery of four moons orbiting Jupiter came hard on the heels of two other dramatic observations of celestial phenomena, the New Stars of 1572 and 1604. Each of these findings showed that cosmic entities undergo change, whereas the Aristotelian World model required that the Heavens be immutable. The Galilean moons also contradicted Aristotelian dogma by showing that Earth was not the only center of motion in the Universe. In addition, the moons disproved the argument invoked by advocates of the Old Astronomy that the Earth would leave the Moon behind if it moved in a heliocentric orbit, for evidently, Jupiter moves in an orbit but carries its moons with it.

FIGURE 1.10. Detail from Galileo's *Sidereus Nuncius* (1610) illustrating Jupiter and its four moons.

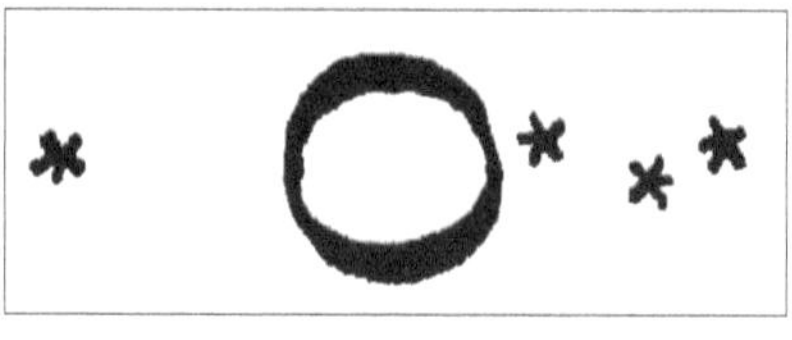

For these reasons among many, cornerstone beliefs of the scholastic worldview fell under suspicion. In 1610 Shakespeare was nearing the end of his writing career but would surely have noticed the public announcements emanating from Italy. I posited in 2003 that Shakespeare lauded the discovery of the Galilean moons in *Cymbeline*, wherein the four ghosts of Posthumus' immediate family circled him and represented the four ghostlike attendants to the planet Jupiter.

Until July of 1610, Saturn was too close to the Sun for Galileo to observe it. Figure 1.11a is a cartoon of what Saturn looked like to him,

FIGURE 1.11. Cartoons illustrating Saturn's image: (a) (Uppermost) as it appeared to Galileo in 1610, (b) (middle) how he interpreted what he saw, and (c) (nethermost) how the planet would have appeared at the time if his spyglass had had better optics.

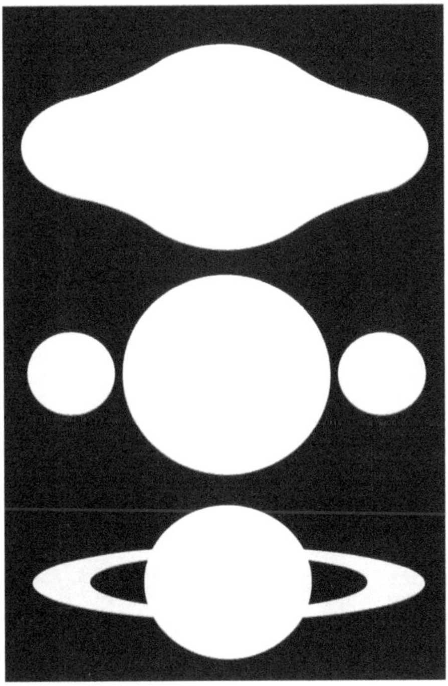

and figure 1.11b is how Galileo interpreted what he saw. Figure 1.11c shows how the planet (with a ring inclined to the line of sight) might have appeared in 1610, if his spyglass had had better resolution. Figure 1.12 shows an Earth-based image comprised of a stack of 504 separate images taken through a small telescope with modern equipment. The eye's capacity to integrate over time is similar to such image superposition, enabling it to distinguish these features through a telescope with good optics. However, the image of Saturn seen through the Digges' telescope reproduction (figure 1.9) shows a ring that is not resolved into two rings. For comparison, figure 1.13 is a space-age image.

Writing to Cosimo's secretary on July 30, Galileo described Saturn as an arrangement of three images "placed on a straight line along the

FIGURE 1.12. Saturn on January 14, 2007, imaged by a Toscano 8-inch (200 mm) telescope. The image is a sum of 504 separate images each of 1/25 second exposure and shows the Cassini gap separating the inner brighter B-ring and the outer fainter A-ring. These rings lie respectively from 92,200 to 117,500 and from 121,000 to 136,200 kilometers (57,300 to 73,000 and 75,200 to 84,600 miles) from the center of Saturn yet are only a few hundred meters (yards) thick. The equatorial radius of Saturn is 60,000 kilometers (37,300 miles). For comparison, the Earth's equatorial radius is 6,400 kilometers (4,000 miles).

Figure 1.13. Saturn and its rings depicted by NASA's Voyager 2 spacecraft on July 21, 1981, when the spacecraft was 34 million kilometers (21 million miles) from the planet.

zodiac, the one in the middle being about three times larger than the other two on the sides" (*Sidereus* 102; figure 1.11b). Galileo planned to publish his result in a new edition of *Sidereus Nuncius* in 1610, but to ensure credit for himself in the meantime, he encoded it into a nonsensical chain of alphabetical letters which he sent to Kepler in Prague and to the Jesuits at the Collegio Romano in Rome in October of 1610. The letters, when properly arranged, announced that he had observed what he believed was the triplicity of Saturn.

Also in October of 1610, Galileo began to observe Venus and discovered that the planet had phases (figure 1.14). This established Venus' orbit as circumsolar and motivated John Donne to bemoan the confusion wrought by Copernican heliocentricism.[7] Galileo buried this latest discovery in an anagram whose solution, "the mother of love [i.e., Venus]

FIGURE 1.14. Illustration of successive phases of the "dark star" Venus seen from Earth as both orbit the Sun (see figure 1.5). (Top left) Venus is just past its farthest point from the Earth (Superior Conjunction), so its image is relatively small and gibbous. (Top middle) The image becomes larger and less gibbous as it approaches the Earth and reaches maximum elongation. (Top right) The image becomes crescent. (Bottom row) Venus appears larger and its phase more crescent as it nears its closest point to the Earth (Inferior Conjunction).

emulates the figures of Cynthia [i.e., the Moon]," meant that Venus has phases like the Moon. He revealed this on New Year's Day, 1611.

To cap off this run of discoveries, in November of 1612 Galileo was astonished to find that Saturn's "satellites" had disappeared. In an allusion to the myth of the god Saturn devouring its children, Galileo wondered whether the eponymous planet had gobbled up its attendants. Only in 1659 did Christiaan Huygens (1629–1695) divulge that a thin, flat ring encircles Saturn. The evolution of images of Saturn's ball and ring from the time of Galileo's observation up to that time is contained in Huygens' *Systema Saturnium* ("Saturn's System") (see Berry, figure 67).

The ring is inclined to the plane of the Earth's orbit, which means that as Saturn revolves about the Sun in 29½ years, observers can sometimes see one face of the ring and sometimes the other, but the ring becomes invisible when it is either backlit, edge on to observers, or edge on to the Sun (see Berry, figure 66). In anticipation of objections from scholastics who still thought that sphericity was the sole celestial figure, and as an echo of Galileo's defense of his telescopic imagery, Huygens added, "I see the ring clearly with my eyes" (Pannekoek 255).

Modern science largely evolved from independent academies and societies that sprang up in the sixteenth and seventeenth centuries. While not wishing to appear competitive with the universities, scholars who were curious about the workings of nature and imbued with a love of learning formed societies in which they could exchange ideas. Florence was the cradle of such activities in Italy, eventually attracting the patronage of the Grand Duke, whereupon the autodidacts transformed one society whimsically titled "Academy of the Moistures" into one with the more respectable name: "The Florentine Academy" (Drake, *Discoveries* 10, 77). In 1603 Federico Cesi (1585–1630) and three friends founded the Academy of the Lynx-eyed, and in April of 1611, Galileo and a few others were feted at one of its banquets. University professors were slighted, however (Drake, *Work* 166–167). At that time, Galileo's spyglass received the name *telescopium* ("telescope"). Under the auspices of the Lincean Academy, Galileo wrote a treatise on solar blemishes titled *Epistolas de Maculis Solaribus* ("Letters on Sunspots"), which brought the essentials of the Copernican Worldview to the attention of literate compatriots.

Francesco Sizzi (1585–1618) was the first author to attack Galileo on theological grounds (Drake, *Work* 162), but after the chief mathematician at the Roman College, Christopher Clavius (1537–1612), confirmed Galileo's discoveries, Pope Paul V (1552–1621) received Galileo cordially when he visited Rome in 1611. The faculty of the Roman College formally recognized Galileo's work, yet its head Cardinal, Robert Bellarmine (1542–1621), secretly began to investigate him. Bellarmine sought to determine whether any link existed between Galileo and Cesare Cremonini

(1550–1631), who headed the philosophy department at Padua and whom the church suspected of heresy (Drake, *Discoveries* 69, 75–76).

After another of Galileo's friends named Maffeo Barberini (1568–1644) became Pope Urban VIII, he allowed Galileo to discuss heliocentricism as long as he did not present it as a description of reality. In 1632, after censors had granted approval and with the encouragement of Urban VIII, Galileo published his *Dialogo* ("Dialogue") on the two chief World systems (Ptolemaic and Copernican), using fictitious characters to argue the cases for and against heliocentricism. *Dialogo* belittled Aristotelian cosmology and made its advocate, Simplicio, look like a simpleton. Galileo wrote in Italian in order to reach as wide an audience as possible, and his enemies could not ignore it. They started to foment trouble for him, and aides led the Pontiff to believe that Galileo had held him up to ridicule. Within a few months, a special tribunal had started to look into the matter. Galileo received a summons to appear before the Roman Inquisition, and "having been shown the instruments of torture" (Drake, *Work* 351n22), he confessed to his belief in a moving Earth. On June 22, 1633, he was compelled on bended knee to abjure Copernicanism, and legend has it that on rising he muttered, '*Eppur si muove*' (Nevertheless, it [the Earth] moves). The church placed Galileo's astronomy works on the *Index Librorum Prohibitorum* ("Index of Prohibited Books"), where they remained until 1835 "pursuant to an 1822 action based upon the Holy Office's 1820 decision not to oppose Copernican views" (Seeger 32–33).[8]

Galileo was nearly 70 years old when he received a sentence of life imprisonment, but Cardinal Francesco Barberini (1597–1679), who was Pope Paul's nephew and "Galileo's ablest and most highly placed protector at Rome" (Drake, *Work* 287n26), commuted Galileo's place of confinement to the Florentine embassy. Later, Galileo was remanded to the custody of the Archbishop of Siena. He spent the rest of his life under house arrest, but that did not stifle his creativity. He returned to his studies and in 1638, after a delay caused by censors, *Due Nuove Scienze* ("Two New Sciences") emerged, which is arguably the bedrock of modern Physics.

Digges' *A Perfit Description*, which appeared in 1576 and became extremely popular in England, was reprinted several times afterward, including in 1592, 1596, and 1605 (Johnson 315–321), and Galileo's discoveries in Astronomy were published and publicized between 1610 and 1612. These dates fall within the span of time from about 1589–1591 to 1613–1615 in which Shakespeare wrote his plays (Crystal and Crystal 592). Surely, the bard did not ignore these transformations in worldview, particularly in light of Shakespeare's attention to Digges' publications (Hotson, *Appoint* 237–259).

The Canon

In 1543 Copernican heliocentricism had set humankind spinning and revolving, and thirty-three years later in 1576, the habitat of life was set adrift in a sea of stars that stretched out seemingly forever. A poet able to grasp the import would surely write about it, but until recently, the evidence from Shakespeare's Canon has been slight and ambiguous. For example, in *Troilus and Cressida*, Ulysses is a Greek commander and political operative who upholds traditional virtues of rank and order. His "degree" speech contains a passage that suggests a planetary arrangement.

> ULYSSES The heavens themselves, the planets, and this centre
> Observe degree, priority, and place…
> And therefore is the glorious planet Sol
> In noble eminence enthroned and sphered
> Amidst the other… (1.3.85–1.3.91)

Concerning the sequences *G* and *H* stated previously, the geocentric arrangement *G* has three Ancient Planets lying on either side of the Sun, which Shakespeare calls "planet Sol," and which is therefore "amidst" the other Ancient Planets. However, the passage also could refer to the modern heliocentric sequence *H* because the royal Sun, like an enthroned King that is the center of his subjects, is at the physical center of the orbits of six (modern) planets and is therefore "amidst" them as well. And it is possible that Shakespeare intended both meanings.

The Digges family was schooled in the classics, and its members would have known that inquisitors had a history of pillorying free thinkers. The Diggeses, as followers of Roger Bacon, would have known of the persecution that the friar underwent for his scientific investigations and beliefs (Clegg 143–144) as well as of Copernicus' fears and the frosty reception that his ideas received in some quarters. In the fifteenth and sixteenth centuries, the concept of the infinitude of space as an abode for deities was a topic of much discussion, and by broadening its scope to include the material World, Thomas Digges had to proceed warily to avoid charges of atheism. It is reasonable to believe that Shakespeare would follow suit for the same reasons and, declining to cater to the convictions of closed minds, would disguise the transformations in Worldview using the many literary devices at his command.

Subsequent events in the first half of the seventeenth century amply justified such caution. Church institutions that had hitherto monopolized interpretation of the Heavens felt threatened by heliocentricism, and matters came to head when in 1609–1610, Galileo built a spyglass and discovered phenomena that imperiled the foundation of Aristotelian cosmology and (indirectly) Christian doctrine.

The mystery deepens when one discovers that Shakespeare's *Hamlet* contains metaphorical descriptions of celestial data that no one could have known without telescopic aid. By the turn of the seventeenth century, therefore, it appears that Shakespeare was the only poet to notice the revolution wrought by the New Astronomy.

The genre of books on Shakespeare's astronomy dates back at least 40 years (see Clark, Meadows), but following the publication of *Hamlet's Universe* (herein *HU*), the criterion of evidence for Shakespeare's knowledge of astronomy has shifted from short passages lifted out of context to the analysis of entire plays. To seek further evidence for Shakespeare's recognition of competing worldviews, I begin by examining *Love's Labour's Lost* in the next chapter.

Chapter 2

Love's Labour's Lost

> The knowledge of man is as the waters, some descending from above, and some springing from beneath; the one informed by the light of nature, the other inspired by divine revelation.
>
> —Francis Bacon

Love's Labour's Lost (LLL) is a romantic comedy, though it is deemed "a distinctly odd and difficult play" with a "hidden meaning which can be recovered only if the right key…is found" (Woudhuysen 1). This chapter suggests that *LLL* exposes the narrowness of the medieval worldview and acts as a searing indictment of religious intolerance. Mythology and astronomical phenomena provide a chronology of events that combine to solve riddles posed by Dull and Holofernes. The Spanish Armada figures prominently in more than just the naming of the character Armado. Astronomical considerations explain the metaphysical significance of the slain deer, and Rosaline's traits indicate that in 1594, Shakespeare was aware of properties of the ancient planets that he could not have known without telescopic aid.

NAVARRE

In 1572 the Protestant Henry of Navarre (1553–1610) ascended the throne of Navarre, which is now a province in northern Spain bordering on France. Shakespeare (*Love's Labour's Lost*) models the King of Navarre after Henry and names his three companions—Berowne, Longaville, and Dumaine—after political operatives prominent in France in the 1590s (Mowat and Werstine, *Love's* 227–228). (In this Chapter, unattributed notes, lineation, and pagination refer to Mowat and Werstine's edition of *Love's Labour's Lost*, and other editions of Shakespeare's plays are also identified by editors' names.)

Navarre was head of the French Protestants, the Huguenots. He was in Paris for his wedding when, toward the end of the postnuptial celebrations, there was an attempt to assassinate the Protestant leader Admiral Gaspard de Coligny (1519–1572). This plot was orchestrated by Catherine de' Medici (1519–1589), mother to the young French King Charles IX (1550–1574). She and Charles' counselors convinced him that the Huguenots were a threat, and he eventually succumbed to this pressure. He ordered, "Kill them! Kill them all…" (Noguères 67), and early in the morning of August 24, 1572 (Saint Bartholomew's Day), the massacre of Huguenots began (Buisseret 7–8). The leader of the assassins was Henri of Lorraine, third Duke of Guise (1550–1588), and one of his first victims was Coligny, who was recuperating from wounds suffered during the previous attempt to kill him. Among many atrocities was the murder (by stabbing) of the anti-Aristotelian Huguenot philosopher Petrus Ramus (1515–1572).

The French court and Navarre himself acted as if nothing untoward had occurred (Buisseret 8), but abroad, the massacre did not pass unnoticed. Some refugees sought asylum in Canterbury in Kent, England, and in about 1594, the Kentishman Christopher Marlowe (1564–1593) wrote *The Massacre at Paris*, a play in which the Duke of Guise orders his followers to commit atrocities. The date of Marlowe's play compares closely to the likely date of the composition of *LLL* (Hibbard *Love's* 45), and topical allusions in *LLL* cluster around 1594 as well (Pendergast 4). Supposedly, *LLL* contains not so much as a hint of the religious wars of the sixteenth century (Hibbard, *Love's* 50), but the present interpretation suggests otherwise.

Stage directions refer to the King of Navarre as "Ferdinand," but like Claudius in *Hamlet*, the name is never part of the dialogue (Hibbard, *Love's* 1.1.0.1), and its omission may be an attention-getting stratagem. A likely candidate for the source of the name is Ferdinand V (1452–1516), King of Castile and Leon, who in 1512 annexed most of the kingdom of Navarre and completed the recapture of Spain from the Moors. With Queen Isabella I (1451–1504), Ferdinand financed Christopher Columbus (1451?–1506) and instituted the religious tribunal known as the Spanish Inquisition.

At the outset, Shakespeare introduces the themes of hunting, the passage of time, and the god of Time, Saturn. The King of Navarre maintains that fame is something "all hunt after in their lives" (1.1.1), and he expresses hope that despite "cormorant devouring time" (1.1.4), the endeavor upon which he and his companions are about to embark shall bate "scythe's keen edge" and make them "heirs of all eternity" (1.1.6–1.1.7). Cormorants are aquatic birds that prey on fish and metaphorically devour time just as Saturn devoured his children. The scythe is a tool of the harvest, and in mythology, it is also the instrument of death. Saturn is god of the harvest, commonly depicted as an old man holding an hourglass to measure time and a scythe to fell those whose time has come (48).

Navarre summarizes the articles of a pact by which he and his companions have sworn to abide. In order to buy honor and prestige after death, they plan to deprive themselves of food and sleep for three years while devoting themselves to study. They also plan to forego the company of women. Longaville and Dumaine sign on without hesitation, but Berowne is skeptical.

In a passage replete with optical imagery, Berowne points out the limitations of book learning.

> BEROWNE As painfully to pore upon a book
> To seek the light of truth, while truth the while
> Doth falsely blind the eyesight of his look.
> Light seeking light doth light of light beguile.
> So, ere you find where light in darkness lies,
> Your light grows dark by losing of your eyes. (1.1.76–1.1.81)

According to primitive theory, eyes emit beams that illuminate the object gazed upon, and Berowne alleges that if light emphasizes canonical wisdom at the expense of the curiosity of an open mind, then it can blind the observer and achieve the opposite of its intended goal.

The contrast of light and dark persists as Berowne assails pedantry.

> BEROWNE Study is like the heaven's glorious sun
> That will not be deep searched with saucy looks.
> Small have continual plodders ever won
> Save base authority from others' books.
> These earthly godfathers of heaven's lights,
> That give a name to every fixed star,
> Have no more profit of their shining nights
> Than those that walk and know not what they are.
> Too much to know is to know naught but fame;
> And every godfather can give a name. (1.1.86–1.1.95)

There is little profit in merely seeing a star without seeking to know more about it. Berowne is an iconoclast who is unafraid to interrogate nature and is therefore likely to aid Navarre in his quest to become a philosopher-king.

Reminded of the pending visit of female royalty, the King decides that he and his companions must dispense with the article of their pact that forbids contact with females. This may be Shakespeare's way of noting that Henry of Navarre was a notorious womanizer. Pledges of fealty meant little to Henry, notwithstanding his religious faith, which he alternately embraced and renounced to suit the situation at hand. Shakespeare brackets the topic of immediate concern to Navarre (the

Princess' visit) with the one that concerns Berowne (epistemology), and as the plot proceeds, both become foci of attention.

Armado

Berowne signs the pact, and Navarre raises the topic of a visiting Spaniard named Armado. The name refers to the ill-fated armada launched in 1588 by the Spanish Sovereign Philip II (1527–1598) (1.1.174n). Philip II became King of Spain upon the abdication of his father, Charles I (1500–1558). He married Queen Mary I (1516–1558) of England in a futile attempt to secure the English crown, and after failing in his attempt, left England and his wife in 1555. In England, the Marian persecution is renowned for its brutality, and on the Continent, the Spanish Inquisition attained the pinnacle of its power under Philip.

Shakespeare casts the braggart Armado as the archetypal pedant. The King announces that he plans to use him as a source of entertainment, and Longaville adds that they can also make sport of Costard, a rustic whose name literally means "head" but which is also the name of a large apple (1.1.183n). This is a case of the pot calling the kettle black, for the prospective autodidacts are as incapable of accurate self-appraisal as the peasant whom they demean. The script has an abundance of I / eye puns, signifying the importance of vision in overcoming self-centeredness and establishing an objective worldview.

Costard enters, escorted by a Constable named Anthony Dull and bearing a letter from Armado to the King. Apparently, Costard had consorted with a maid, Jaquenetta, and Armado has arrested him. The King remands the yokel to the custody of Armado, and the next scene (1.2) reveals the injustice of this treatment—when Armado (who is in love with Jaquenetta as well) asks the boy called Moth, "what sign is it when a man of great spirit grows melancholy?" (1.2.1–1.2.2).[1] "Melancholy" is a reference to Saturn, the planet of sadness and death, which prepares the way for future drama.

PRINCESS

The Princess arrives, escorted by Boyet and other Lords. Accompanying her are three ladies-in-waiting: Rosaline, Katherine, and Maria. The ostensible reason for the royal mission is to lay claim to the province of Aquitaine, which Navarre holds as security in lieu of cash that he claims the French owe him.

The Princess has heard that the King plans to commit to a monkish existence, so she dispatches Boyet to ascertain the manner of her reception. Boyet returns with the news that Navarre cannot receive ladies at court and will lodge them in adjacent fields. The King greets the visitors there, accompanied by his three associates. The Princess scolds the King for his inhospitality and questions the wisdom of the oath he has sworn.

Navarre offers to return Aquitaine upon payment of 100,000 crowns, but the Princess avers that France has paid the sum already and that she has the papers to prove it. Boyet promises, "Tomorrow you shall have a sight of them" (2.1.171). Inexplicably, the Princess has set out without evidence vital to her mission. Instead, the proof is entrusted to an independent courier who never arrives.

The King and two of his companions depart, and Berowne remains to commend his heart to Rosaline. She greets his overture impolitely, and Berowne leaves. Dumaine enters and asks Boyet for Katherine's name, which Boyet supplies, although he skillfully declines to give Maria's name to Longaville. Berowne reenters and, in turn, asks Boyet for Rosaline's name, which he supplies. Boyet opines that Navarre is in love with the Princess, which completes the pairings of affection.

In Act 3, Armado releases Costard and pays him three farthings to deliver a love letter to the peasant girl, Jaquenetta, while Berowne pays him a shilling to deliver a love letter to Rosaline. Berowne wants his letter delivered that very afternoon because the Princess plans to hunt deer the next day.

Act 4 begins the next day with the Princess inquiring of a forester, "Was that the king that spurred his horse so hard / Against the steep uprising of the hill?" (4.1.1–4.1.2) The woodsman answers that he does not know, but thinks the equestrian was not Navarre. "Whoe'er he was," she adds, "he showed a mounting mind" (4.1.4). Although this may refer to the desire to copulate, the script never divulges who the equestrian is or why he is the first topic of Act 4. Perhaps the rider stands for Shakespeare himself mounting a campaign against pedantry and prejudice, and one can well imagine that (*c.* 1594) this would be an uphill struggle.

Using the royal "we," the Princess tells the forester "today we shall have our dispatch. / On Saturday we will return to France" (4.1.5–4.1.6). Shakespeare uses the noun "dispatch" to mean "settlement of business," which depends upon evidence that Boyet had assured the Princess would arrive this very day (2.1.171). If the evidence were to arrive as Boyet had promised, then it is reasonable to suppose that the Princess would settle her business and depart the next day, particularly because Navarre received her inhospitably. If Saturday is the next day, the current day must be Friday, so the royal party must have arrived on Thursday. In addition, the verb "dispatch" means "to kill," so with great economy of words, the Princess makes her expectations clear: today (Friday) she plans to accomplish her mission and bag a buck.

The Princess harbors no ill will toward her prey ("my heart means no ill," 4.1.37), but intends to show herself the equal of any hunter. This is unsurprising, as in matters of wit, she consistently bests every male she meets. *LLL* is laced with puns, and in this case, she twice refers to "heart" (4.1.35, 4.1.37), which puns on "hart," a male deer, particularly one in its fifth year (*OED*). Apparently, the Princess aims to kill a 5-year-old buck. It is appropriate for this occur on a Friday, which takes its name from Fria, the Nordic goddess of love and the only female to lend a name to a day of the week.

Costard arrives and locates the Princess and her retinue in the forest, but instead of giving Rosaline the letter from Berowne, he mistakenly delivers the letter that Armado wrote to Jaquenetta. Armado's letter has the correct address but the Princess orders it to be opened anyway. Perhaps this intrusion is a prerogative stemming from the divine rights of royalty. In the letter, Armado misquotes the dictum *veni, vidi, vici*, which means, "I came, I saw, I conquered," by translating it as, "He came, see, and overcame" (4.1.75n, 4.1.77n). Perhaps the error refers to medals, struck in England after the defeat of the Armada, that bore the inscription *Venit, Vidit, Fugit* ("it came, it saw, it fled"; Camden, Section 32). Armado's error is a parody on the "dog-Latin of the Schoolmen" (Dampier 98), as is for example, "hig, hag, hog" in *The Merry Wives of Windsor* (4.1.39), and is indirectly an admonition for the English language to replace it. Scene 4.1 ends with the shout, "Sola, sola," and the hunt is on. "Sola" is a cry that accompanies the blowing of horns on a hunt (4.1.178n).

The next scene 4.2 opens with the Princess having already bagged her buck. A pedant Holofernes offers a description of the animal that refers to "earth" and "sky."

> HOLOFERNES The deer was, as you know, in blood, sanguis, ripe as the pomewater, who now hangeth like a jewel in the ear of caelo, the sky, the welkin, the heaven, and anon falleth like a crab on the face of terra, the soil, the land, the earth. (4.2.3–4.2.7)

This is the opening salvo of a barrage of bombast. The curate Sir Nathaniel compliments Holofernes on the quality of the epithets, which he feels are scholarly and "sweetly varied" (4.2.9). Perhaps they are as sweet and varied as pomewaters and crabs, where a "pomewater" is a type of large, juicy apple and a "crab" is a crabapple. One can expect varied meanings for the epithets as well.

Holofernes likens the deer to a pomewater that is like a "jewel" hanging in the sky, which means that he regards the pome as silhouetted against "the sky, the welkin, the heaven," where "welkin" refers to the "starry welkin" as in *A Midsummer Night's Dream* (3.2.356; *OED*). The Latin *caelo* in "of caelo" was originally 'celo' (Hibbard, *Love's* 4.2.5n) and is in the ablative case, whereas if it had been in the genitive case as the preceding "of" might imply, it would have meant simply 'of the sky' (*OED*). The combined English-genitive and Latin-adverbial construction suggests that a straightforward genitive assignation needs qualification. The pomewater and the deer may appear to be 'of the sky,' but in reality they are foreground objects belonging to the space between *caelum* (the firmament) and *terra* (the ground). In Aristotle's World-view, Air and Aether fill this space, the former containing such things as apples, and the latter, planets.

According to Holofernes, the deer resembles a pomewater that will soon ("anon") fall to ground "like a crab." In this temporal sequence, the pomewater will follow the course of the crabapple, which falls first. Something is awry, however, for where large ripe apples regularly fall to ground, crabapples remain on the tree and provide food for deer and other wildlife during the winter. Thus, the analogy must have another meaning.

Chronology

The naming of Navarre, the context of hunting and butchering, and Shakespeare's parody of the practitioners of peripatetic philosophy suggest that the date of the deer's demise is near Saint Bartholomew's Day—August 24, 1572. This is a Sunday. The Princess killed the buck on a Friday, which hypothetically would be August 22, 1572, so she would have arrived on Thursday, August 21, 1572. Then, having killed the deer on Friday, she plans to depart on Saturday August 23, 1572—provided of course that the necessary documents arrive so that she can complete her mission.

Just as stars are believed to host the spirits of the dead, so on this particular Saturday does the planet Saturn host the spirit of the slain deer. Saturday is "Saturn's day," and this particular Saturday is Saint Bartholomew's Eve, when plans for the Paris massacre were set into motion. The chronology recalls the association from the opening lines of the play (1.1.1–1.1.6), between hunting and Saturn whose symbol is a stylized scythe (♄), the mythological instrument of the Grim Reaper.

Astronomy supports the position. On the afternoon of Friday August 22, 1572, after the Princess has slain the deer, the Sun sets at about 19:30 local time (7:30 p.m.). The nearly Full Moon had risen a few minutes earlier; Venus, Mars, and Saturn have not yet set; and Mercury is less than 10° from the Sun and is about to set. Thus, six of the seven Ancient

Planets are in the sky just before sunset, the exception being Jupiter, which rises at about 20:25 hours (8:25 p.m.). In the nearly two-hour interval after the deer dies, all seven Ancient Planets have been or have become visible, but the script names only six of them—Sun (1.1.86), Moon (4.2.46), Mercury (5.2.1002), Venus (2.1.270), Mars (5.2.723), and Jupiter (4.2.139). Saturn is conspicuously absent from the list.

Shakespeare confirms that Saturn is the object of interest because when it sets (about 25 minutes after Jupiter rises), the constellation Cancer (the Crab) has set completely. In accordance with the postulate, the deer that is like the pomewater, and which is associated with Saturn, "anon falleth like a crab on the face of terra"—in other words, the Crab constellation sets before Saturn. This resolves the anomaly noted previously, that crab (apples) fall to ground before pomewaters.

The appearance of the sky on the next day (Saturn's day, Saint Bartholomew's Eve, August 23, 1572) is not much different from that just described. The Moon rises slightly after the Sun sets, and is essentially Full, the better to light the way for wise man *homo sapiens* to kill his kind.

THE BUCK

Immediately after Holofernes has described the state of the sky, Nathaniel describes the slain deer. "I assure you, it was a buck of the first head" (4.2.9–4.2.10) he says, which means that the slain buck had its first full head of horns and was therefore 5 years old. Apparently, the Princess has slain a "hart," but Dull says it was a pricket, which is a buck only in its second year (4.2.12n). Holofernes calls Dull's pronouncement a "barbarous" intimation (4.2.13), which is apt in light of the pending massacre. After Dull reasserts his opinion, Holofernes quotes others who believe it was a "sore" (4.2.69), which is a four-year-old buck, and although a "sore" is also a young hawk (Hulme 55), the context points to the former meaning. Whereupon, Holofernes confuses matters even further by calling it a "sorel," which is a three-year-old buck.

It is hard to believe that anyone could confuse bucks of ages 2, 3, 4, and 5 years, yet Holofernes admits that Dull's pronouncement has merit.

> HOLOFERNES Yet [it is] a kind of insinuation, as it were, in via, in way, of explication; facere, as it were, replication, or rather, ostentare, to show, as it were, his inclination, after his undressed, unpolished, uneducated, unpruned, untrained, or rather unlettered, or ratherest, unconfirmed fashion, to insert again my haud credo for a deer. (4.2.13–4.2.19)

Holofernes utters "as it were" three times, implying that Shakespeare is transferring information on another front, the nature of which must lie in the bilingual blather. It is incumbent upon the audience to discover these insinuations.

No sooner has Holofernes ended his harangue than Dull reaffirms his belief that the deer was a pricket, and Holofernes immediately confirms Dull's meaning. After Nathaniel's report that the deer was a "full head," and after Dull's two injections of "pricket," Holofernes calls Dull "monster ignorance" and says he looks "deformed" (4.2.23–4.2.24). Dull is not described as being physically deformed, but it is well known that *LLL* is renowned for lack of stage directions, so perhaps Holofernes is addressing the deer carcass that lies lifeless on the ground.

Riddles

The clergyman, Nathaniel, demeans Dull's intellect and scholarship, but the Constable puts Christian teaching into practice and turns the other cheek. He compliments the bookmen on their learning and poses a riddle, "What was a month old at Cain's birth that's not five weeks old as yet?" (4.2.41–4.2.42) Holofernes answers promptly, "Dictynna" (4.2.43–4.2.44). Nathaniel knows pagan mythology and explains that this is one of many names for the Moon. Depending on culture, location, and epoch, the Goddess of the Moon is known variously as Artemis, Astarte, Cynthia, Diana, Luna, Phoebe, or Selene; but in eastern Crete her name is Britomartis, and in western Crete, Dictynna, where "dictyon" means a net used for hunting or fishing (Graves, section 89, 2–4). Like Artemis and the others in her league, the Princess is a huntress, and it becomes clear that Shakespeare associates her with the Moon.

Dull has phrased his riddle in terms of a coincidence between the end of a lunation and the birth of "Cain," but his phraseology is deceptive. If the "Cain" in question refers to the jealous offspring of the first man, Adam, then no one can know whether such a coincidence actually occurred because *Genesis* gives no specific information on when Cain was born. Nevertheless, the solution is still readily apparent because the Moon never does attain an "age" of 5 weeks (35 days), regardless of whose birthday it is. The lunar "age" starts afresh every synodic period of 29½ days, or as Dull says, "the moon is never but a month old" (4.2.57), where a calendar month is on average 30½ days long.

Holofernes elaborates on his answer, which is a riddle in its own right.

> HOLOFERNES The moon was a month old when Adam was
> no more.
> And raught not to five weeks when he came to fivescore.
> Th' allusion holds in the exchange. (4.2.47–4.2.51)

The first line of this passage implies that the Moon "was a month old when Adam was no more [than a month old]" (4.2.47–4.2.48n). Mosaic chronology confirms this, as in *Genesis* it is stated that on the fourth day of creation, God made two great lights—the greater light, the Sun, to rule the day, and the lesser light, the Moon, to rule the night—and He set them in the firmament to give light upon the Earth. Then, it states further, on the sixth day, God created Adam. In this earliest time of recorded history, the Moon is 2 days older than Adam—in other words, when the Moon attained an age of one month, Adam was younger by about 2 days. However, does this hold "in the exchange" (i.e., does it apply to both Cain and Adam)? *Genesis* is silent on Cain's time of birth, but if the words "was no more" are interpreted to mean a postmortem state (i.e., if Adam or Cain were no longer alive), then the Moon will continue periodically to attain an age of a month after the death of either one.

The first part of the second line ("And raught not to five weeks") is independent of the rest of the line ("when he came to fivescore") because, as Dull says, "the moon is never but a month old," where "never but" means "on no occasion" or "at no time" (*OED*). Thus, the second line is true regardless of how old anyone is. This assumes that "fivescore" (i.e., 5 × 20 = 100) refers to age in years, although Holofernes never states so explicitly. Perhaps this is a hint that if one were to encounter the value of one hundred again, then one could again assign suitable units as before. The third line of the quoted passage means that the solution to the riddle applies to Adam as well as to Cain (Hibbard, *Love's* 4.2.41n).

Dull segues from Adam's age to the deer's age and, after a hiatus of 37 lines, reaffirms that "'twas a pricket that the Princess killed" (4.2.58). Holofernes decides to ignore the information imparted by Nathaniel that the deer was 5 years old and "to humor the ignorant" by calling "the deer the Princess killed a pricket" (4.2.60–4.2.62). So far, Dull has told the audience three times that the deer is a pricket, and Holofernes has added a fourth instance. Holofernes then reels off a seemingly inane doggerel (4.2.67–4.2.78) in which he refers to prickets, sorels, and sores and plays word games using the letter "L" in both upper and lower case, the former being the Roman symbol for 50 and the latter to transform a "sore" to "sorel."[2] In sum, Holofernes' explanation seems learned and profound, but in truth, it is trite and shallow as befits scholastic sophistry.

Nathaniel congratulates Holofernes on the excellence of his verse. "A rare talent," he declares (4.2.79); whereupon Dull—whose name belies his sharpness—mumbles aside, "If a talent be a claw, look how he claws him with a talent" (4.2.80–4.2.81). One spelling of "talon" is "talent," therefore Dull likens Nathaniel's fawning over Holofernes to clawing by a bird of prey. The reason for this oblique reference to falconry becomes clear as similar disparagements accumulate.

The irony continues as Holofernes explains, "This is a gift that I have, simple, simple—a foolish extravagant spirit, full of forms, figures, shapes, objects, ideas, apprehensions, motions, revolutions" (4.2.82–4.2.85). Holofernes speaks better than he knows because all eight of these could pertain to geocentric models of the Universe. In particular, they could refer to Plato's "forms" as representing "ideas" that include use of circles and spheres and the manner of their motions. Most persuasive is the eighth and final item in the list—"revolution"—which may well refer to Copernicus' *De Revolutionibus* and its plural form to subsequent advances. This would not be the first time that Shakespeare places seminal ideas in the mouths of buffoons.

Holofernes tells how he generates his bright ideas.

> HOLOFERNES These are begot in the ventricle of memory, nourished in the womb of pia mater [the brain], and delivered upon the mellowing of occasion. But the gift is good in those in whom it is acute, and I am thankful for it. (4.2.85–4.2.89)

He might as well be speaking as an empiricist adept at observing, remembering, and synthesizing data, and delivering it when the coast is clear. Especially striking is the last sentence, which sounds like an expression of humble gratitude unbecoming the pretentiousness of pedantry.

The curate Nathaniel invokes the Almighty, "I praise the Lord for you, and so may my parishioners, for their sons are well tutored by you, and their daughters profit very greatly under you" (4.2.90–4.2.93). The irony here is that the last thing that religious leaders wanted to do was expose their own eyes and those of their pupils to evidence of the New Astronomy. Concerning the curate's claim that young women would profit "under" Holofernes, the pedant replies that, "if their daughters be capable, I will put it to them" (4.2.96–4.2.97). The sexual innuendos further cast the curate and the schoolman in a bad light.

Jaquenetta and Costard enter and ask Nathaniel to read a letter that they mistakenly think is addressed to the illiterate Jaquenetta. However, this is Berowne's love letter addressed to Rosaline. Upon learning of its intended recipient, Holofernes orders the two to take it to the King.

Pupil and Sky

Berowne admires Rosaline as if she were a creature of the heavens. He calls her "celestial" (4.2.142), and in the next scene (4.3) her subtextual identity is revealed. Before the schoolmen part company, Holofernes invites Nathaniel and Dull to join him and partake of victuals at the family home of one of his pupils. There, Holofernes will undertake his *ben venuto* ("welcome" in Italian) and "prove" that the verses Berowne wrote in his letter to Rosaline are "very unlearned, neither savoring of poetry, wit, nor invention" (4.2.183–4.2.185).

Who is this pupil? In the sixteenth century, the primary meaning of "pupil" was "an orphan under the care of a guardian" (*OED*), although the meaning "student" was also relevant. It happens that Thomas Digges was both a ward and a student of John Dee (*DNB*; Johnson 157), and the joke is that Thomas gathered information telescopically through the pupil of his eye. The identity of the hosts in scene 4.2 is a mystery, but if the persistent irony in *LLL* is any gauge, then they are likely to be surreptitious supporters of the New Astronomy.

Navarre himself is not immune to Cupid's arrow and is unaware of Berowne's presence. He recites a sad sonnet bemoaning his lot, and upon the arrival of Longaville, he takes cover as Longaville starts to read a line of a poem to "sweet Maria." He calls her the "empress" of his love (4.3.54), but dissatisfied with his effort, he tears up the product of his pen and tries again, this time elevating Maria to a "goddess" (4.3.64). This provokes Berowne to mutter aside that Longaville makes a "green goose" (meaning either a gosling or a giddy girl) into a deity, which he states is "pure idolatry" (4.3.75). Longaville, in turn, takes cover as Dumaine happens upon the scene and reads an ode to "divine Kate" (4.3.84). Dumaine is unaware that his companions are in love, but wishes that they were so that he alone among the four would not be tempted to break his vow.

Misery loves company, so Longaville comes forward and announces that he is similarly stricken, whereupon Navarre, thinking that he alone has heard the two confessions of love, steps out of the shadows and asks scathingly (4.3.152), "What will Berowne say?" Berowne then reveals himself and, knowing of Navarre's lovelorn lament, reveals the King's hypocrisy.

Berowne thinks that his expression of love in the letter he sent Rosaline is safely in her hands, so feels that he can afford some hypocrisy of his own. However, the irony is that Costard and Jaquenetta are on their way to deliver Berowne's love letter to Navarre, and sure enough, the two enter at that moment. Berowne recognizes the letter they carry as the one he wrote to Rosaline. He attempts a getaway, which Navarre forestalls, and then he tries to tear up the letter, but Navarre reassembles the pieces and reveals that Berowne had authored a love letter of his own. This in turn reveals Berowne's hypocrisy.

With his love for Rosaline now known, Berowne waxes overly eloquent in his praise of her and likens her to the Sun. He asks rhetorically,

> BEROWNE What peremptory eagle-sighted eye
> Dares look upon the heaven of her brow
> That is not blinded by her majesty? (4.3.246–4.3.249)

Supposedly, only eagles could stare at the Sun (Harting 24–25); otherwise, Berowne has misspoken because the Sun is the province of male royalty and Rosaline's supposed "majesty" does not blind anyone. Navarre corrects Berowne by telling him that Rosaline's mistress, the Princess, is "a gracious moon" (4.3.250), and because she herself is a lesser luminary than the Princess is, she is not nearly the equal of the Sun.

In the pyramidal structure of early modern society, the Princess is appropriately associated with the goddess of chastity and of the hunt. This association is remarkable, as by the mooted chronology, on the morning of Saturday August 23 (when she plans to leave Navarre for Paris) the Moon is Full at about midday.

According to the King, Rosaline is "an attending star scarce seen [as] a light" (4.3.251). No star or satellite is physically associated with the Moon, so the King probably refers to a planet (which in olden days, as has been noted, could be called a "star"). It must be fainter than the Moon and an attendant of some sort. Mercury and Venus never stray far from the direction of the Sun and are normally fainter than the Moon, and of these two, Venus is female. This implies an association between Rosaline and Venus. In addition, the Latin *venus* means "grace" or "charm," and when the King calls the Princess "a gracious moon," he associates with her both *venus* (grace and charm) and Venus (Rosaline as an attendant).

At a masque in scene 5.2, the Princess and Rosaline disguise themselves each as the other, inviting confusion between "Venus" and the "Moon," as if each had properties of the other. In 1610–1611, Galileo revealed that Venus reflects sunlight and shows a full range of phases from Full to New (figure 1.14). If Shakespeare intended to ascribe the lunar property of phases to the planet Venus (i.e., to imply that Venus is not self-luminous and has phases like the Moon), then it would be expected that he would remove doubt because when he wrote *LLL*, the fact would become news to everyone.

Berowne persists in misattributing solar properties to Rosaline. He insists that if it were not for her, "day would turn to night" (4.3.253), and he extols her beauty by likening it to the duration of the Sun above the horizon (and hence to its apparent daily cycle). Because Shakespeare makes it clear that Berowne errs in associating Rosaline with the Sun, he probably refers to some other cycle of shining, and again Venus is a possibility. The Latin *venustas* is closely related to *venus* and means charm, grace, and beauty. In later Roman times, "beauty" became a property of the goddess Venus that she shared with her Greek equivalent, Aphrodite. This further supports the association of Rosaline with Venus.

When Berowne adds, "O, 'tis the sun that maketh all things shine!" (4.3.266), he gainsays the Aristotelian position that the Sun illuminates only the Earth and Moon and that everything else shines by the element Fire. Unwittingly, he eliminates this element as the source of light for all other Ancient Planets. This means that the remaining six are inherently "dark," or non-self-luminous. If this is Shakespeare's meaning, there should be textual confirmation here as well.

Unexpectedly, Navarre tells Berowne, "By heaven, thy love is black as ebony" (4.3.267). Navarre asserts that Berowne's love, Rosaline, is like a heavenly body that is as dark as ebony is black, which confirms that the Sun illuminates Venus, goddess of love. In the script that follows, "black" occurs six more times, interspersed with four mentions of "fair." It seems contradictory to claim that something that shines can be both black and fair, and indeed Berowne ponders how someone with a "fair cheek" (4.3.255) can be "black as ebony" (4.3.267). Berowne's pronouncement that "No face is fair that is not full so black" (4.3.273) denies that a dark object can shine and still be "fair," or perfect, which prompts the King to call the apparent contradiction a "paradox" (4.3.274).

Perfection is a unique condition, and to call an object "fair" when it is both light and dark is to beg the question of the proportions needed to attain that state. Shakespeare trusts that the audience sees the resolution of the "paradox," which of course is to dispense with the Aristotelian notion of heavenly "perfection" and to allow the illuminated hemisphere of Venus to be "fair" while the other is "black" (figure 1.14). It is known today that Venus is cloud covered and featureless when seen in the yellow-orange light of the Sun, so to a sixteenth-century telescopic observer, the illuminated hemisphere would look "fair."[3] Navarre's utterance, "beauty's crest becomes the heavens well" (4.3.276), may allude to beauty's "*cres*[cen]*t*" and be a veiled reference to Venus in its crescent phase.

In a hotly debated passage (Hibbard, *Love's* 4.3.251–4.3.252n), Navarre says that black is "The hue of dungeons and the school of night, and beauty's crest becomes the heavens well" (4.3.275–4.3.276). The "school of night" may refer to a shadowy group of poets and playwrights that included the "Wizard Earl" Henry Percy (1564–1632), 9th Earl of Northumberland, and the mathematician Thomas Harriot (1560–1621), who tutored the members in astronomy (Hall 242). The group was sometimes labeled the "School of Atheism," perhaps because in the early modern age, curiosity about Nature hazarded charges of godlessness.

Nevertheless, Berowne still insists that his Rosaline is as fair as the Sun. "I'll prove her fair, or talk till doomsday," he says (4.3.294).

When Berowne discovers that he is wrong (4.3.295), Navarre says, "No devil will fright thee then so much as she." Berowne errs because, as discussed earlier, Rosaline is not associated with the Sun but rather with Venus, which puts him in peril of having to talk until kingdom come. Evidently, those who misappropriate godly properties and who are ignorant of the New Astronomy belong in the Devil's house. Through a process of logical inference from empirical evidence, Navarre has disputed Berowne's violation of canonical myth and his doctrinaire devotion to received wisdom, and has delivered significant facts in support of the New Astronomy.

At the close of Act 4, the would-be suitors gather their wits and resolve to entertain their inamoratas in "the afternoon" of that day (4.3.370), which is Friday, August 22, 1572.

Fit for a Queen

Act 5 opens with Armado telling Holofernes and Dull that Navarre has asked them to stage a show for the Princess that same afternoon (5.1.90), and Holofernes decides that they will present the pageant of the Nine Worthies. The last scene (5.2) opens with the Princess showing off a diamond that Navarre sent her along with screeds of poetry proclaiming his love. The topic of Rosaline's subtextual identification resurfaces as she converses with Katherine and announces in reference to Navarre, "You'll ne'er be friends with him. He killed your sister" (5.2.13–5.2.14). From the fourteenth century, "sister" can signify a mythological or imaginary being (*OED*), so the comment could mean that Navarre "killed" a metaphorical "sister" of Katherine. Maria and Rosaline are sisters, but they are not dead. Instead, as revealed previously, Navarre killed what Berowne imagined was Rosaline's representation, namely her identification with the Sun.

In the next line, Katherine explains that the King turned Rosaline's celestial persona into something "melancholy, sad, and heavy," following which, "she died" (5.2.15). "Melancholy" means dark and gloomy as in "shade of melancholy boughs" (Bevington, *As You Like It* 2.7.110).

"Sad" means dark colored or gloomy as in "sad habiliment" (Bevington, *Titus Andronicus* 5.2.1), and "heavy" means dark and gloomy as in "heavy night" (Bevington, *Othello* 5.1.43). Melancholy is a Saturnian trait, and of all the planets, Saturn receives the least amount of sunlight. Here in the same vein, the descriptors apply to the dark hemisphere of Venus, and what has died is Rosaline's erroneous association with the Sun.

Katherine tries to define Rosaline further using an amalgam of literal and figurative conceits.

> KATHERINE Had she been light like you,
> Of such a merry, nimble, stirring spirit,
> She might ha' been a grandam ere she died.
> And so might you, for a light heart lives long. (5.2.16–5.2.19)

Rosaline is suspicious and asks, "What's your dark meaning, mouse, of this light word?" (5.2.20–5.2.21) The cattiness continues as Katherine contends that Rosaline is, "A light condition in a beauty dark" (5.2.22), as if light were value added by the Sun. Rosaline calls Katherine a "mouse," which can be a term of endearment but also implies weakness and timidity (*OED*), and Katherine has retaliated by calling Rosaline wanton and unchaste (Hibbard, *Love's* 5.2.20n). To this Rosaline responds, "We need more light to find your meaning out" (5.2.23). Katherine offers to snuff out the altercation by ending the argument "darkly" (5.2.25), but the clawing persists for two more rounds (5.2.26–30) before the Princess terminates it. Shakespeare again contrasts "light" and "dark" in order to emphasize that Rosaline and planet Venus do not meet the standards of the old ideal of heavenly perfection.

Rosaline too has received a gift from her adorer along with verses and a sketch that she states is "fair as a text B in a copy book." Rosaline hints at her state of celestial presumption by proclaiming "B" to be her "red dominical" and "golden letter" (5.2.47). A dominical letter B establishes that Sunday is the second day of the year, which makes Saturday the first day of Rosaline's year. Thus, her "golden letter" leads to the day named for Saturn and its glyph ♄, where gold or yellow is the color of Saturn. Rosaline continues in her role of planetary claimant, but now she is staking a claim on sad Saturn. Surely, a rebuke is in order.

Rosaline continues as if still addressing Katherine, "O, that your face were not so full of O's!" (5.2.48). The exclamation "O" is "a signifier in search of something signified" (Woudhuysen 21). Possibly, Rosaline describes Katherine's face as pockmarked, but stage directions are sparse in *LLL*, and it is reasonable to suppose that she is addressing the Princess instead. In fact, it is the Princess who responds. "A pox of that jest!" she says (5.2.49). It appears that the Princess takes the remark personally. She is "a gracious moon" (4.3.250) and it is plausible that the O's signify pockmarks on her face, connoting craters on the Moon's face.

Figure 2.1 is an image of the Moon recorded by Gainer using his reconstruction of the Digges telescope, showing craters on the surface.

FIGURE 2.1. The Moon as seen through Gainer's reconstruction of the Digges telescope (exposure 1/20 sec, ISO 100, 5-megapixel camera, good seeing).

It is likely, therefore, that Shakespeare had access to this knowledge. In *LLL*, Rosaline is the means by which Shakespeare describes telescopic results on the appearance of Venus and the Moon, fifteen to sixteen years before Galileo's announcements thereof.

Boyet declares that the four Navarrean gallants have disguised themselves as Russians and will arrive soon, whereupon the ladies swap favors in order to confuse them. The suitors arrive and the ladies mask their faces. Berowne asks the Princess to remove her mask and "show the sunshine of [her] face" (5.2.212), referencing the illuminated lunar hemisphere. The Princess and Rosaline have swapped identities, so when Berowne attributes sunshine to the Princess, he is in effect repeating the error of scene 4.3 by attributing sunshine to Rosaline. Ironically, Rosaline is sufficiently astute to know that she is supposed to represent the Princess who represents the Moon, and she responds, "My face is but a moon, and clouded too" (5.2.214). She speaks ironically for herself too, as she represents Venus, which has phases like the Moon.[3] Rosaline confirms that she represents Venus when she tells Navarre, "Thus change I like the moon" (5.2.224), meaning that Venus has phases like the Moon.

Boyet warns the ladies that the nimrods of Navarre will return to try again. When they arrive, the Princess reveals that the ladies saw through the men's disguises, whereas they did not penetrate theirs. The exchange of favors deceived the suitors, and they mistakenly wooed the wrong ladies, the moral of the deceit being (one supposes) that appearances are deceptive.

Armado and his entourage attempt to stage the pageant of the Nine Worthies by dividing the nine roles between "the pedant Armado, the braggart Holofernes, the hedge-priest Nathaniel, Costard the fool and Moth the boy" (5.2.596–5.2.597). This prompts Berowne to allude to the game of dice, *novem quinque* ("nine five") (5.2.598LN). "Abate throw at novum, and the whole world again / Cannot pick out five such, take each one in his vein," he says, because five players will play nine roles.[4]

Mercury

In scene 2.1, which occurs on Thursday August 21, 1572, Boyet had assured the King of Navarre that the acquittances necessary to prove the French claim to Aquitaine would arrive the following day, but it is now Friday evening and nary a document is in sight. A messenger from Paris does arrive, but not with the requisite proof. Instead, he brings news of the death of the French King. The courier's name is Marcade, which is probably a conflation of "macabre" and "Markedy"—a mispronunciation of "Mercury," the mythical messenger of the Olympian gods (Graziani 392–394).

The death occurred after the Princess left, yet this messenger has traveled faster than the one who is supposedly carrying the documentary evidence. The latter's absence may be due to any number of factors, but the circumstances of this muddle are worth a second look. The Princess is highly intelligent, and when she gave Navarre the letter requesting transfer of Aquitaine to France, she may have believed that it was sufficient to make the case because when Navarre asked her for proof of payment, she does not reply. If she had had a ready answer, she would have given it; instead, the diplomat Boyet chimes in with a lame excuse that a carrier will soon arrive bearing the proof. The delinquent courier may be a figment of Boyet's imagination, which raises the possibility that the proof of payment never existed to begin with.

The death of the French King apparently resolves the French party's dilemma, although Shakespeare never explains how—"supposing he ever knew" (Hibbard, *Love's* 5.2.727n). Saturn is the *deus ex machina* that with one artful sweep of his scythe has resolved the dispute, and Shakespeare does not need to explain further because from the outset he has emphasized the role of the god of Time.

To the disappointment of the suitors, the Princess resolves to depart that very night, but she promises Navarre that if he wishes, he may resume his courtship after she has observed the customary one-year period of mourning (5.2.884–5.2.887n). In the meantime, she advises him to repair to "some forlorn and naked hermitage" (5.2.872) to repent his perjuries. Rosaline imposes a similar stricture on Berowne, and the other two nobles are equally outdone.

SONGS

The play ends with two songs, one in praise of the cuckoo (representing the season of spring) and the other of the owl (representing winter). Through the agency of the cuckoo, spring's song "mocks married men" (5.2.973). The cuckoo lays its eggs in other's nests, and lends its name to the "cuckold," who is someone with horns supposedly protruding from the forehead and suffering from disaffection. The gender allusion is to the Princess' skill at pointing and shooting a stag with a full head of antlers. She is not about to make a snap decision on a "world-without-end bargain" (5.2.866) like matrimony, and she especially wishes to ensure that Navarre has wised up before she admits his overtures.

By contrast, winter's song concerns the cosmic subtext. It has the owl calmly watching and enduring the rigors of the cold, singing "Tu-whit to-whoo" on a "merry note" (5.2.992, 5.2.1000). The owl is the emblem of wisdom and is the bird associated with Athena and her Latin equivalent, Minerva, goddess of wisdom. The relevance of the owl's happy refrain becomes clear upon hearing the last lines (which appear in the quarto edition of *LLL*), "The words of Mercury are harsh after the songs of Apollo" (5.2.1002–5.2.1003) because *a priori* the messages of Mercury can clash with Apollonian ideals.

Apollo plays a central (albeit invisible) role in the plays analyzed in this present book. The name 'Apollo' may derive from *abol* meaning "apple," the symbol of wisdom. This would be appropriate both because of the role of crabapple and the pomewater in *LLL*, and because Apollo is the enemy of ignorance and barbarity, and champions "music, poetry, philosophy, astronomy, mathematics, medicine" (Graves, section 21.10), and moderation in all things. The new Queen intends to promote these causes, and thus it is she and not Navarre who will become a philosopher-monarch. In contrast to this ideal are the harsh realities of mundane existence against which all ideals must compete, making the apparently farcical *LLL* a tale of the struggle of knowledge over ignorance.

The Folio edition adds at the end, "You that way. We this way," but there are no stage directions telling players which way to go (5.2.1003n). Arguments presented here suggest that Armado, Holofernes, Nathaniel, and Dull follow the cuckoo while the new Queen and her retinue follow the owl, trailed perhaps by Moth. The tragic aspect in this romantic comedy lies in the audience's knowledge that the god of Time can only do so much, for even if the Princess departs on Friday evening, she and her Apollonian spirit will not arrive in Paris in time to forestall the fratricide that began in the early morning of Saint Bartholomew's Day, 1572.

Last Laugh

As for Navarre, the Queen expressly forbade him from following her, and it would be undignified for him to follow the foolish five, so for comedic effect, Navarre and his nobles are left dithering on stage trying to decide which way to go. Yet the audience cannot wait "twelvemonth and a day" for the worthies to decide what to do because (as Berowne opines) a year is "too long for a play" (5.2.950–5.2.952). Suitably acted, the suitors' confusion could provide the last laugh of the play.

Navarre himself is on the horns of a dilemma, yet has little choice but to follow the Queen's suggestion and repair to a hermitage. He had intended to lead a Spartan existence anyway, and lest he and his hangers-on should lose their labors of love, they must carry on as they had previously planned, though wiser now thanks to the influence of the Princess. Navarre's plan to make his court "the wonder of the world" and a "little academe" (1.1.12–1.1.13) has shifted emphasis away from pedantry. Berowne seems likely to provide the local impetus for progress because he at least "forswears linguistic affectation…in favor of plain speech" (260) and has opposed uncritical assimilation of other's thoughts (1.1.74–1.1.95). The Huguenot martyr Petrus Ramus would approve.

Members of the audience who see beyond appearances would have mixed feelings because they too must decide which group to follow.

Their quandary and the indecisive ending of the play express "an unfinished process" (Bevington 33), perhaps because Shakespeare intended to pose the question of whether the struggle for Knowledge over Ignorance and of Good against Evil will ever end, or whether, like the cycles of the celestial machinery, it will persist forever. The lack of a resounding climax is itself a crux of the play, as it forewarns playgoers to anticipate further developments.

Shakespeare emphasizes the crux of time by making the last scene the longest of the Canon. The emphasis is consistent with the thesis that, as his writing career progresses, Shakespeare returns to some of the topics previously broached. For example, the tension between winter and spring recurs in *The Winter's Tale*, which also contains the Canon's second-longest scene and which, like *LLL*, is concerned with Saturn and time. Another crux is Shakespeare's apparent knowledge of craters on the Moon and phases of Venus, which points to telescopic observations in or before 1594. These data agree with those in *Hamlet*, which I study next.

Chapter 3

Hamlet

Science, and her sister Poesy,
Shall clothe in light the fields and cities of the free!

—Percy Bysshe Shelley

In *Hamlet's Universe* (*HU*), I argued that Shakespeare's most famous play is a cosmic allegory that relates the competition between the four chief cosmological models extant at the turn of the seventeenth century. Shakespeare (*Hamlet*) inserted multiple and hitherto unexplained conceits that describe properties of the Sun, Moon, planets, and stars that Shakespeare could not have known without telescopic aid. This chapter follows the script of *Hamlet* and presents the cosmic subtext by addressing matters more or less as they arise in the script, and additionally by expanding on the cosmic interpretation in *HU* in ways that I had not discovered between 1996 and 2006. These further developments include a chronology of Act 1 and the significance of the most famous skull in literature, the skull of Yorick.

AMLETH AND HAMLET

It is useful first to summarize the chief literary source of *Hamlet* and then follow the parallel development of the textual and subtextual plots. In *Historia Danica* ("History of Denmark"), which was written circa 1200 (Saxo Grammaticus 1: 208), Saxo Grammaticus (*fl.*1188–1201) chronicled the exploits of the legendary Prince Amleth, who affected a mental disorder as a means to survive the malice of his King. Shakespeare modeled Hamlet on Amleth; in particular, he contrasted Hamlet's feigned madness with the deceitful methods of the King and his court.

The relevant part of *Historia Danica* begins when Amleth's father, Horvendile, King of the Jutes, duels with Koll, King of Norway. Prior to the duel, they had agreed that if either were maimed then he should perish, as he could never live with the humiliation. Horvendile severs Koll's foot and, true to their pact, Horvendile kills him. Later, for good measure, he kills Koll's sister too, only to die at the hands of his brother, Feng. "Incest" caps "unnatural murder" when Feng marries his murdered brother's widow, Geruth.

Amleth feels threatened by Feng, who is now both his uncle and his stepfather. Amleth feigns madness and resorts to doublespeak in order to protect himself, but Feng suspects him of cunning and hatches a plot to lure him into the presence of a "fair woman." Amleth, alerted to the scheme, plans accordingly. As luck would have it, the temptress had been Amleth's childhood companion and is still fond of him, so they make love. To keep Feng from learning of this, Amleth swears the damsel to secrecy and later describes their meeting equivocally.

In a second test, a spy watches as Amleth visits his mother in her chambers, but Amleth kills the spy and chastises his mother. Feng is convinced of Amleth's guile, so he sends him to Britain with two companions who carry a message ordering Amleth's execution. Amleth alters the message that the guards carry, and the British monarch cautiously decides to await developments. Amleth demonstrates his worth by perceiving correctly that something is rotten both in the comestibles of a royal banquet and in the royal lineage. When the King discovers that

Amleth is right on both counts, he accepts Amleth's word as if it were divinely inspired. Amleth proves to have such exceptional qualities that the King decides to follow through with the altered message and put Amleth's guards to death, whereupon Amleth returns to his native land and liquidates his uncle's henchmen and then Feng.

Table 3.1 shows the relationship between Saxo's and Shakespeare's characters. The death of Koll at the hands of Horvendile is the basis for Old Fortinbras' death at the hands of Old Hamlet. The deaths of Rosencrantz and Guildenstern duplicate the execution of Amleth's guards. Feng and Geruth are associated with the Danish royal couple Claudius and Gertrude. For reasons soon to be explained, Shakespeare drops the name Feng in favor of Claudius but retains a variant of Geruth.

Amleth's killing of Feng corresponds to Hamlet killing Claudius. Saxo's British King believes in divine direction, which matches the

Table 3.1. Character identifications in *Hamlet*.

Shakespeare	Saxo	Identification
Claudius	Feng	Ptolemy
Old Hamlet	Horvendile	Leonard Digges
Hamlet	Amleth	Thomas Digges
Rosencrantz, Guildenstern	guards	Tycho Brahe
Gertrude	Geruth	—
Ophelia	a fair maiden	—
Horatio	—	Horace
Barnardo	—	Bernardus Silvestris
Francisco	—	Francesco Petrarca
Marcellus	—	Marcellus Palingenius
Polonius	—	Robert Pullen
Reynaldo	—	John Reynolds
Laertes	—	Thomas Harriot
Osric	—	Walter Raleigh

proclamation of Hamlet's friend Horatio that "Heaven will direct it" (Edwards, *Hamlet* 1.4.91). (In this Chapter, unattributed notes, lineation, and pagination refer to Edwards' *Hamlet*, and other editions of Shakespeare's plays are also identified by editors' names.) In the literal storyline, Claudius kills his brother Old Hamlet and usurps the throne, and shortly thereafter violates the laws against incest by wedding Old Hamlet's widow, Gertrude. From the new King's perspective, this gives him license to bring her son Hamlet under his sway.

The usurper King's name is Claudius, but the script refers to him simply as "king," perhaps because Shakespeare wanted the audience to attend especially to the name. I posit that Claudius represents his namesake Claudius Ptolemy, who was the last Greco-Roman astronomer of note and avatar of the bounded geocentric model of the Universe. Table 3.1 lists this and other relationships between the characters in the play, and table 3.2 lists characters who personify the chief models of the Universe in contention at the turn of the seventeenth century.

Wittenberg was the first academic center to include a Sun-centered planetary system in its teaching, a fact that prompted speculation that its mention in *Hamlet* means that the play is astronomically significant (Payne-Gaposchkin 162). In the play, the King of Denmark, Claudius, seeks to counter the radicalism of Wittenberg and thus preserve Elsinore as the seat of bounded geocentricism. Polonius is a pedant and defender of Aristotelianism whose death foreshadows the end of scholasticism. The death of the twin courtiers is Shakespeare's way of killing the

TABLE 3.2. Model personifications in *Hamlet.*

Role	Model	Builders
Claudius	Bounded geocentric	Claudius Ptolemy
Rosencrantz Guildenstern	Bounded hybrid	Tycho Brahe
Hamlet	Heliocentric unbounded	Nicholas Copernicus Thomas Digges

Tychonic model, and the slaying of Claudius represents the triumph of the New Astronomy over the Old.

Francisco, Barnardo

The opening scene of *Hamlet* is set on a high terrace of Elsinore (i.e., Kronborg) Castle, which lies in the town of Helsingor (Elsenor). The play begins just before midnight. Francisco is the sentry on duty, and when Barnardo arrives to relieve him of his watch, the two have trouble identifying one another. It will be revealed that the reason for this difficulty is that there is no Moon in the sky. Barnardo asks, "Who's there?" but Francisco demands that the newcomer announce himself first. "Nay answer me" (1.1.1–1.1.2), he commands. These words introduce the issue of identity from the start.

Barnardo's name is a variant of "Bernardo" (Andrews 1.1.0SD). I have argued that Barnardo represents Bernardus Silvestris and that Francisco represents Francesco Petrarca (Petrarch) (1304–1374). Bernardus was an early proponent of a moving Earth and his major work, *Cosmographia*, is an excellent fit to the *Hamlet* subtext. Petrarch was a humanist with a literary style rivaling that of Dante (1265–1321) but whose works are comparatively free of medieval scholasticism (Dampier 98).

Francisco is on duty when Barnardo enters (Hibbard, *Hamlet* 1.1.0n), which establishes the joint literary, humanistic, and scientific contexts. Barnardo is the first to speak. He notes that it is midnight and suggests that Francisco get himself to bed. This effectively prepares the way for the entrance of Marcellus, another herald of the New Astronomy who represents the literary figure Pietro Angelo Manzoli (*c.*1500–1543), better known as Marcellus Palingenius Stellatus, the Stellified Poet (table 3.1).

Marcellus

Zodiacus Vitae ("The Zodiac of Life") by Marcellus Palingenius first appeared in 1531. The epic poem is a compendium of contemporary knowledge of the World in which each of its twelve books bears the name of a Zodiacal constellation. After Barnaby Googe (1540–1594) had translated it, the poem became popular throughout England and helped pave the way for challenges to the old Worldview (Johnson 145). In 1559 the poem made it onto the first "Index of Prohibited Books," but luckily, Palingenius had died sixteen years earlier.

In Book VIII of *Zodiacus Vitae*, subtitled *Scorpius*, Palingenius wrote,

> Whatsoever Aristotle saith, or any of them all,
> I pass not for: since from the truth they many times do fall.
> Oft prudent, grave, and famous men, in errors chance to slide,
> And many wits with them deceive when they themselves go wide.
> (as cited in Johnson 147)

Aristotle was certain of the inerrancy of his thinking, a trait that Palingenius denies in his own.

> Let no man judge me arrogant, for reason ruleth mee,
> She faithful guide of wisemen is: let him who seeks to find
> The truth, love her, and follow her with all his might and mind.
> (as cited in Johnson 147)

Book XI, subtitled *Aquarius*, deals primarily with astronomy. According to Gabriel Harvey (*c.*1545–1630), Thomas Digges memorized the whole of it and took "mutch delight to repeate it often" (Johnson 162–163). Palingenius wrote, "Where reason is away, / No faith nor credit must we give, to words that men do say," (Palingenius 213) and he continued in a polemical vein.

> What store of fond Foolosophers and such as hunt for praise,
> The earth brings forth, it is not good to credit all he says,
> Though great his estimation be in the mouths of many men,
> Though many reams of paper he hath scribbled with his pen,
> For famous men do oftentimes make great and famous lies.
> And often men do miss the truth though they be ever so wise.
> (Palingenius 215–216)

He reiterated the need for reason.

> Therefore must reason first be sought. For in such doubtful things,
> More credit reason ought to have than men's imaginings,
> For such are often proved false. (Palingenius 215–216)

In effect, Marcellus Palingenius recognizes the need for empirical verification of theory no matter how esteemed the theoretician. In *A Perfit Description*, however, Thomas Digges gives Aristotle the benefit of the doubt because he believed that the Stagirite philosophized to the best of his ability; instead, he faults his uncritical followers.

Digges admired Palingenius also because he contrasted celestial appearance and reality. In *Aquarius*, Palingenius wrote,

> All stars are not of bigness like, for many less there be,
> And in such sort, as comprehend no men may them we see.
> Some are again of larger size, in number few and fine,
> That in clear nights amidst the skies with gorgeous light do shine.
> (Palingenius 214)

Palingenius allows stars to have various sizes, which is the standard contemporary explanation for why brighter stars seem larger to the naked eye. It is known today that this perception ignores the characteristics of the receptor (the eye) and the circumstances of the observations. Nevertheless, Palingenius is right in the sense that the larger the star, the greater the surface area from which it radiates, and therefore the more light it can emit. He concludes that some stars are too small to be visible, which is a remarkable insight as it foresees the revelation that some parts of the macroscopic World are inaccessible to direct sensory perception.

Palingenius had a qualitative grasp of the concept of radiant power flowing from the surfaces of stars, but he went further.

> Some do in compass far exceed both seas, and earth, and all,
> And bigger are their shining globes, though they do seem so small
> Because so far from us they be. For everything else beside,

> The farther it is from our eyes, the less in sight is spied,
> And doth deceive the lookers on. (Palingenius 214)

He shifts attention from the surface of a star to the aperture of the eye, and concludes correctly that the further away a star is, the dimmer it seems. This is a qualitative assertion of the law of flux that Kepler enunciated in the seventeenth century (Caspar 145). In addition, Palingenius says humans are deceived into thinking that differences in stellar brightness are solely intrinsic (i.e., due to their size) and independent of their distance.

In effect, the history of astronomy relates the struggle to transform appearances on the two-dimensional surface of the sky into the three-dimensional picture of stellar space. Although Palingenius asserts the finiteness of the physical Universe and fills the rest of space with divine light, he prepares the way for the Diggesian World model and the feature of the New Astronomy that embraces the idea of an infinite Universe of stars by positing that stars can lie at different distances.

HORATIO

Returning to the play, on the previous two mornings, Marcellus and Barnardo had seen a ghost hovering over the battlements, and Marcellus had left to seek out Horatio in order that he too might witness the spectacle. On his return, Marcellus calls out, "Holla, Barnardo!" (1.1.18). Barnardo asks, "What, is Horatio there?" and Horatio answers, "A piece of him" (1.1.19). It turns out that Horatio has arrived from Wittenberg, Germany—home to the Lutheran reformation and the first academic home of the Sun-centered cosmology of Nicholas Copernicus. Heliocentricism received its first major impetus at the end of 1541 when Rheticus returned to Wittenberg from Frauenburg, Poland, where he had studied under Copernicus, and had arranged for the publication of Copernicus' seminal work, *De Revolutionibus*. The "piece" of Horatio that arrives from Wittenberg is the heliocentric component of the New Astronomy (figure 1.4).

Shakespeare makes Marcellus part of the action before the play starts, but introduces Horatio only after it starts. This is the same order in which the corresponding key publications—*Zodiacus Vitae* (1531) and *De Revolutionibus* (1543)—appeared. Together, Marcellus and Horatio represent the germinal components of the New Astronomy: an infinite Universe and a heliocentric planetary system (figure 1.7).

On the previous two nights, the specter hove into view at the same hour (1:00 a.m.), and now on the third night, Marcellus arrives with Horatio shortly before that time, as if confident of the hour. Marcellus has asked Horatio to keep watch with them so that if the phenomenon recurs, Horatio can elicit a response from it, but Horatio is skeptical. "Tush, tush, 'twill not appear," he says (1.1.30). Ensuing events prove Marcellus correct and Horatio wrong.

Ghost

Horatio asks Barnardo to describe the phenomenon, and Shakespeare emphasized the significance of time by commencing the description with an incomplete line:

> BARNARDO Last night of all,
> When yond same star that's westward from the pole
> Had made his course t'illume that part of heaven
> Where now it burns, Marcellus and myself
> The bell then beating one — (1.1.35–1.1.39)

At this moment the Ghost appears, rendering Barnardo speechless.

The Ghost walks on Barnardo's watch and reappears from the direction of a bright star. The play concerns natural and preternatural existence, and the fact that three apparitions occur from a precise direction in the night sky at a precise time ("jump at this dead hour," 1.1.65n) suggests an examination of events from a calendrical and astronomical perspective.

The "pole" to which Barnardo refers is an imaginary point in the sky called the North Celestial Pole, toward which the rotational axis of the Earth points. To all observers in the Northern Hemisphere, the sky appears to rotate about this point. The north-south meridian connects the observer's zenith to the North Celestial Pole and drops straight down upon the northern and southern horizons. Thus, facing north, any star that lies to the left of the north-south meridian is reasonably "westward of the pole." Furthermore, when Shakespeare writes of a "star that's westward from the pole," he refers directly to Thomas Digges, who was a star mathematician in England.

Barnardo identifies the star in question in such a casual way as to suggest that it is conspicuous compared to other stars that lie westward of the pole. Payne-Gaposchkin notes that Shakespeare must have seen the "New Star" of November 1572 (162), and Olson, Olson, and Doescher have suggested that the passage quoted earlier refers to it. This New Star erupted suddenly in the direction of the constellation Cassiopeia, which lies about 30° from the North Celestial Pole and is westward of the meridian in the early morning in November. Other suggestions for the identification of the star exist in their own interpretive framework.

The New Star of 1572 is now known popularly as "Tycho's supernova" after Tycho Brahe, who published his research in *De Stella Nova* ("On the New Star") in 1573.[1] Tycho's book is better known than Thomas Digges' publication of the same year, which has a quaint title—*Alae seu Scalae Mathematicae* ("Mathematical Wings or Ladders")—that gives no hint that it contains data on the same topic. The apparent brightness of the New Star rivaled that of the planet Venus at its brightest, at which time it was ten times brighter than Sirius, the brightest star in the sky. It was then effectively the third brightest object in the sky after the Sun and the Moon, and was even visible during the day. Its advent was evidence that the heavens undergo change, when previously under the Old Astronomy, astrologers, philosophers, and theologians believed that the heavens were immutable.

Marcellus believes that the Ghost is more likely to speak to Horatio than to a common soldier because Horatio is a scholar from the University of Wittenberg. Identity is an issue for Horatio too as he interrogates the apparition and notes similarities to Old Hamlet.

> HORATIO What art thou that usurp'st this time of night,
> Together with that fair and warlike form

In which the majesty of buried Denmark
Did sometimes march? (1.1.46–1.1.49)

Horatio charges the Ghost to speak, but it stalks silently away. Two earlier lines (1.1.19, 1.1.21) also begin with "What" and emphasize the question of identity. Under the hypothesis of cosmic allegory, the Ghost represents the spirit of Leonard Digges (table 3.1).

The spectacle puzzles Horatio.

HORATIO In what particular thought to work I know not,
But in the gross and scope of mine opinion
This bodes some strange eruption to our state. (1.1.67–1.1.69)

Horatio does not know in which particular area to concentrate his thoughts, but on "taking a wide view" and allowing all possibilities, he supposes that the apparition serves as an omen of change (1.1.67–1.1.68n). This change will affect both political and celestial outcomes.

Marcellus and Horatio agree that the Ghost resembles the recently deceased King, Old Hamlet, and Horatio identifies the spirit's armor as the very same that Old Hamlet wore when he combated the King of Norway. He connects the apparition with what he knows about the current ambition of Young Fortinbras, who has resolved to seek the return of lands forfeited to Denmark upon the death of his father. Thus, Horatio correctly foresees parallel upheavals in politics, particularly the replacement of the false King by a true heir to the throne, which translates in the cosmic allegory to the replacement of the false center of the planetary system with the rightful one—the royal Sun.

One of the differences between F1 and Q2 is the passage 1.1.108–1.1.125, which appears only in Q2 and is one of many that are closely associated with the cosmic subtext.[2] Horatio declares that the apparition is "a mote to trouble the mind's eye," meaning that just as an irritant in the eye prevents one from seeing clearly, so the image of the Ghost "perplexes the mind" and confuses insight (1.1.112n). It is a symptom of the ontological challenge implicit in the Diggesian Worldview, which is that of reconciling physical and metaphysical infinities.

Horatio contributes to the subtextual storyline by naming portents of change.

> HORATIO As stars with trains of fire, and dews of blood,
> Disasters in the sun; and the moist star,
> Upon whose influence Neptune's empire stands,
> Was sick almost to doomsday with eclipse.
> And even the like precurse of feared events,
> As harbingers preceding still the fates
> And prologue to the omen coming on,
> Have heaven and earth together demonstrated
> Unto our climatures and countrymen. (1.1.117–1.1.125)

The "moist star" is the Moon, and its eclipse is as ominous as comets with their "trains of fire." The passage "Dews of blood, / Disasters in the sun" form an incomplete sentence, but the plural "dews" and "disasters" occurring in a solar context suggest a multiplicity of phenomena. Thus, the reference is probably not to solar eclipses because the parallel adumbration is to the singular event of a lunar eclipse ("the moist star… sick…with eclipse").

A possible explanation is that the plural "Disasters in the Sun" refers to sunspots. Owing to laws governing thermal equilibrium between matter and radiation, they look like dark holes in the Sun's skin (figure 3.1), so much so that William Herschel believed that they were actual holes through which observers could see a dark solar interior. In mythology, puncture wounds to the Sun would qualify as hazardous to a monarch's well-being because the Sun is the embodiment of royalty, and in fact, in the end all three aspirants to the throne—Hamlet, Laertes, and Claudius—suffer puncture wounds.

The Q2 and F1 texts resume with the reentry of the Ghost. Horatio orders it to stay and speak, and it seems about to do so when a crowing cock startles it. Horatio commands Marcellus to use force to prevent it from leaving, but Marcellus flails at the phantom without effect. In ordering Marcellus to arrest the Ghost's flight, Horatio has assumed that it is susceptible to physical force. This and Horatio's earlier prediction that the specter would not reappear are elementary examples of

FIGURE 3.1. Image of the Sun showing sunspots.

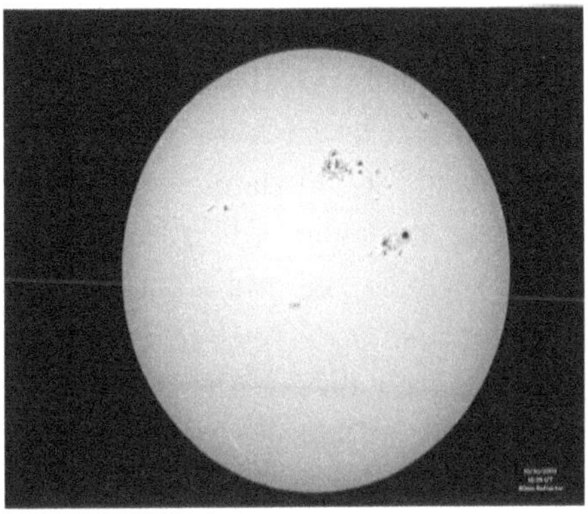

testable hypotheses, exemplifying the need for evidence-based inquiry and showing the vulnerability of untested pronouncements. Thomas Digges' first scientific publication, *Alae seu*, stressed the importance of evidence-based inquiry in testing hypotheses (Johnson 160), which he reiterated three years later in *A Perfit Description*. Shakespeare introduces the topic of epistemology through a discarnate entity with connections to a superior intelligence, and the two-fold occurrence of failed hypotheses so early in the play stresses the importance of wastebaskets in rational inquiry.

It occurs to Horatio that a Ghost resembling Old Hamlet will most likely speak to Young Hamlet, a conclusion that Barnardo and Marcellus could have reached after the first or second apparition. Shakespeare never writes but to a purpose, and it may be that he delays bringing Hamlet on stage in order first to introduce precursors to the New Astronomy.

The first scene ends with Marcellus announcing that he knows where to find Hamlet "most conveniently" (1.1.175). The scene began at about midnight and progressed toward dawn. This anticipates the rising of Phoebus Apollo, "god of day" (1.1.152n), who in turn presages the dawn of a new age of political and scientific enlightenment.

CLAUDIUS

Claudius holds court later the same day. He is in an expansive mood and seems unconcerned that he has violated ecclesiastical law by marrying his brother's widow, and that he has done so with impolitic speed. He expresses regret at the death of Old Hamlet and vows to carry on as the new regent. He dismisses the topic of Old Hamlet with a haughty "So much for him" (1.2.25), which is reminiscent of Thomas Digges' crisp and contemptuous sentence in *A Perfit Description*, "Thus much *Aristotle*" (Johnson and Larkey 90; italics original).

Claudius dispatches the ambassadors Cornelius and Voltemand to Norway to request Old Norway to intercede with his nephew, Young Fortinbras, who is planning to attack Denmark. Cornelius is the name of a Roman centurion to whom an angel of God appeared, leading him to become among the earliest Gentile converts to Christianity; moreover, Voltemand literally means a "turned" or a "changed man" (Sohmer, *Mystery* 242n8). One can sense that the times are changing, but it is not yet certain whether those changes are for good or ill.

Claudius announces that Fortinbras plans to attack because he believes that the Danish state is "disjoint and out of frame" (1.2.20). Enough time must have passed since the death of Old Hamlet for word to have filtered through to Denmark that Fortinbras takes a dim view of the new Danish regime. Fortinbras is correct to think that the figurative center of geocentricism is "out of frame," or disordered and out of shape like the Aristotelian Worldview. Fortinbras is both an enemy of the Claudian state and its cosmology, and an assault on Denmark is therefore understandable from both a political and a cosmological standpoint.

Claudius receives a request from Laertes to return to France. Only later is it revealed that Paris is his specific destination (2.1.7). Claudius ascertains that Laertes' father and his chief minister, Polonius, does not oppose the trip, so he approves it in short order. Next, Claudius addresses his newfound and grief-stricken "son," Hamlet. "How is it that the clouds still hang on you?" he asks, but Hamlet replies, "Not so my lord, I am too much i'th'sun" (1.2.66–1.2.67). As the son of the deceased sovereign and a legitimate contender for the throne, Hamlet is entitled to bask in the radiance of the Sun, which is the emblem of kingship, but in accordance with Horatio's ominous warning, Hamlet says also that he is *too much* in the Sun.

Hamlet's mother scolds him too. "Why seems it so particular with thee?" she asks, but Hamlet has a suitable response, "Seems madam? Nay it is, I know not seems" (1.2.75–1.2.76). Hamlet likens his pall of grief to an inky cloak and a suit of solemn black. He explains that his weeds of woe have seams that seem but that he has within him that "which passes show" (1.2.85). Like a book, Hamlet is not to be judged by his cover. The King calls Hamlet's grief a "mourning" duty (1.2.88), a pun that anticipates the early morning meeting between Hamlet and his father's spirit.

Claudius wishes that Hamlet would accept him as his new father and urges the Prince not to prolong his sorrow.

> CLAUDIUS But you must know, your father lost a father,
> That father lost, lost his. (1.2.89–1.2.90)

Unwittingly, Claudius alludes to the theories of past geokineticists and atomists whose theories have faded into obscurity and who are all but forgotten.

> CLAUDIUS It shows a will most incorrect to heaven
> A heart unfortified, a mind impatient,
> An understanding simple and unschooled. (1.2.95–1.2.97)

The false monarch is keen to correct Hamlet's disposition and blames his prolonged grieving on a lack of a scholastic education.

The geocentric monarch asserts the reality of appearances.

> CLAUDIUS For what we know must be, and is as common
> As any the most vulgar thing to sense. (1.2.98–1.2.99)

He blames nature for apparent celestial anomalies and says that those who seek to explain them are wasting their time.

> CLAUDIUS Why should we in our peevish opposition
> Take it to heart? Fie, it is a fault to heaven,
> A fault against the dead, a fault to nature,
> To reason most absurd, whose common theme
> Is death of fathers, and who still hath cried,
> From the first corse till he who died today,
> 'This must be so.' (1.2.100–1.2.106)

He thinks that argument carried to extremes is "reason most absurd" whose common theme is "death of fathers," all of whose arguments for change in the standard model were short lived.

By the customary count, *Hamlet* speaks to the death of four fathers—Old Hamlet, Old Fortinbras, Polonius, and the father of Pyrrhus in the poem of scene 2.2—but in the cosmic interpretation one must add also fathers of the New Astronomy (chapter 1). Leading these ancestral ranks is Pythagoras, the "father of Greek philosophy" and "the progenitor of astronomy as a science" (Heninger 28–29). Equally significant, I believe, is the deceased father who is the chief cause of the false King's anxiety—Old Hamlet—who is allegorically Leonard Digges, father of Thomas Digges and the father of optical telescopy.

Claudius opposes Hamlet's return to school in Wittenberg. "It is most retrograde to our desire" (1.2.114), he says. He refers to the planetary phenomenon of retrogression, which occurs over periods of weeks and months and is readily observed for planets whose orbits lie outside the Earth's orbit. It occurs around the time of Opposition, when planets are aligned with Sun and Earth, and Shakespeare emphasizes its significance by placing "retrograde" 14 lines after "opposition" (1.2.100; figures 1.2, 1.3).

Retrograde motion violates the Aristotelian tenet of uniform motion around the Earth, which devout geocentricists dismissed as a quirk of nature and not a fault of understanding. The sixteenth century was a time when received wisdom was valued and original thought was suspect. Therefore, supporters of the status quo saw any step forward as "a movement backwards" (Koyré, *Astronomical* 92n13). Like the retrograde motion of planets, Hamlet's return to Wittenberg would entail motion in the wrong direction—to the very place where heliocentricism had begun to take root. Gertrude chimes in and asks Hamlet not to go to Wittenberg, and her son's resolve dissolves as he promises, "I shall in all my best obey you madam" (1.2.120). Hamlet has a higher social standing than Laertes, yet his quest for education is stymied whereas Laertes has free rein to make a seemingly purposeless visit to Paris.

Hamlet has pleased his mother but remains disgruntled.

> HAMLET O that this too too solid flesh would melt,
> Thaw and resolve itself into a dew,
> Or that the Everlasting had not fixed
> His canon against self-slaughter. (1.2.129–1.2.132)

Given the customary injunction against suicide, Hamlet wishes that his body would dissolve itself into "a dew" so he could bid his woes adieu. If he could evaporate in the morning sun as easily as the Ghost evanesced in the morning twilight, he would feel as if he were one with his father. "Dew" is part of the ominous tone set previously by Horatio (1.1.117) when he spoke of "dews of blood" and "disasters in the Sun," and it signifies Hamlet's premonition about the manner of his death.

Hamlet can hardly believe that his father is dead or that, within two months, his mother married his uncle. The Sun imagery continues as Hamlet compares his father to Hyperion (the Sun god) and Claudius to a lecherous satyr. In the meantime, he resolves, "I must hold my tongue" (1.2.159).

Wittenberg

Horatio enters with Marcellus and Barnardo. Hamlet is surprised to see his fellow student from Wittenberg and twice asks Horatio why he has come to Elsinore (1.2.164, 1.2.168). Horatio replies that he came to see the funeral of Hamlet's father; however, if Horatio had been at Elsinore for several weeks, he surely would have sought out his good friend earlier. Tongue in cheek, Hamlet suggests that Horatio came to see the new

King's wedding, which followed hard upon his father's funeral, but again the two would surely have encountered one another sooner. Because Claudius now claims to be Hamlet's father, perhaps Horatio came to see the funeral of the *new* King, but how would Horatio know of the imminent death of Claudius? The question is not resolved, and the audience is left to wonder whether Shakespeare, having already introduced a discarnate being, hints that the events follow a paranormal script.

Horatio tells Hamlet how Barnardo and Marcellus had seen the spirit of his father and that he was present for the third apparition.

> HORATIO And I with them the third night kept watch,
> Where, as they had delivered, both in time,
> Form of the thing, each word made true and good,
> The apparition comes. (1.2.208–1.2.211)

To "deliver" means to "report" or "describe." Horatio establishes that his observation of the third apparition matched Barnardo's description of the previous two "both in time" and in the "Form of the thing." In other words, on all three occasions the spirit appeared at the same time (1 o'clock in the morning) and from the same direction (of the star that is westward from the pole).

"Perchance 'twill walk again," Hamlet says, and once again Horatio makes a prediction, "I warrant it will" (1.2.242), where to "warrant" means to "guarantee as true" or "promise or predict as certain" (*OED*). This time he is certain, and he will be right.

The Prince announces that he will meet his companions that very night, but he breaks from the pattern of the past. "Upon the platform 'twixt eleven and twelve / I'll visit you" (1.2.251–1.2.252), he says. Hamlet knows that Marcellus had had good luck on the third occasion by assuming that the Ghost would walk at the same time as on the previous two occasions, yet on this fourth occasion he knowingly opts to start his watch an hour or two earlier. If he intended to ignore precedent, why would he not start watching at nightfall?

It seems that the timing of the arrivals of Barnardo, Marcellus, Horatio, and Hamlet has historical significance. In this scene (1.2), Barnardo utters only a few words and then only in unison with his companions, whereupon in scene 1.4 when the time comes for the group to reassemble, Barnardo is absent, never again to grace the stage. This signifies that the New Astronomy has progressed beyond the *Cosmographia* of Bernardus Silvestris to embrace the sixteenth-century works, *Zodiacus Vitae* of Marcellus Palingenius, and *De Revolutionibus* of Nicholas Copernicus whose influence Horatio has delivered from Wittenberg.

LAERTES AND OPHELIA

Saxo does not name Amleth's seducer, so Shakespeare is free to improvise. Ophelia's name may derive from the Latin prefix *op-* (which can mean either "opposite," "completely facing," or "in the direction of") and the Greek *helios* (the Sun), converted to a feminine ending with the suffix *-a*. Thus, "Ophelia" refers to something lying either opposite to or in the same direction as the Sun (figure 3.2). When the Moon is directly opposite the Sun, it is in Opposition, and when it is directly between the Earth and the Sun, it is in Conjunction. In either case, the Moon is at

FIGURE 3.2. Lunar Opposition and Conjunction with the Sun, illustrating the meaning of *Op-heli-a.*

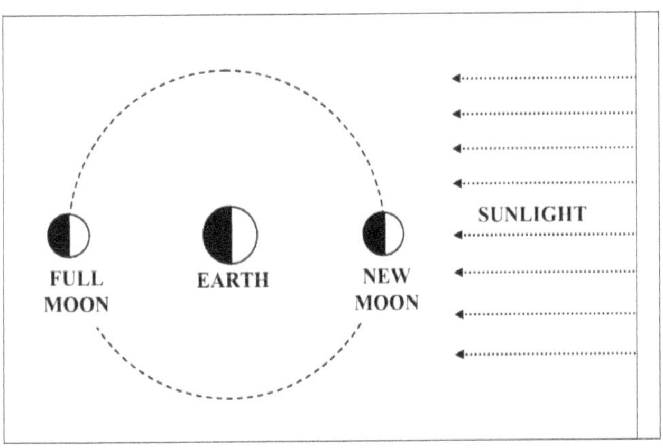

Syzygy.[3] As a prospective queen, Ophelia would play opposite Hamlet who would rule Denmark in conjunction with her. When Polonius and his missing wife named their daughter Ophelia, little did they know that someday she might qualify for such a role, or was the preternatural puppeteer in action even then?

The royal Sun and the chaste Moon rule the heavens as they light "The little O, the earth" (Bevington, *Antony and Cleopatra* 5.2.80), just as Hamlet and Ophelia seem destined to rule Denmark. However, according to Horatio, the omens are not propitious. Eclipses of the Moon can occur when the phase of the Moon is Full, at which time it is directly opposite the Sun and in the Earth's shadow. Spring tides occur at both Full Moon and New Moon when the moon goddess and governess of the flood ensures that high tides are deepest and most likely to swamp the unwary. In scene 1.3, Laertes warns Ophelia of this state, and in scene 4.7, it happens that Ophelia does in fact perish from an excess of water. Not by accident are the lunar deities—Diana and Artemis—also the goddesses of childbirth. The ominous lunar eclipse of which Horatio speaks refers to Opposition when the Moon is Full and opposite Helios, and in a discussion that follows, it becomes apparent that the script verifies the present explanation by referring also to Ophelia's Conjunction when the Moon lies in the same direction as Helios, and thus when its phase is New.

Before setting sail for France, Laertes inundates his sister Ophelia with advice replete with astronomical connotations (Usher, "New Astronomy" 137–141). Laertes knows of the involvement between Ophelia and Hamlet. The couple's potential alignment touches on the issues of love and chastity, which Laertes deals with in that order. First, on the subject of love, he declares that Hamlet is a youth whose amorous interests are false because they are "not permanent" and "not lasting" (1.3.8). He states his reasons.

> LAERTES For nature crescent does not grow alone
> In thews and bulk, but as this temple waxes
> The inward service of the mind and soul
> Grows wide withal. (1.3.11–1.3.14)

Many believe that these lines are fair game for a cut in order to "eliminate difficult words and phrases" (Kliman 63), but in reality they divulge a phenomenon important to the subtext and to the history of Renaissance science.

Most editors refrain from identifying the crescent in 1.3.11 as lunar because the Moon is the goddess of chastity, and because the topic under discussion is love, which is the province of Venus. Laertes describes maturation toward genuine love as a waxing or maturing like a phase that is crescent but widening. A few reports from antiquity do describe Venus as a crescent, but there are no reports of a progressive change in her shape and particularly not as a function of apparent size and elongation angle from the Sun. Even the possibility of a Cytherean crescent was not common knowledge in 1610–1611; otherwise, Galileo would not have made such a fuss when he first observed it. In the subtext, Shakespeare does not use the term "gibbous" (which describes the phase when more than half its visible disk is illuminated; see figure 1.14) because the term entered the English language only in 1690 (*OED*). Instead, Shakespeare describes the complete waxing phase from crescent to gibbous as a "widening." These results on Venus have important ramifications for the chronology of Act 1, as is shown subsequently in this chapter.

Laertes turns next to the topic of chastity and warns Ophelia not to besmirch her honor by opening her chaste treasure to Hamlet's unmastered importunity. Laertes advises Ophelia to stay in the "rear of her affection" and "Out of the shot and anger of desire" (1.3.34–1.3.35). For emphasis, he rolls out two adages and a botanical conceit.

> LAERTES 'The chariest maid is prodigal enough
> If she unmask her beauty to the moon.'
> 'Virtue itself scapes not calumnious strokes.'
> The canker galls the infants of the spring
> Too oft before their buttons be disclosed,
> And in the morn and liquid dew of youth
> Contagious blastments are most imminent. (1.3.36–1.3.42)

The word "blastment" suggests an explosion of some kind and because military conceits precede it ("shot and danger"), the image that springs to mind is ballistic impact. Gunnery was an established art in the sixteenth century, and Shakespeare would know of skin lesions caused by the syphilitic pox and their resemblance to craters. In the sixteenth century, chastity was the only sure defense against venereal disease, which was widespread and much to be feared. "Fear it Ophelia, fear it my dear sister… best safety lies in fear," says Laertes (1.3.33, 1.3.43).

Botany sustains the imagery because plant cankers are lesions that result from injury to plants, allowing fungi and bacteria to assail the tissue. Diseased leaves and fruit have sunken craters and raised margins that resemble the lunar surface (figure 2.1). Lunar craters are indiscernible to the naked eye, yet the several conceits suggest that Shakespeare's astronomer saw a contagion of them covering the Moon's face. All told, Laertes' sermon on love and chastity doubles as a description of the properties of the Moon and Venus that were unknown before the advent of telescopes. These repeat those uncovered in chapter 2 from *Love's Labour's Lost*.

Even prior to the time of Ptolemy, mathematicians knew the Moon's distance with fair accuracy, so Shakespeare and his astronomer would have realized that lunar craters are "most imminent" (1.3.42) and that their diameters can exceed 100 kilometers (60 miles). The imagery involves violence and death. Shakespeare has already implied that Ophelia risks an unwelcome widening in the womb if she loses her chaste state, and now it is revealed that Ophelia must beware becoming like the moon goddess Artemis, whose name derives from a "high source of water" (Graves 2: 382).

Laertes worries that he is dillydallying, and his father arrives and confirms his concern. "Yet here Laertes? Aboard, aboard for shame!" (1.3.55), he says. The rebuke suggests that Laertes ends his harangue of Ophelia comparatively late in the day, and Polonius delays Laertes even longer with 26 lines of his own advice, ending with,

> POLONIUS This above all, to thine own self be true,
> And it must follow, as the night the day,
> Thou canst then not be false to any man. (1.3.78–1.3.80)

On Earth, there are no absolutes, and despite appearances, Polonius has redeeming qualities.

After Laertes departs, Polonius interrogates Ophelia on her conversation with Hamlet.

> POLONIUS 'Tis told me he hath very oft of late
> Given private time to you. (1.3.91–1.3.92)

Someone in the spymaster's thrall has been keeping an eye on the Prince. Ophelia protests that Hamlet has importuned her with love "in honorable fashion" and "with almost all the holy vows of heaven" (1.3.111, 1.3.114), but it appears he has stopped short of actually proposing marriage.

Polonius likens Hamlet's expressions of interest to "blazes" that give "more light than heat" (1.3.117–1.3.118). "Blazes" are short bursts of flame (Hibbard, *Hamlet* 1.3.117n), which, says Polonius, "You must not take for fire" (1.3.120). Perhaps Polonius refers to shooting stars (meteors) or to the New Star whose blaze has come and will go, unlike genuine stars that are steadfast in their light. In Aristotelian teleology, everything in creation serves a purpose, and evidently in Polonius' mind, ephemeral celestial phenomena serve as omens warning maidens of fickleness in love. Polonius cautions Ophelia that the glint in Hamlet's eye is valueless, a mere flash in the pan, and he forbids her to have further contact with him. This quashes the last vestige of her free will. "I shall obey, my lord" (1.3.136), she says.

The Dram of Eale

Scene 1.4 opens on the battlements with the entrance of Hamlet, Horatio, and Marcellus, but Barnardo is absent for reasons mentioned. Cannons fire and Hamlet explains that Claudius is throwing a party. Claudius is in his cups and "drains his draughts of Rhenish down" (1.4.10). This is the first of two mentions of Rhine wine in *Hamlet*, the other occurring in the context of an attribute of Yorick, as shall be shown.

Hamlet addresses Danish overindulgence and his own noncompliance with the cultural norm. The context of wine prompts him to launch his "dram of eale" speech (1.4.17–1.4.38), which is not in F1. A dram of eale is a portion of old liquor added to new stock to establish its enological heritage, as a father begets a son. In Hamlet's mind, a defective dram mars the ensuing product.

> HAMLET So, oft it chances in particular men,
> That for some vicious mole of nature in them,
> As in their birth, wherein they are not guilty,
> Since nature cannot choose his origin… (1.4.23–1.4.26)

Hamlet refers to imbalance in the four humors, which supposedly accounts for unstable temperaments and mental breakdowns. For the time being, he is an apologist for human weakness, as in referring to a "vicious mole of nature" he describes many in the world who are afflicted. He claims that a stamp of a defect corrupts a person's reputation, albeit not the person *per se*.

> HAMLET The dram of eale
> Doth all the noble substance of a doubt
> To his own scandal. (Jenkins 1.4.35n, 1.4.36–1.4.38)

Sohmer (*Mystery* 236) calculated that Hamlet was born at least 53 days before his parents' wedding, and because "vicious" can mean "immoral" or "bad," one may wonder whether the "vicious mole of nature" in *Hamlet* is Old Hamlet's imprint of bastardy.

No sooner has Hamlet finished his speech than the figurative representation of the defective dram—the spirit of Old Hamlet—appears. Forasmuch as Hamlet personifies Thomas Digges, Shakespeare may wish to draw attention to his illegitimacy, in which case one would expect Leonard and his wife to have attempted to inoculate their child against the stigma. Perhaps they purposefully forgot the child's year of birth, and it happens that Thomas Digges' date of birth was neither chiseled on his tomb nor recorded otherwise. When sources do venture to state his year of birth, most often they cite the value simply as *circa* 1546. This value is close to the mark, as the birth date of Tycho Brahe was December 14, 1546, and when Shakespeare states that Tycho's dual representation,

Rosencrantz and Guildenstern, grew up with Hamlet and that the two were "neighboured to his youth and haviour" (2.2.12), he implies that Thomas was a close contemporary of Tycho.

In Q2, the Ghost's entrance occurs at line 1.4.38, but F1 does not contain the dram of eale speech, so the spirit enters effectively after line 1.4.16. Because the disagreement between Marcellus and Horatio over whether or not midnight has struck occurs near the scene's beginning (1.4.3–1.4.4), in Q2 the time of the Ghost's entrance is 34 lines later than the disputed time, and in F1 it is about 12 lines later. In either case, it is reasonable to believe that the Ghost enters shortly after midnight and not at 1 a.m. as it had on each of the previous three occasions. It appears that Hamlet commenced watching at just the right time! Yet how did he know when the Ghost would appear? Perhaps the Ghost was lying low, waiting for Hamlet to arrive first; but then how did Hamlet know that he could rendezvous at a time of his own choosing?

Hamlet marvels that the sepulcher holding Old Hamlet's corpse has cast up his spirit, and he seeks an explanation.

> HAMLET Why thy canonised bones, hearsèd in death,
> Have burst their cerements; why the sepulchre,
> Wherein we saw thee quietly enurned,
> Hath oped his ponderous and marble jaws
> To cast thee up again. What may this mean,
> That thou, dead corse, again in complete steel
> Revisits thus the glimpses of the moon…? (1.4.47–1.4.53)

According to Thomas Digges' preface to *Pantometria* in 1571, the world's first telescopist died about a year prior to one of the most dramatic celestial events of the time, the discovery of the New Star of 1572. To counter the bad timing, Shakespeare lets the dramatized spirit of Leonard Digges burst from its sepulcher like a New Star erupting in the Firmament; furthermore, he allows it to visit its old haunts as if it were alive and able to enjoy the spectacle in that state. Hamlet marvels that the spirit "Revisits thus the glimpses of the moon" (1.4.53) and the chronology of Act 1 (in table 3.3 that follows) explains why.

DRAWN TO MADNESS

The spirit beckons Hamlet to follow but his companions demur.

> HORATIO What if it tempt you toward the flood my lord,
> Or the dreadful summit of the cliff
> That beetles o'er his base into the sea…
> Which might deprive your sovereignty of reason,
> And draw you into madness? (1.4.69–1.4.74)

Of Horatio's two fears, the first is unwarranted because it is Ophelia who is at odds with Neptune's empire, not Hamlet. His second concern is that the spirit might deprive Hamlet of reason and draw him "into madness," which does occur, except that Hamlet's "madness" refers to science, which translates to "sanity" as it presages the dawn of a new age.

Marcellus anticipates the coming struggle. "Something is rotten in the state of Denmark" (1.4.90) he says, and Horatio tacitly agrees. "Heaven will direct it" (1.4.91) he says, meaning that events are under supernatural control. Marcellus adds, "Nay let's follow him" (1.4.91), where "nay" means "let us not leave it to Heaven, but do something ourselves" (Jenkins 1.4.91n). A prevalent belief is that humans have free will, but that a commanding otherworldly presence directs events as if with the happenstance of mundane life. Humans act, but unassisted their efforts are limited, so when it comes to such important matters as the overthrow of corrupt regimes and the end of false cosmology, a beneficent power may lend a hand. This agrees with the Neoplatonic belief in a gradation of being from the supreme One to humankind via a series of intermediaries who act to bring Good to the world.

On this the fourth occasion of the Ghost's appearance, all presynthetic components of the New Astronomy are on stage. Marcellus represents the spatial distribution of stars, Horatio represents heliocentricism from Wittenberg, and the ephemeral specter points to the mutability of the starry skies. Together, they infuse Hamlet with the components of Diggesian cosmology, in which capacity he will wage his struggle on behalf of the New Astronomy.

The next scene (1.5) opens with the spirit identifying itself to Hamlet.

> GHOST I am thy father's spirit,
> Doomed for a certain time to walk the night,
> And for the day confined to fast in fires,
> Till the foul crimes done in my days of nature
> Are burnt and purged away. (1.5.9–1.5.13)

It has appeared for three nights running, and on the fourth, it prepares for penance in Purgatory. Apparently, the transgressions of Old Hamlet were not so egregious as to damn him outright (Jenkins 1.5.12n), which according to table 3.1 implies the same for the Ghost's counterpart Leonard Digges. Leonard may have fathered a child out of wedlock (although Elizabethans would hardly consider this a "foul crime"), or Shakespeare may have looked askance at Leonard's participation in Wyatt's Rebellion of 1554, and because Queen Mary was Catholic, it is appropriate that his spirit atone in accordance with Catholic doctrine. Shakespeare's writing is syncretic, and he uses a Catholic tenet for the purpose of his dramatic construction rather than out of religious conviction.

The Ghost tells Hamlet that it could relate a tale that would make Hamlet's two eyes "start from their spheres" (1.5.17) like stars falling from the Firmament, but that it must not divulge the secrets of its prison house. The Ghost, having crossed into natural space from space beyond the pale of human experience, is not entitled to tell of conditions on the other side. Instead, the specter tells Hamlet that if he loved his father, he must "Revenge his foul and most unnatural murder" (1.5.25).

From the fifteenth century, "unnatural" means "not in accordance or conformity with the physical nature of persons or animals" and "not in accordance or agreement with the usual course of nature" (*OED*), but "unnatural" may also mean "supernatural." Shakespeare refers to an evil spirit that has promoted the murder of Old Hamlet, operating unseen in order to usurp power and capture Elsinore.

At the time of the murder, Young Hamlet had already enrolled at Wittenberg, and it may be surmised that the Hamlets were on track to develop the New Astronomy. Leonard Digges, anterior to his death reported in 1571, was on a similar course. Thomas and Hamlet are both compelled by the spirits of their deceased fathers to finish the job. To that end, the phantom divulges the identity of the murderer, calling Claudius an incestuous, adulterate beast who "won to his shameful lust" his "most seeming virtuous queen" (1.5.45–1.5.46). Hamlet will remember the latter phrase with its connotation of appearance over reality, and will use the play-within-the-play "to try conclusions" (3.4.196).

Hamlet has become a puppet in the hands of a superior power, yet for all the strings attached, he is still a human with a conscience. It is important therefore, that the Ghost address his mundane concerns.

> GHOST But howsomever thou pursues this act
> Taint not thy mind, nor let thy soul contrive
> Against thy mother aught. Leave her to heaven. (1.5.84–1.5.86)

Gertrude is not part of the subtextual script and Hamlet must not judge her. He is partly a denizen of the supernatural world who must feel no guilt for acting on behalf of the puppeteer.

Claudius too is beholden to supernatural influence, except that in his case the spirit is the epitome of evil. Like Hamlet, Claudius is human, and his conscience is vulnerable, as evident from his spasms of remorse (3.1.49–3.1.50, 3.3.36–3.3.38), the lesson being (one supposes) that supernormal agencies can act for good or ill.

The Ghost senses the morning air and knows it must repair to the sulfurous and tormenting flames. "Adieu, adieu, adieu," it mutters as the temperature falls to the dew point, "Remember me" (1.5.91), whereupon it departs the stage presumably through a trapdoor (Jenkins 1.5.91SDn). The specter changes phase like a watery mist. Instead of evanescing as on previous mornings, it retains its spectral form and simply drops out of sight, destined for a place warmer than Denmark in autumn.

Far be it for the Prince to forget these momentous events.

> HAMLET Remember thee?
> Ay thou poor ghost, whiles memory holds a seat
> In this distracted globe. (1.5.95–1.5.97)

"Globe" is a triple pun (1.5.97n): on the sorry state of the World, on Hamlet's head with its now discombobulated thoughts, and on the Globe Theatre wherein appearances distract patrons to the detriment of their understanding of the subtext. Ambiguity persists.

> HAMLET Remember thee?
> Yea, from the table of my memory
> I'll wipe away all trivial fond records,
> All saws of books, all forms, all pressures past,
> That youth and observation copied there. (1.5.97–1.5.101)

Hamlet will erase his impressions of trivial and foolish ("fond") memories. He will remove set and formulistic ideas and all imprints and impressions ("pressures") that were inculcated by "deferential, even obsequious, attention to his superiors" as well as "imitation of them or obedience to them" (1.5.100nn, 1.5.101n). Hamlet must revamp the political world and reconstruct the cosmic World. Horatio and Marcellus catch up with Hamlet and the first thing Horatio says is, "Heavens secure him" (1.5.113). Hamlet's response, "So be it" (1.5.114) suggests that he is resigned to his fate.

Falconers

Marcellus and Hamlet use birdcalls to hail one another. Marcellus shouts, "Illo, ho, ho, my lord," and Hamlet mimics the cry of a hawker, "Hillo, ho, ho, boy! Come bird, come," (1.5.115–1.5.116). On such a momentous occasion as dialogue with an evanescent being, Marcellus and Hamlet communicate as if they were falconers. In Hamlet, this "begins a train of imagery associated with hawks" (Hibbard, *Hamlet* 1.5.119n), but what purpose could Shakespeare have in mind to associate this notable occasion with a bird of prey?

According to the printed *OED*, in 1550 a "leonard" or "leonarde" was a type of hawk, as in "hawkes of the towre, as leonardes, leonerettes, fawcons..." (In the online *OED*, this definition is found under "hawk, *n*[1].") From early times, a hawk meant any diurnal bird of prey used in falconry, or any bird of the family *Falconidae*, which includes falcons (*OED*). By this definition, leonards and falcons are types of hawks. "Leonard" is a variant on "lannard" or lanner, which is a species of falcon. In the class-conscious society of the early modern age, each species received a rank appropriate to the station of the falconer. Accordingly, squires were supposed to fly Lanner Falcons (Berners and Gryndall, n.p.).

A leonard is a large bird of prey with an eagle eye that enables it to see and resolve targets from afar, just as Leonard Digges "Esquire" (Loades 253) had an instrument able to resolve distant objects with unparalleled clarity. A hawk's eye is commonly known to possess a high enough resolution to spot a mouse at a distance of two miles, making the visual acuity of an eagle is nine times better than that of the most farsighted person (Harting 24). These specifications correspond to a resolution of a few arc seconds or less, which is comparable to the clarity with which a telescope with good optics and an aperture greater than a few centimeters can discern objects.

With this identification, the "bird" in 1.5.116 is presumably a reference to the just-departed spirit of Old Hamlet, here also known as Leonard Digges. Thereby, Shakespeare reaffirms that the deceased Old Hamlet and his son are equivalent to the deceased Leonard Digges and his son. The Ghost has lifted Hamlet from a state of innocence, and it would be unwise for the Prince to divulge details to his companions lest they unwittingly betray the confidence. Hamlet asks them to keep secret what they know, which the specter encourages by uttering "Swear" (1.5.149) from below stage.

The Prince proposes an oath, and the disembodied voice again cries "Swear" (1.5.155). According to Jenkins, "The shifting locality of the voice adds to the impression of a subterranean demon" (1.5.157LN). Hamlet reinforces the impression by asking, "*Hic et ubique*?" ("Here and everywhere?") and adding, "then we'll shift our ground" (1.5.156). Hamlet gets the impression that the demon in the cellarage moves by shifting ground, which prompts the announcement that he and his companions will shift their ground as well. For a third time, the ghost utters, "Swear" (1.5.161). Shakespeare emphasizes the identity of the Ghost as

Hamlet replies, "Well said old mole, canst work i'th'earth so fast? / A worthy pioneer" (1.5.162–1.5.163). A "pioneer" is "a soldier responsible for excavations and tunneling" (1.5.163n), and the Diggeses are both diggers of sorts because they are pioneers who undermine the old paradigm entrenched at Elsinore.

Shortly after Thomas Digges had reported the death of his father in 1572, Ludwig Lavater (1527–1586) published a confirmation, *Of ghostes and sprites walking by nyght*, which critics regard as one of Shakespeare's sources. The reading discussed previously is supported by the passage, "Pioneers or diggers for mettal, do affirme, that in many mines, there apeare straunge shapes and sprites, who are apparelled like unto other laborers in the pit. These wander up and down in caves and underminings" (Wilson 81). In the same vein, the Spirit of Old Hamlet has come down from on high and has holed up below like a mole in the subterranean chambers of its habitat. The spirit of Old Hamlet moves the *earth*, whereas the spirit of Young Hamlet moves the *Earth*.

Hamlet is of the opinion that there are more things "in heaven and earth" (1.5.166) than are dreamt of in Horatio's philosophy. Thanks to the paranormal visitation, Hamlet knows that there is more to cosmology than is known at Wittenberg. By the end of the play, the future historian of the scientific revolution, Horatio, will know more about the spirit who spent time in heaven and in the earth.

In order to protect himself, Hamlet will emulate Amleth and act in a "fantastic and foolish manner" (1.5.172n). His companions must pledge secrecy too, and the mole mutters "Swear" for the fourth time (1.5.182). The four-fold exhortation suggests that events are occurring for the good, yet Hamlet is apprehensive. "The time is out of joint," he complains, "O cursèd spite / That ever I was born to set it right" (1.5.189–1.5.190).

By the end of Act 1, Shakespeare has developed the revenge motive that in the subtext translates to a tension between the two chief World models, the Ptolemaic and the unbounded Copernican-Diggesian. The Ghost has actuated the contest and toward the end of scene 1.5, it is free to begin a stint in Purgatory.

CHRONOLOGY OF ACT 1

The events of Act 1 have infused Hamlet with properties of the New Astronomy, and their correspondence with phenomena in the heavens leads to the chronology shown in table 3.3. The chronology builds on Olson, Olson, and Doescher's identification of the star westward of the Pole (1.1.35–1.1.39) with the New Star of 1572.

Examining celestial ephemerides of the Sun, Moon, and planets will show whether they correspond to events in the play. The quest is apposite given Shakespeare's interest in eclipses (Levy, "Eclipses"). In table 3.3, the first column identifies scenes from Act 1 of *Hamlet*. Column 2 gives the corresponding Julian (Old Style) calendar dates of events that appear in column three, where italics distinguish events in the play from those in the sky. Column 3 gives lunar and planetary ephemerides, column 4 gives the angle *A* of the Moon above the horizon, and column 5 gives the fraction *f* of the Moon's disk that is illuminated as seen from Elsinore. The last column gives pertinent Christian religious holidays. A standard atmospheric model accounts for refraction. All times quoted pertain to Elsinore. The New Style (Gregorian) calendar began 10 years after the dates considered, so dates listed are according to the Julian calendar, whereas phenomena are of course proleptic.

Two constraints bracket the celestial events of column 3 (Dreyer 61–63).

First, on the evening of Sunday, November 2, 1572, Hieronymus Munosius (*fl.* 1572) and his students were viewing constellations from Valencia, Spain, and the professor was certain that the New Star had not appeared, which agrees with the reports of Spanish shepherds. In the absence of a planetarium, Professor Munosius perforce had to hold classes at night, not during the day, in order to teach students about the stars and constellations. At Valencia in early November, observing begins about when astronomical twilight ends at 18:30 hours (6:30 p.m.). As November 2 is a Sunday, it is reasonable to expect that the professor would have started his class early and sent his students home well before midnight. This constitutes the early limit on the appearance of the

TABLE 3.3. Chronology of Act 1 of *Hamlet*: astronomical events and dates at Elsinore in November 1572.

Scene	Day	Event	*A*	*f*	Day
	Sat 1				All Saints'
	Sun 2	New Star not seen from Valencia, Spain			
	Mon 3	*Apparition on day 1 at 01:00.* Moon, Venus in Virgo			All Souls'
		Moon rises 04:16	0°	0.067	
		Civil twilight begins 06:52	18°	0.062	
	Tue 4	*Apparition on day 2 at 01:00.* Moon in Libra			
		Moon rises 05:33	0°	0.025	
		Civil twilight begins 06:54	9.5°	0.023	
1.1	Wed 5	*Apparition on day 3 at 01:00*			
		Moon rises 06:50	0°	0.003	
		Civil twilight begins 06:56	0.9°	0.003	
1.2		*Hamlet plans to watch between 23:00–24:00*			
1.3		*Laertes lectures Ophelia on chastity*			
		Moon is New at 19:01		0.000	
1.4	Thu 6	*Hamlet sees apparition shortly after 00:00*			
1.5		*Ghost in the cellarage* New Star seen about 06:00 from Wittenberg			

New Star. The fact that the precise time of appearance of the New Star is unknown leaves Shakespeare the flexibility to weave a tale that accommodates the uncertainty.[4]

A later constraint follows from its first known sighting on November 6, 1572, in the morning twilight at about 06:00 hours (6:00 a.m.), when a professor of mathematics at Wittenberg University, Wolfgang Schuler (d.1575), spied it above the north-northwestern horizon.

One must question whether the Elsinore clock ("The bell then beating one," 1.1.39) is an apparent time or a mean time, the former set by the times of transit of the Sun across the meridian that occur at unequal intervals, the latter being uniform and set by the rotation of the Earth relative to the stars. I argue that Shakespeare uses the latter.[5] Values listed in column 3 correspond to the Elsinore locale, derived from the relative longitudes of Valencia (Spain), Wittenberg (Germany), Elsinore (Denmark), and Greenwich (England), the last required to convert Greenwich Mean Time supplied by modern ephemerides into other local mean times.

It suffices to estimate longitudes from contemporary maps. This is justified because Shakespeare was familiar with cartographic advances, as evidenced by *Twelfth Night* in about 1601–1602. There, Maria tells Sir Toby that Malvolio "does smile his face into more lines than are in the new map"—referring to the new projection by Gerardus Mercator (1512–1594) that appeared in 1599 as a part of *Principal Navigations* by Richard Hakluyt (1552–1616)—where Malvolio's smile lines refer to its crisscrossing rhumb lines (Bevington, *Twelfth Night* 3.2.77n; Shirley, *Mapping* xv). In England in the 1540s, before being attainted for treason in connection with Wyatt's Rebellion, Leonard Digges was a principal contributor to the use of mathematics and surveying in cartography and marine navigation. By the 1560s, the government did not need to rely on military engineers because they could turn to "well-educated and patriotic squires like Thomas Digges" for guidance in the application of the latest techniques (Barber 59–60).

A map from 1570 shows a longitude difference between Greenwich and Helsingor (Elsinore) which I estimate at about 13° 30′ (corresponding to 54 minutes in time) (Oertel map numbers 5, 21). Longitude values collected by Tycho Brahe before his death in 1601 put the value at 13° 35′ (54⅓ min). Maps on the Mercator projection, compiled at the end of the sixteenth century (Bricker 69–71), show a value of about 12° 30′ (50 min) compared to the modern value of 12° 44.4′ (51 min). The chief results are that contemporary sources show a time difference between London and Elsinore of less than an hour and that they disagree by only a matter of minutes in time, so for convenience let the time difference be expressed in round numbers at 50 minutes. Further, Valencia and Greenwich have virtually the same longitude, thus the early limit of November 2 at Valencia translates to essentially the same limit at Greenwich; Elsinore and Wittenberg lie close in longitude too, so the late limit of November 6 at Wittenberg corresponds to the same limit at Elsinore (Oertel map numbers 5, 7, 13).

The key to the chronology of table 3.3 is the supposition that the early morning of Thursday November 6, 1572, when Hamlet meets the Ghost for the first time, corresponds to the morning when Schuler first saw the New Star. At that instant, Urania infuses Hamlet with the muse of the New Astronomy and sets the stage for subtextual revelations to come. Working backward in time reveals the correspondence between events on stage and in the heavens. Of course, in the end the independent evidence must be tested to see whether it confirms the chronology.

The Munosius and Schuler constraints bracket the start and end of the chronology, but apply at sites with different longitudes. Valencia and Greenwich have similar longitudes, so clocks at the two sites keep roughly the same time. The Munosius constraint from Valencia is that, seen from Greenwich, the New Star had also not appeared before midnight on November 2. The corresponding time at Elsinore is later, about 00:50 hours on November 3. Thus, the null datum from Valencia (that the New Star had not appeared on the evening of November 2) is compatible with the time of the first apparition of the Ghost at Elsinore at 01:00 hour (1 a.m.) on November 3. Concerning the Schuler constraint, there is slightly less than an approximately six-hour delay between Hamlet's first sighting of the Ghost at Elsinore, which occurred just after midnight on Thursday November 6, 1572, and the first real-time observation of the New Star at about 06:00 hours (6 a.m.) the same morning at Wittenberg. Thus, the script of *Hamlet* satisfies both constraints.

Table 3.3 indicates that scene 1.1 opens at about midnight at Elsinore Castle and progresses toward dawn. At 01:00 hours, the Ghost appears from the direction of the same star whence it had appeared on the previous two nights. It is doomed "for a certain time to walk the night" (1.5.10). It is safe to assume that nighttime lasts to the start of civil twilight when by definition, the Sun is 6° below the horizon. Scene 1.2 begins later the same day with Claudius holding court. After the court adjourns, Hamlet agrees to meet his companions that same night between 23:00 (11 p.m.) and 24:00 hours (midnight). In scene 1.3, Laertes sets sail late on the same day. In scene 1.4, Hamlet, Horatio, and Marcellus enter at about midnight, and shortly thereafter, in the early morning of the next day, the Ghost enters for the fourth time. As posited, this is soon after the start of November 6, 1572, and less than about 6 hours before Schuler first spies the New Star. At that instant, the Ghost turns over responsibility to Hamlet for the promulgation of the New Astronomy. Shakespeare probably does not know exactly when the New Star actually became visible, but from November 2–4 he lets the Ghost prepare for the event. Violations of causality are no hindrance to supernatural beings, and there is a slim chance that the Ghost, who is here and everywhere, knew that the event is about to occur (or in modern terms, that light from the event is soon to arrive at Earth).

VALIDATIONS

As demanded, additional information from the script of *Hamlet* validates the chronology in table 3.3.

First, on November 5, 1572, the waning Moon lies close to the direction of the Sun and is not above the horizon until the start of twilight. Thus, there is no Moon on the night of the opening scene, which places fact behind what has hitherto been conjecture, that in the opening scene, darkness hinders the soldiers in recognizing one another.

Second, through the duration covered, the illuminated fraction of Venus increases from 0.24 to 0.27, and thus Laertes' description of Venus as a waxing crescent (1.3.11–1.3.14) matches the actual phase of the planet at the time. On the days listed in table 3.3, Venus lies in the direction of the constellation of Virgo, which matches what Laertes supposes Ophelia's chaste state to be. It so happens also that on November 3 when the Ghost first appears, the Moon too is in Virgo, although she leaves the sign later that same evening.

Third, in scene 1.1, Horatio warns of Ophelia's fate by citing an eclipse of the Moon, which can only occur when the Moon is Full (i.e., in Opposition; figure 3.2). In scene 1.3, Laertes devotes the second part of his sermonizing to warning Ophelia to guard her chastity, and within a matter of hours, the phase of the icon of chastity, the Moon, becomes New (i.e., in Conjunction). Because it has been demonstrated that Ophelia's name means, "aligned with the Sun in Opposition or Conjunction" or "aligned in Syzygy," the script itself supports both cases from which Ophelia takes her name.

Fourth, the Moon sets in the midafternoon of Sunday, November 2, 1572. Thus, the sky is dark that evening, which may be one reason why Munosius chose that time to teach students about the stars.

Fifth, the occurrence of the New phase of the Moon on the evening of November 5, 1572 coincides roughly with the times in scene 1.3 when Laertes advises Ophelia to beware of Hamlet's amorous designs, and when Polonius waxes apoplectic about his daughter's dalliance. In several history plays, Shakespeare uses "conjunction" to describe "union," so the Moon's conjunction with the royal Sun would be a matter of concern to an overly protective father like Polonius.

Sixth, Hamlet marvels that the sepulcher holding Old Hamlet's corpse has cast up his spirit, and he wonders why it "Revisits thus the glimpses of the moon." Monday November 3, 1572 corresponds to the time when the Moon lies about 18° above the southeastern horizon. It is then a waning crescent with an illuminated disk of 6%. According to Barnardo, the same scenario plays out the next morning, at which time, at the onset of civil twilight, the Moon is a thin sliver 9.5° above the horizon. When the Ghost appears for a third time, it is Wednesday, November 5. At the start of civil twilight that morning, the Moon is barely above the horizon and only 12 hours away from the New phase. Thus, on November 3 and 4, 1572, the Ghost could glimpse the Moon before fleeing the dawn. Brahe, Dee, and Digges showed that the New Star lay well beyond the Moon, and therefore above the clouds surrounding the Earth, so even with an overcast sky, the specter could glimpse the Moon from on high. Depending on local weather conditions, it might even catch sight of the Moon on November 5 as it prepares to descend into Purgatory.

Seventh, in 1610, the news travel time from southern Germany to England was about a fortnight (van Helden, *Sidereus* 87), and because the straight-line distance from Wittenberg to Elsinore is about half that, Horatio would have taken a few days to make the journey. Horatio does not convey Schuler's information about the New Star when he journeys from Wittenberg to Elsinore because he left before Schuler saw it.

Eighth, the dominical letters for the leap year 1572 are FE, so All Souls' Day falls nominally on Sunday November 2, but in the traditional rite Christians observe it on the following Monday. All Souls' Day celebrates the bond between the living and the dead, and is pertinent therefore to the relationship between Hamlet and his deceased father. By the present reckoning, the Ghost appears at Elsinore at 1 o'clock in the morning of All Souls' Day, 1572 (table 3.3), but why did Shakespeare not choose to have the Ghost herald the start of that holy day at Elsinore by making it appear at midnight? Time in England is about 50 minutes earlier than at Elsinore, so in effect (and within the margin of errors stated), the Ghost's first apparition at Elsinore occurs almost immediately after the start of All Souls' Day in England. By implying a secondary temporal zero point appropriate to England, Shakespeare not only draws attention to the role of Englishman in promoting the New Astronomy but also directs commentary to a coterie of cognoscenti such as might frequent the Inns of Court. This answer assumes that England had clocks and clock keepers every bit as good as Denmark's, which does not strain credulity if one assumes that the Diggeses, at least, were capable craftsmen.

Ninth, All Saints' Day occurs on Saturday November 1, but Shakespeare is careful to omit any association of the Hamlets with that day as, by evidence of the script (e.g., 1.5.10–1.5.13, 3.1.120), neither father nor son is singularly saintly.

Tenth, looking ahead to scene 2.2, Polonius enters to announce the arrival of the thespians whose task it is to cure Hamlet's melancholy.

By then the political battle lines are drawn, with the King's side attempting to convert Hamlet from his mourning duties and Hamlet aiming to demonstrate the guilt of the usurper King. Hamlet makes a seemingly gratuitous reference to a certain Monday, "You say right sir, a Monday morning, 'twas then indeed" (2.2.355). This refers to the morning of Monday, November 3, 1572, when the Ghost first appears and when both sides gird themselves for the contest ahead.

Paris, France

Act 2 opens as Polonius dispatches his manservant, Reynaldo, to Paris to spy on his son and other Danes. The spymaster advocates information acquisition through deceit. "By indirections find directions out" (2.1.64) he declares. Shakespeare could well have created Polonius as a caricature of Robert Pullen (d.*c.* 1147), a medieval schoolman whose Latinized name is Polenius, and Reynaldo may take his name from John Reynolds (1549–1607) (table 3.1), who was a contemporary of Shakespeare and an inveterate enemy of the theatre. The juxtaposition of the names Polonius and Reynaldo in scene 2.1 is likely a slight upon these Oxford schoolmen, which is plausible given the suspicion that at a performance of *Hamlet* at Oxford, someone changed their names to avoid trouble for the players (Hibbard, *Hamlet* 75).

It is only in scene 2.1 that Paris, France is revealed to be Laertes' destination, which is apt because Thomas Aquinas had studied at the University of Paris under Albertus Magnus (1193–1280) and later became a Professor of Theology there. Paris University became a citadel of scholasticism whose docents regarded Ptolemy's calculus as the one true system of the World, which Laertes will eventually embrace.

Infatuation

Ophelia bursts in upon her father and complains that Hamlet has frightened her. He asks, "Mad for thy love?" and she replies truthfully, "My lord I do not know" (2.1.83). She tells her father that she obeyed his command and denied Hamlet access to her, and at this early stage in her life's journey, there is no reason to suspect her of duplicity. Polonius leaps to the conclusion that the Prince is infatuated. "That hath made him mad," he concludes (2.1.108). By "mad," Shakespeare may mean any or all of the meanings from the eleventh and thirteenth centuries, like enthusiasm (as in *Jeremiah* 1:38, "they are madde upon their idols"), anger, or mental illness. Hamlet is not an idolater, but a cool-headed empiricist. By contrast, Polonius suffers from the delusion of infallibility and does not doubt conclusions reached through his own supposedly inerrant thinking. Once Polonius has pondered an issue, his mind becomes like a Möbius strip: mono-faceted and perpetually self-validating.[6] In *Hudibras*, Samuel Butler (1612–1680) satirized hypocrites because they know "what's what and that's as high as metaphysic wit can fly" (Part I, Canto I, 149–150). Previously, Hamlet too had jumped to an unwarranted conclusion when he assumed "It is an honest Ghost" (1.5.138), but in Act 3, he will seize an opportunity to test his assumption.

Polonius tells Ophelia, "Come, go with me, I will go seek the king" (2.1.99), and again, "Come, go we to the king…Come" (2.1.115–2.1.118). Three times he bids her to accompany him, but in the next scene when he arrives to address Claudius, he arrives alone.

HAMLET'S TRANSFORMATION

Claudius seeks to restore the errant Prince to health, which amounts to redeeming him from his supposed misconceptions. To that end, he has summoned two courtiers, Rosencrantz and Guildenstern, and upon their arrival, he explains why he needs their help.

> CLAUDIUS Something have you heard
> Of Hamlet's transformation — so call it,
> Sith nor th'exterior nor the inward man
> Resembles that it was. (2.2.4–2.2.7)

Claudius is concerned about Hamlet's "inward" and "exterior man," which allegorically are his personification of the New Astronomy, heliocentricism, and the infinite Universe of stars. He calls on the twin sycophants to bolster his position as the defender of geocentricism, which is the basis of the Tychonic hybrid (table 3.2).

Polonius enters without Ophelia and announces the return of the Danish ambassadors from Norway, and eight lines later announces that he has found "the very cause of Hamlet's lunacy" (2.2.49). When he withdraws briefly to summon the ambassadors, the Queen offers the opinion that the causes of Hamlet's melancholy and apparent affective disorder are due to his father's sudden death and her hasty remarriage. Claudius feels that they should consult Polonius, which shows that he has a slightly better grasp of methodology than Polonius because he takes Gertrude's doubt seriously and plans to seek testimony and examine evidence. By the second century AD, Claudius Ptolemy had progressed slightly beyond most of the philosophers of ancient Athens because he gathered new data and incorporated them into his work.

Polonius and the ambassadors enter, and Voltemand announces that Old Norway had been under the impression that Fortinbras was staging a *coup de main* against Poland, but that upon further inquiry discovered that he was really planning to attack Denmark. He reports further that Old Norway intervened and persuaded his nephew to turn pretense into practice by unleashing his troops against "the Polack" (2.2.75) and "never more / To give th'assay of arms against [Claudius]" (2.2.70–2.2.71). However, to reach Poland overland, Fortinbras and his army must pass through Denmark, so the ambassadors report that part of the deal is for Claudius to permit this.

Old Norway is a heroic figure under the direction of a supreme being. He acts as an intermediary to improve the lot of humankind by channeling the energy of Fortinbras against Poland, possibly knowing that Hamlet will take care of business in Denmark. Later in scene 4.4, which is marked by a significant cut from Q2, Fortinbras' army passes through Denmark without incident, and in the end, he will return to Denmark as the beneficiary of the paranormal orchestration. The irony is that by dispatching ambassadors to Old Norway, Claudius unwittingly helps promote the demise of the very cosmology he personifies. The further irony is that the apparent success of the diplomatic mission pleases Claudius and prompts Polonius to celebrate: "This business is well ended" (2.2.85), he thinks.

LOVE LETTER

Polonius returns to the subject of Hamlet's health, claiming on the evidence of a letter that Hamlet supposedly sent to Ophelia and which he claims she gave him, that Hamlet is mad. The spymaster reads the letter.

> To the celestial, and my soul's idol, the most beautified Ophelia in her excellent white bosom, these, et cetera.
>
> Doubt thou the stars are fire,
> Doubt that the sun doth move,
> Doubt truth to be a liar,
> But never doubt that I love.
>
> O dear Ophelia, I am ill at these numbers, I have not art to reckon my groans; but that I love thee best, O most best, believe it. Adieu. Thine evermore, most dear lady, whilst this machine is to him, Hamlet. (2.2.109–2.2.122)

He had not even finished reading the first sentence before Gertrude asks in disbelief, "Came this from Hamlet to her?" By the letter's end, there can be little doubt that it dismays her, and one might also wonder why Hamlet would burden his inamorata with such inanities.

Stars in the sky serve the interests of romance quite well, but so does the Moon, yet Hamlet is so gauche that he ignores it in favor of the Sun. By calling Ophelia "celestial," the earnest swain oversteps his bounds because it is ridiculous, even blasphemous, to think that any terrestrial existent can aspire to celestial perfection. The double superlative, "I love thee best, O most best, believe it," emphasizes the extremity of his infatuation. His letter abounds with adjectives and convoluted grammatical constructions. The writing makes Hamlet appear so enamored of Ophelia that he cannot write from the heart. The four lines of doggerel suggest a puerile versifier. Indeed, he seemingly admits to literary incompetence ("ill at these numbers"), and his admiration of Ophelia's qualities is so excessive as even to warrant censorship ("et cetera").

The first two lines of the verse advise Ophelia to doubt two tenets of the Aristotelian Worldview, that the stars shine by fire and that the Sun moves around the Earth.[7] Hamlet would have to be a trifle touched to think that urging Ophelia to abandon articles of cosmic faith would benefit her and promote his romantic interests. Line three of the verse sounds as if he urges her to doubt that truth is falsity (i.e., to accept that truth is truth), which is a vacuous truism unworthy of the utterance unless perhaps to someone crazy enough to doubt the obvious. This hardly promotes romance either. In line four, Hamlet makes his confession of love, whereupon he compares his body to a machine. A sane admirer would not stoop to such vulgarity. Apparently, Polonius is right—Hamlet is mad and madly in love. However, Polonius' entrance without Ophelia is cause for suspicion, and his sudden devotion to empirical evidence (i.e., the letter) is out of keeping with his character.

Shakespeare draws attention to the contradiction in the spymaster's presentation in two passages of similar construction that occur at approximately equal distances into their respective scenes.

> OPHELIA No my good lord; but as you did command,
> I did repel his letters and denied
> His access to me. (2.1.106–2.1.108)
>
> POLONIUS I have a daughter — have while she is mine —
> Who in her duty and obedience, mark,
> Hath given me this. (2.2.106–2.2.108)

The two passages are contradictory. In the first, Ophelia exemplifies obedience to her father's command by citing her rejection of Hamlet's letters, yet in the second Polonius claims that Ophelia received a letter from Hamlet. This casts doubt upon the provenance of the letter (2.2.109–2.2.122n).

Goddard, in an article rescued from oblivion by his daughter and published posthumously, noted Ophelia's absence and Polonius' perversity and deceit and suggested that the jealous father forged the letter during the interval between scenes 2.1 and 2.2. Polonius has even more time because he enters after scene 2.2 has begun. Polonius is no stranger to forgery because he had instructed Reynaldo to put on his son "What forgeries you please" (2.1.20). He argued further that the style of writing seems calculated to cast Hamlet in an unfavorable light, and that the letter is replete with platitudes and superlatives that a schoolman might devise to impugn the sanity of a heterodox thinker.

If Polonius authored the letter in order to make Hamlet seem mad, it would make sense that lines 1–2 of the letter's verse would urge Ophelia to doubt the received wisdom of the Old Astronomy, which of course the pedant thinks everyone in their right mind accepts. Line 3 is a banal truism that needs no repeating except possibly by someone out of touch with reality, and line 4 establishes the source of Hamlet's alleged insanity.

Scientific thinking was often mistaken for insanity. Robert Greene (1558–1592) accused Marlowe of blaspheming with "the mad priest of the sun" (Nicholl 206), the Nolan Giordano Bruno (1548–1600), who on philosophical and theological grounds advocated a materially infinite Universe in 1584. After Tycho fell into disfavor with the Danish court, he too was deemed mad, and Copernicus' protégé, Rheticus, was similarly regarded (Danielson 123–124). In Act 5, the audience encounters a gravedigger who fantasizes that all Englishmen are as mad as Hamlet, but in the subtext, Hamlet's supposed madness is a reflection of Thomas Digges' progressively minded use of empiricism in scientific inquiry. If what the gravedigger claims is true, there must exist other late sixteenth-century scholars in England who advocate a fresh look at the World.

Remarkably, the subtextual meaning of the letter follows from its literal interpretation. If celestial *Op-heli-a* were to become a partner to Hamlet, she would have to doubt two tenets of geocentricism, that "the stars are fire" and that "the Sun doth move." The former requires a repudiation of the ancient element Fire, and the latter denies that the Sun orbits a centrally placed and immobile Earth. Thus, lines 1–2 of the doggerel refer to the Copernican component of the New Astronomy, whereas line 3 suggests the language of logical argument and is a pejorative reference to methodological principles prevalent among Aristotelians. Not incidentally, line 3 has a peculiar construction because "truth" is an abstract notion whereas "liar" is a category of human. The imperative is ambiguous to the extent that "liar" may simply connote falsehood, but the lack of parallelism suggests that Shakespeare had in mind a particular prevaricator, and the three lines suggest that the liar of reference is Aristotle. The anaphora "doubt" in lines 1–3 indicates that Aristotle did not doubt the correctness of his thinking. He is a falsifier for his failures of method and for his neglect even of the possibility of a moving Earth.

Line 3 of the verse means that Ophelia should doubt that the truth of the New Astronomy is wrong, or equivalently that she should accept that the empiricism inherent in celestial observation is correct. As one who accepts appearances at face value, Aristotle is the opposite of a doubting Thomas, who in this case is Thomas Digges. The fact that the third line of the verse is so contorted suggests that it is a parody of peripatetic philosophy. Perhaps Shakespeare had in mind scholars like Bernard of Verdun (*fl.*13th century) and his Parisian contemporaries whose methodology was so deficient that they brooked no alternative to Ptolemaic astronomy. Thus, lines 1–3 of doggerel indict Aristotelian philosophy, and line 4 establishes that Hamlet has a both a literal and figurative interest in Ophelia.

In the remaining text of the letter, Hamlet alleges that he is "ill at these numbers," but he does not divulge what these numbers are. The only plural antecedent to "numbers" is "stars." If stars are in fact the referent, the groan-provoking number cannot refer to those visible to the unaided eye because, as will be shown in the following discussion of scene 3.3, these stars number about 10,000, which is a conceptually manageable quantity. In addition, Hamlet has the ability to comprehend numbers like 20,000 (4.4.60) and 40,000 (5.1.236), so it seems that the number in question is far larger.

A likely possibility is the number of stars visible through a telescope. There are about 120,000 stars to an apparent visual magnitude of nine, which is about ten times fainter than good eyesight can register on a dark night. One magnitude fainter than that, the number jumps to about 400,000, which is ten times more than the value of 40,000 by which

Hamlet measures his love for Ophelia. Shakespeare ensures that counting is an issue when Horatio tells Hamlet that the duration of the Ghost's presence was as long as "one with moderate haste may [count to] a hundred" (1.2.237). Counting continuously to 400,000 would take ten or more days, a task that would exasperate the most dedicated arithmetician. The passage immediately following the verse suggests that it refers to the Diggesian vision of a Universe of stars stretching out into space, and if so, Shakespeare uses the love letter to introduce all three major components of the New Philosophy extant in the sixteenth century—the New Organon of evidence-based knowledge, heliocentricism, and the seemingly unending distribution of stars.

Hamlet signs off his letter by referring to his body as "this machine." The *OED* cites the example of a human or animal frame seen "as a combination of several parts" each performing a function, like a mechanical device. "Machine" is also a structure of any kind, material or immaterial, a fabric, or an erection. The term is apt because the Ptolemaic model, with all its rods and wheels, resembles an intricate machine. Similarly, Digges refers to the form or frame of the Universe, and in a passage from the same scene, Hamlet refers to cosmic existents using architectural terminology, such as "this goodly frame, the earth," "this canopy the air," "this brave o'erhanging firmament," and "this majestical roof fretted with golden fire" (2.2.282–2.2.285). Vulgarity aside, Hamlet's "machine" could be the new frame or structure of the Universe.

The meaning of Hamlet's proviso, "while this machine is to him," is that "there is a great deal to him" (Jenkins 2.2.122–2.2.123n), which is true because the components of the New Astronomy are an integral part of his dramatic construction. Certainly, the infinite Universe contains a "great deal" (of stars). Altogether, Hamlet's closing sentence means that he will remain faithful to Ophelia for as long as she remains true to him and for as long as he is alive to personify the New Astronomy.

The forgery is fraught with dramatic irony because, in attempting to besmirch Hamlet, the archetypal pedant unwittingly advances the cause of the New Astronomy under the nose of its staunchest detractor, the false King. To bolster his case, Polonius explains that he told Ophelia to have no dealings with Hamlet. "Lord Hamlet is a prince out of thy star" (2.2.139), he explains, the irony here being that he correctly classifies the pair's figurative relationship, with Ophelia representing the geocentric Moon and Hamlet claiming association with the royal and central Sun.

More Doubt

Polonius states that he had ordered Ophelia to sequester herself, admit no messengers and, receive no tokens. Although the royal couple harbors doubts, they fail to ask how Hamlet managed to deliver a letter to her if she were isolated. It is unlikely that she received a letter from Hamlet prior to the proscription because then Hamlet had no need to resort to that mode of communication when he was free to see her. The royal couple's doubt spurs Polonius to defend his thinking, which he does indignantly, as if protesting too much.

> POLONIUS Hath there been such a time, I'ld fain know that,
> That I have positively said, 'tis so,
> When it proved otherwise?…
> If circumstances lead me, I will find
> Where truth is hid, though it were hid indeed
> Within the centre. (2.2.151–2.2.157)

If Polonius thinks something is true then it is true, and if he thinks something is false then it is false. The eristic seeks the truth at the core of his beliefs, the Earth, or equivalently in his own mind where of course he always finds it.

To his credit, Claudius is unconvinced. "How may we try it further?" he asks (2.2.157). It is hard to fathom why Claudius does not summon Hamlet to ask him directly about the letter or for a sample of his handwriting. Instead, Polonius suggests that he "loose" his daughter to the Prince as he walks in the lobby (2.2.160) while he and Claudius hide behind an arras to mark the encounter. The spymaster plumbs the depths of scurrility, as he lacks empathy even for his own children. Claudius agrees and the deceit plays out in the next scene.

At that moment, Hamlet enters, prompting Polonius to hustle the royal couple away. Hamlet calls Polonius a fishmonger, which puzzles Polonius, but Hamlet wishes that he were because then he would be honest and be "one man picked out of ten thousand" (2.2.176–2.2.177). The reference is to book 3 of Aristotle's *Politics*, in which the philosopher characterizes the ideal city as comprised of various professions ("carpenter…husbandman…shoemaker, and so on") who have sound family connections and are in alliance against evildoers. Hamlet imputes a lack of these qualities to Polonius, and then warns him that Ophelia may conceive.

Hamlet's equivocation is part of his strategy for survival, provoking Polonius to comment, "Though this be madness, yet there is method in't" (2.2.200–2.2.201). In a further irony, he defines Hamlet's "madness" as "method." Polonius mutters further that Hamlet's replies are "pregnant" (2.2.203). Perhaps Ophelia's condition pricks her father's conscience—if he has one. Alternatively, "pregnant" is an appropriate term for a Prince who basks in the light of the royal Sun, as when Copernicus (26) wrote, "The Earth…is fertilized by the sun and conceives offspring every year."

Polonius needs to leave in order to contrive a meeting between the Prince and Ophelia, and he tells Hamlet, "I will…take my leave of you" (2.2.208–2.2.209). The Prince reinforces his need to survive by making it clear that Polonius cannot take from him anything that he will more willingly part with, except his life. For emphasis, Hamlet repeats "except my life" three times (2.2.210).

COURTIERS

The Sovereign's sycophants, Rosencrantz and Guildenstern, enter and greet Hamlet like long-lost friends. Hamlet asks them, "What news" (2.2.227), to which Rosencrantz replies that the world has grown honest. A passage of 28 lines follows, which editorial opinion ascribes to a cut in Q2 rather than an addition to F1. In the restored passage, Hamlet asks why Fortune "sends you to prison hither?" (2.2.231–2.2.232). He opines that Denmark is one of the world's worst prisons having many confines, wards, and dungeons. Tycho's castle had cells for detaining peasant debtors, and Kronborg Castle was renowned for its dungeons in which, it was rumored, prisoners were tortured.

The cut may have assuaged the sensibilities of Anne of Denmark (1574–1619), who was deceased when F1 appeared. It relates political restrictions in Denmark to bounded space in the Aristotelian model, to which Rosencrantz replies that Denmark is too narrow for Hamlet's mind. This prompts the response (2.2.243–2.2.244), "O God, I could be bounded in a nutshell and count myself a king of infinite space, were it not that I have bad dreams." In the fifteenth and sixteenth centuries, theologians thought of space as infinite only if it referred to the space associated with the deity, and it is therefore unlikely that Hamlet sees himself as a "king" of that sort of space. Probably, he has some sort of physical space in mind, in which case the referents "nutshell," "infinite space," and "bad dreams" require explanation.

"Nutshell" refers to the rigid shell of stars in Tycho's model, as well as to Nut—the ancient Egyptian sky goddess who arches over the heads of observers. The meaning could also refer to Tycho's model, which is a generalization of an ancient Egyptian model. In addition, a "nut" has a hard shell, and during the fourteenth century it symbolized "something of trifling value" (*OED*). The conceit refers to Tycho's *minutum*, which (with its shell of stars) is the most compact of the three bounded models. It could refer also to Ptolemy's bounding sphere, which is barely larger than Tycho's.[8]

"Infinite space" was a hotly debated topic in the sixteenth century, and the reference is not surprising in the present context. The word "infinite" originated in theology, and from 1413, it had the chiefly theological property of an unassignable end, whether real or imaginary. This is the meaning intended by Nicholas of Cusa and Palingenius to describe the abode of the Almighty. In 1660 the word entered the language of mathematics through the geometry of Euclid (*fl.c.* 300 BC) and later acquired its current spatial and temporal senses (*OED*). From about 1385, however, there existed "loose or hyperbolic" meanings for quantities indefinitely or exceedingly great, immense, vast, or immeasurably and incalculably large, as in *The Merchant of Venice*, "an infinite deal of nothing" (1.1.114). In *Hamlet*, Shakespeare writes of virtues (1.4.33–1.4.34), faculties (2.2.287), and jest (5.1.157) that are "infinite," and it is reasonable to expect that the fourth instance ("infinite" space) is hyperbolic too.

Digges used "infinite" to describe the incidence of things that are arguably finite such as "infinite absurdities" and "infinite multitude of absurd imaginations" (as cited in Johnson and Larkey 79, 86). When he wrote, "This orbe of starres fixed infinitely up extendeth hit self in altitude" (Johnson 166; see figure 1.7), he referred to a limit that is exceedingly great but uncertain or incalculable, as is the meaning of the symbol "∞" in physics. The induction bears comparison to a crucial step on Plato's ladder of understanding, in which objects of perception evolve into objects of intellect. In this sense, the idea of an infinite Universe becomes a Platonic Form—a new image of the World. An "infinite" distribution of stars would extend beyond existent ones accessible to the telescopically aided eye to include those that exist only in the mind's eye and as such would have an unobservable and incalculable bound. Thomas Digges made such an intuitive leap by distributing stars uniformly in "infinite space," and it is reasonable to believe that Shakespeare did so too.

"Bad dreams" refers to both the oppressiveness of Elsinore and the threat of persecution because within a few lines Hamlet remarks, "by my fay, I cannot reason" (2.2.251–2.2.252), meaning that free inquiry about the Universe is proscribed. The Diggesian model is antithetical to the confining space of the Aristotelian World which, to a devotee of the New Astronomy, feels figuratively like a prison. By dictate of the false monarch, Hamlet is heir to the Old Astronomy, but he feels incarcerated at Elsinore. Just as Digges tore down the walls of the bounding Firmament, Hamlet must free himself from the old paradigm. However, Hamlet cannot speak openly about this for fear of repercussions, the mere anticipation of which would cause the most redoubtable citizens to have bad dreams.

Hamlet anticipates the contest between the New Astronomy and the Tychonic Worldview, and he asks the King's myrmidons pointedly, "what make you at Elsinore?" (2.2.256–2.2.257). Rosencrantz tries to dodge the question, but Hamlet knows at least part of the answer. "I know the good king and queen have sent for you," (2.2.266–2.2.267) he says. Anticipating their mission, Hamlet explains that he does not know why he has lost all his mirth. He is, on the one hand, "most dreadfully attended" (2.2.255–2.2.256) by a murderous regime and a ghost who has sworn him to secrecy. He knows that his life is not his own, so he must speak equivocally.

> HAMLET…it goes so heavily with my disposition that this goodly frame, the earth, seems to me a sterile promontory; this most excellent canopy the air, look you, this brave o'erhanging firmament, this majestical roof fretted with golden fire…appeareth no other thing to me but a foul and pestilent congregation of vapours. (2.2.281–2.2.286)

The Prince suffers from "world-weariness" (2.2.280–2.2.290n), like Portia in *The Merchant of Venice* (1.2.1–1.2.2). When it comes to Worldviews, Hamlet is weary of Elsinore's, and Claudius is wary of Wittenberg's.

In this passage, "promontory" means a point of land that juts out like Jutland or anything that resembles it, and coincidentally, the "Promontory of Noses" is where Tycho went for a prosthetic nose (Crompton 502). Tycho had two artificial noses, which he secured with an adhesive salve. The one he used on important occasions was made of gold and silver blended to a flesh tone, and the other was made of a lighter copper alloy and was reserved for everyday use as well as for his burial (Thoren 26). The passage connects Tycho's disfigurement to the element Air and conjures up images of nasal stuffiness, oppressiveness, and the crowded Tychonic model. Mention of Air and an infinite Universe in the same breath refers to the fanciful Pythagorean notion that space must be infinite for it to "breathe," a notion manifest in Digges' "famous diagram," figure 1.7 (Heath xxix; Heninger 42n50). In the entire passage,

Shakespeare conflates air with Tycho's disfigurement and his theory of the Heavens.

Hamlet expresses his belief that humankind can do better than the sorry effort to which he has just alluded.

> HAMLET What a piece of work is a man! How noble in reason, how infinite in faculties, in form and moving how express and admirable, in action how like an angel, in apprehension how like a god! The beauty of the world, the paragon of animals — and yet to me, what is this quintessence of dust? Man delights not me — no, nor woman neither, though by your smiling you seem to say so. (2.2.286–2.2.292)

Quintessence is the fifth element that is supposed to constitute the heavenly bodies and be latent in all things, including humans (Hibbard, *Hamlet* 2.2.306n), but the courtiers disregard Hamlet's noble sentiments and instead smirk at the remark "Man delights not me." Rosencrantz denies misinterpreting it but Hamlet presses him, "Why did ye laugh then, when I said man delights not me?" (2.2.294).

In the sixteenth century, death by dueling was only a slight misunderstanding away and Rosencrantz had better think fast (2.2.295–2.2.297). "To think, my lord, if you delight not in man, what lenten entertainment the players shall receive from you," he says, "we coted them on the way, and hither are they coming to offer you service." "Lenten" means austere or meager, so Rosencrantz claims that he laughed because sixteenth-century actors were male and he expects that Hamlet will give them a lukewarm reception. By changing the subject and introducing the brand new topic of traveling thespians, Rosencrantz dodges a confrontation, and it is not worth Hamlet's time to pursue the matter. Whether by supernatural influence or native ability, the Prince can think with blinding speed. He anticipates that one of the thespians will play a monarch. "He that plays the king shall be welcome, his majesty shall have tribute of me," he announces, meaning that he will honor one pseudo-king with as much seriousness as another (2.2.298–2.2.299n). He is under no illusion that the tragedians will do him any favors.

The irony in Rosencrantz's chuckle is now apparent, for indeed Hamlet will not delight in these actors, not because of their gender but because they are puppets of the Elsinore regime. He asks, "What players are they?" (2.2.303), and the courtier explains that they are the tragedians of the city, the ones that Hamlet was "wont to take such delight in" (2.2.304). Hamlet wonders why they are on the road when, in theory, they should fare better in residence, and Rosencrantz explains, "their inhibition comes by the means of the late innovation" (2.2.308–2.2.309).

"Innovation" means primarily "revolution" or "the alteration of what is established by the introduction of new elements or forms" (*OED*). Hamlet's cosmic "transformation" (2.2.5) and his means of observing the "revolution" (5.1.75–5.1.76) fit the definition. A second meaning is "political revolution…rebellion or insurrection." Thus, the revolution that Shakespeare has in mind applies equally to the two facets that predominate in *Hamlet*—the political world and the scientific World. The lines that follow (2.2.313–2.2.333) strikingly exemplify text omitted from Q2. Hamlet wonders whether the thespians are touring because they grow rusty, but Rosencrantz claims they left because child actors are now the fashion. Perhaps the "late innovation" refers to this fad and to the "war of the theatres" that erupted around 1601. Shakespeare goes on to compare child players to nestlings—"an eyrie of children, little eyases" (2.2.315–2.2.316)—which is appropriate because some young male birds of prey are known to oust older birds of the same species from their nests. This comparison prepares the way for Hamlet's perennially puzzling remark on hawks and wind directions, to be addressed shortly.

Hamlet asks whether the boys carry it off ("carry it away," 2.2.332), and Rosencrantz agrees, "Ay that they do my lord, Hercules and his load too"—a reference to the emblem of the Globe Theater, which was "supposed to have been Hercules carrying the celestial globe on his shoulders" (2.2.333n).[9] Between them, Atlas and Hercules support the finite, rigid structure symbolic of a closed Universe. The fact that the boy actors "carry it away" signifies that they dispose of it symbolically (if not physically), which qualifies them as supporters of the New Astronomy, whereas their superannuated elders cling to outmoded ideas and are no longer welcome in the city. It is not surprising that the Claudian court has invited this touring troupe to its Elsinore redoubt.

PORTRAITS

The Q2 script resumes with what sounds like a rueful segue.

> HAMLET It is not very strange, for my uncle is king of Denmark, and those that make mouths at him while my father lived give twenty, forty, fifty, a hundred ducats apiece for his picture in little. (2.2.334–2.2.346)

This could mean that people who formerly derided his uncle Claudius are now willing to pay a high price for a miniature portrait of him. It is like the unreasoning adulation that the child actors enjoy (Jenkins 2.2.359–2.2.362n), but the phrase "his picture" could also mean a portrait *belonging* to Claudius, which in turn could refer to one or other of Tycho's portraits (figures 3.3, 3.4). Hamlet's next sentence supports the conjecture. "'Sblood, there is something in this more than natural, if philosophy could find it out," meaning "there is something abnormal about the miniature as scientific investigation would show" (2.2.337–2.2.338n). The remark has meaning even beyond the loss of Tycho's nose and the artificial nose that replaced it, as will be explained.

A flourish announces the arrival of the actors. Hamlet's disputatious interaction with the twin courtiers has lasted over 120 lines, and because the two votaries personify Tycho's model, it is likely that the "picture in little" has something to do with them. It will soon become apparent that this segue is further preparation for understanding Hamlet's enigmatic remark on wind directions.

Rosencrantz did not expect Hamlet to welcome them, but Hamlet does so even before they are on stage. His next words are, "Your hands, come then" (2.2.340–2.2.341), which many believe is Hamlet's way of encouraging applause for the troupe, but his greeting is genuine only in the sense that his thinking has progressed far beyond the appearance of a warm welcome. Rosencrantz is unaware that Hamlet plans to use the

FIGURE 3.3. First de Gheyn engraving of Tycho Brahe for his 40th birthday (1586), from Huizinga.

FIGURE 3.4. Second de Gheyn engraving of Tycho Brahe (1586).

players to further his own ends, and that this remark is the next salvo of a thinly disguised parody on Tycho's portraits. Figures 3.3 and 3.4 show these engravings made in Amsterdam by the renowned artist Jacob de Gheyn II (1565–1629).

In 1590 Tycho Brahe sent two copies of *Liber Secundus* to the English savant, Thomas Savile (*c.* 1560–1593), younger brother of the founder of the Savilian professorships (Dreyer 182). He enclosed four copies of the portrait depicted in figure 3.4. In his cover letter, Tycho sent regards to John Dee and Thomas Digges, both of whom (like Tycho) had written books on the New Star of 1572. By 1590, Thomas Digges had authored or coauthored a number of treatises on mathematics, military science, politics, and astronomy, so it is likely that Savile honored the astronomical side of Tycho's request by forwarding Tycho's book to him.

Tycho also requested approbation from English poets, but it appears that no one took up the challenge. Savile would have known that a foreign dignitary of Tycho Brahe's stature deserved the best poesy that the kingdom could offer. Shakespeare's *Sonnets* and at least six plays of the Canon including an early version of *Hamlet* had appeared possibly as early as 1589–1590, and even if Savile was unaware of Shakespeare, the Digges family was not so deprived and word may have reached Shakespeare by that means (Hotson, *Appoint* 97–140). In fact, a copy of Tycho's portrait ended up in the possession of Thomas' oldest son, Dudley Digges (1583–1639) (Rowse, *Man* 226).

The portraits depict escutcheons of Tycho's sixteen great, great grandparents. In 1910 Huizinga suggested that Shakespeare named the twin courtiers in *Hamlet* for two of them, Erik Rosencrantz and Sophie Guildenstern. He believed this was more probable than other suggestions made in 1908 that attributed these names to reports brought back by touring actors or to the rosters of foreign universities where Danish students had enrolled given how these surnames were common.

The shields of Rosencrantz and Guildenstern depicted in the engravings are both in favored positions on Tycho's right-hand side. Erik's shield leads the procession of male ancestors across the sky, and Sophie's is closest to the ground.

De Gheyn's first engraving has two genealogical errors (figure 3.3), which he corrected (figure 3.4). In the later picture, Tycho has donned his plumed beret, *Instrumentorumque* replaces *Machinarumque*, facial features are redrawn, and there is a mirror reversal of the torso but not of the head. Yet with the opportunity to make corrections, one would think that an artist of de Gheyn's stature would correct all errors and avoid making new mistakes.

One blunder in particular attracts attention. When a hand rests palm down on a flat surface, the thumb is the only finger that can bend nearly in the plane of the surface, but in both engravings, Tycho's fifth ("little") finger has this property. It looks as if it might have been broken at the knuckle and twisted through about 90°, causing it to resemble a thumb. No evidence exists that Tycho was handicapped in this way, and no malformed hands were reported in 1901 when his remains were exhumed. Thus, it appears as if each engraving shows Tycho with swapped hands, so that what appears to be a little finger is in fact a thumb. The other hand is none too prettily depicted either, which justifies Shakespeare's reference to plural hands ("Your hands, come then," 2.2.340–2.2.341). In line 2.2.337, "S'blood" refers to imaginary bleeding from the graven swap, and real bleeding from the facial wound.

There are other travesties. The lighting in figure 3.4 is highly selective, with that from the left illuminating the facets and jambs of the framing architecture, and that from the right illuminating the subject. Only a contrived placement of artificial lighting could produce this effect, one consequence of which is that Tycho's dominant right side is in the dark.

Hamlet's parody continues. "Th'appurtenance of welcome is fashion" (2.2.341) refers to the extravagant garb and complementary appurtenances depicted, such as a plume on the bonnet, a gold chain, and a medallion, and remarkably, the word "garb" occurs in the next line (2.2.342). The welcoming speech concludes with "You are welcome—but my uncle-father and aunt-mother are deceived" (2.2.344–2.2.345). Hamlet's uncle is his self-proclaimed father who, in a sense, has turned his mother into his aunt, but he speaks also of Tycho's uncle and aunt who abducted Tycho when he was an infant and served in effect as his father and mother. On the one hand, Tycho's mother and father are his parents, and on the other hand, so are his uncle and aunt. The couples are interchangeable, like the hands in the portraits.

HAWK AND HANDSAW

Thus, on the subject of parentage, Shakespeare finds it convenient to draw a parallel between the fictional Hamlet and the real-life Tycho Brahe. This temporary duality prepares the way for understanding Hamlet's enigmatic termination of interaction with the courtiers, "I am but mad north-north-west. When the wind is southerly, I know a hawk from a handsaw" (2.2.347–2.2.348). Shakespeare adapted the last sentence in this passage from a proverb, "He knows not a hawk from a hand-saw," wherein "hand-saw" passes as a corruption of "hernshaw" or "heron" (Clarke and Cowden 726). In effect, Hamlet is saying that he knows the difference between a bird of prey and the prey itself (Hulme 54–55). In other words, he wonders whether the courtiers are the predators and he the prey, or vice versa.

The coupling of the aphorism with compass directions has generated more heat than light, but the cosmic allegory enlightens the audience. In classical antiquity, the term "wind" was synonymous with compass bearings because of the predictability of wind directions in the Mediterranean. The term applies to 32 headings 11¼° apart (32 × 11¼° = 360°) of which north-northwest (NNW) is one. Tycho himself determined that Elsinore lay NNW from Hven, and by the end of the sixteenth century, cartographers knew that Wittenberg was approximately south of Hven, which makes it approximately south of Elsinore too because Hven and Elsinore are close together compared to distant Wittenberg.

As remarked, at this point in the script Hamlet and Tycho are temporarily interchangeable like the hands in the portraits, so the passage in question can be interpreted as if each were on Hven. In Hamlet's case, the winds confirm the Worldview that follows the advent of the Ghost. When the wind is southerly (from the direction of Wittenberg), Hamlet on Hven would be discerning (i.e., he knows a hawk from a handsaw).

However, when the wind is NNW (from the direction of Elsinore), he must feign madness. Then, on swapping Hamlet with the rightful observer, a southerly breeze imbues Tycho with heliocentric correctness, whereas the NNW breeze affects him with geocentricism. These twin influences account for the twin attributes of Tycho's model (as personified by the twin courtiers).

By referring both to Tycho's astronomy and to Hamlet's power of discernment, Shakespeare kills two birds with one stone. Hamlet is discerning because he can tell the difference between a hawk and a heron, yet this is hardly a notable achievement for a Prince who is likely familiar with the royal sport of falconry. If Hamlet's welcoming speech concerns Tycho's portrait with its swapped hands, perhaps Hamlet compares a hawk not just to a hernshaw, but also to a tool suitable for severing hands, like a "handsaw." Then if "handsaw" refers to Tycho, to whom might "hawk" refer?

The answer should satisfy two constraints, namely that the context is discernment and that the person in question has qualities opposite to Tycho's. I have argued (*HU*) that the distinction is between the astronomer with artistically interchanged hands who can discern stars to the limit of the naked eye, and one with superior visual power like a hawk. Leonard Digges is the appropriate referent because his invention of the perspective glass vastly improved the capacity of the human eye to detect distant objects in fine detail (i.e., with resolution better than that of the human eye). The identification is further justified by the fact already mentioned that leonards are birds of prey closely related to hawks.

The meaning of an "eyrie" of "eyases" (2.2.315) is now apparent. The association of young hawks with the first telescopist, Leonard Digges, suggests that a new generation of observational astronomers will oust their elders. The passage refers to child actors who warrant applause, literally for their performance and figuratively for promoting the New Astronomy. Thus in *Hamlet*, Shakespeare predicts the manner by which future empirical research into the mysteries of the heavens will occur. Birds of prey will endure into the future, as will telescopists. The fact that Rosencrantz states that the "eyases" are "tyrannically clapped" (2.2.316) suggests that the new generation of fledgling telescopists might abuse the newfangled device, a possibility that is credible because this entire episode involving the courtiers and the thespians is sandwiched between an exit and an entrance of the spymaster Polonius. In keeping with the overall tenor of the play, Shakespeare sees that technological advances have both deleterious and beneficial uses, a sentiment echoed by Winston Churchill (1874–1965) who opined in 1946 that primitive times may return on the gleaming wings of science.

Shakespeare's fondness for sustained imagery explains Rosencrantz's assurance that the "endeavour" of the traveling thespians "keeps in the wonted pace" (2.2.314). In the sixteenth and seventeenth centuries, "pace" could mean a company or herd of asses, and forty-eight lines later Hamlet remarks that each actor came on his ass, so it seems that Rosencrantz inadvertently associates the thespians with creatures renowned for slowness of wit. Immediately thereafter he calls the heliocentric child actors who displaced them "little eyases" (2.2.315), which, suitably slurred, sounds as if Shakespeare is referring to a company of foals.

TYRANNICAL PLAYERS

Polonius enters and greets Hamlet and the courtiers civilly, but Hamlet likens him to a baby not yet out of swaddling clothes. Rosencrantz explains that Polonius is "the second time come to them, for they say an old man is twice a child" (2.2.352–2.2.353). By speaking proverbially, Rosencrantz alerts Hamlet that Polonius is meeting the players for a second time, which implies that the courtiers already know about the summons for them to perform at Elsinore.

Hamlet foresees that Polonius has come to tell Hamlet that the actors have arrived. "I will prophesy," he tells Rosencrantz *sotto voce*, "[that] he comes to tell me of the players." Whereupon, pretending to ignore Polonius, he adds, "You say right sir, a Monday morning, 'twas then indeed" (2.2.354–2.2.355). With reference to the chronology of Act 1 (table 3.3), Polonius hired the troupe on the morning of All Souls' Day (Monday, November 3, 1572), immediately following the first apparition of the Ghost because the need to counter its influence suddenly became a matter of great urgency for the Elsinore establishment.

Polonius does indeed announce that "The actors are come hither my lord" (2.2.359). Polonius has not been privy to Hamlet's previous discussion with Rosencrantz, yet he announces their arrival using the definite article ("the" actors). Similarly, without any prior mention of actors, Rosencrantz had said that "the" players are on their way (2.2.296). Use of the definite articles denotes a particular band of players, which further suggests that Polonius and the courtiers knew about the plan beforehand. The inference is that Claudius is the original source of the summons, and Hamlet's suspicion of their motives is justified.

Polonius praises the players as, "The best actors in the world, either for tragedy, comedy, history, pastoral, pastoral-comical, historical-pastoral, tragical-historical, tragical-comical-historical-pastoral…" (2.2.363–2.2.365). Ridicule aside, the category "pastoral" can refer to the care of souls, and thus perhaps to the thespians' goal to save Hamlet's soul by converting him to the doctrine of bounded geocentricism. As for the remaining categories, "tragical" and "historical" speak to the genre of *Hamlet* and its subtext, and "comical" is an apt description of the geocentric Weltanschauung.

The players arrive on stage and Hamlet welcomes them again, but this time he inquires whether they have come to challenge him. He asks them to recite a speech immediately as if they were "French falconers" who fly at the first quarry they see (2.2.391–2.2.392). When trained, falcons hunt particular kinds of prey and not anything that happens to come along. Once again, Hamlet is uncomplimentary to the French, who so far have proven to be doctrinaire allies of Elsinore.

Player I recites a poem about the savagery of Pyrrhus that is so moving as to engender sympathy in the gods, "unless things mortal move them not at all" (2.2.474). If the gods were capable of caring, their passion would "make milch the burning eyes of heaven" (2.2.475). Whereupon, Polonius comments that Player I has tears in his own eyes, which just goes to show how heartrending the recitation is. Polonius' observation suggests that if the gods were to have seen this brutality, they too would have become bleary eyed. Tears blur vision, effectively causing a loss of optical resolution. As a result, discrete images spread out and overlap, giving a milky appearance. Conversely, when the tears dry up, the opposite occurs. "Milch" refers to the Milky Way Galaxy (2.2.475n) which appears milky (*gala*, Greek for milk) to the naked eye because human vision cannot resolve the myriad of faint stars that comprise it. The implication is that improved optical quality resolves nebulous patches of the Milky Way into stars. Galileo reported this in 1610, but Shakespeare implies *circa* 1601 that the stars are not resolved because he gives no parallel description of stars themselves being nebulous.

GONZAGO

The troupe has been invited to Elsinore to stage a play, and apparently Hamlet has the freedom to choose any play he likes on the theory that any entertainment (whether cheerful or tearful) is bound to lift his spirits. Hamlet asks Player I if the troupe can play *The Murder of Gonzago*. Luckily, they can, and he asks if they would insert twelve to sixteen lines of his devising. Player I agrees and the actors leave, followed by Rosencrantz and Guildenstern. The courtiers were present when Hamlet made his request, and one wonders whether they and the players are simply humoring him.

When alone, Hamlet speculates that, "The spirit that I have seen / May be a devil" (2.2.551–2.2.552). In the spirit of the scientific renaissance, he plans to correct his earlier error, which is the unwarranted assumption that the Ghost is honest. He will amend the Gonzago play in order to see whether Claudius incriminates himself. "I'll have grounds / More relative than this," he puns (2.2.556–2.2.557), referring to epistemology and his father's burial. Act 2 ends with Hamlet's pronouncement, "The play's the thing / Wherein I'll catch the conscience of the king," (2.2.557–2.2.558). To him, the stage is like a laboratory in which he will conduct a test.

Act 2 confirms that Hamlet has undergone a transformation in his inward and outward self, resulting in personification of heliocentricism and an infinite Universe. His inward man transcends anthropocentricism by advancing heliocentricism, and his exterior man promotes the cosmic vision of a universally free state. Like Thomas Digges in 1576, Hamlet is in the flower of his youth and is the fountain from which will spring books on new visions of the World.

A Kind of Joy

To start Act 3, Claudius inquires of the courtiers whether they have discovered why Hamlet feigns a "dangerous lunacy" and puts on a "confusion" (3.1.2–3.1.4). Guildenstern complains that, whenever they try to elicit a confession of his "true state" he stays aloof "with a crafty madness" (3.1.8–3.1.10). Gertrude inquires, "Did you assay him / To any pastime?" and Rosencrantz replies that on their way to Elsinore they passed "certain players" (3.1.14–3.1.16). This accounts for the fact that on the previous day the courtiers had entered ahead of the troupe (2.2.493, 3.1.21). The word "certain" means "reliable" or "trustworthy," signifying that the troupe is of the right political and cosmic persuasion and thus suited to perform at court. Claudius takes news of the troupe's arrival in stride, as if he expects them, for neither Polonius nor the courtiers would have risked his ire by withholding that information until the next day.

Rosencrantz describes Hamlet's reaction to the arrival of the actors as "a kind of joy" (3.1.18). This derives from Hamlet's intuition that they present an opportunity for him to turn disadvantage to advantage. Claudius does not hesitate to accept Hamlet's invitation to attend the players' play because, as a devoted husband, he will do what he can to return his newly acquired son to the geocentric fold.

Lawful Espials

In the meantime, Claudius and Polonius set up the experiment to test the theory that they had formulated earlier (2.2.160–2.2.165). They will arrange for Hamlet to "affront" Ophelia "as 'twere by accident" while they, the self-appointed "lawful espials," will lurk sight unseen behind an arras (3.1.30–3.1.37). The idea is to contrive an event that is plausibly attributable to chance but which, when seen in a temporal context of a string of similar events, amounts to persecution of the purveyor of the new paradigm.

Polonius hands Ophelia a prayer book and utters a platitude laced with irony.

> POLONIUS 'Tis too much proved, that with devotion's visage,
> And pious action, we do sugar o'er
> The devil himself. (3.1.47–3.1.49)

Aside, Claudius whispers his agreement, "Oh, 'tis too true. / How smart a lash that speech doth give my conscience!" (3.1.49–3.1.50). In the scene immediately following the players' play, Claudius will admit prayerfully that his lese majesty "is rank" and "smells to heaven" (3.3.36). Despite his malign influence, Claudius has a human side to him. His two admissions of conscience bracket *The Murder of Gonzago* and help to explain Hamlet's tactics during the performance.

The principals are in place when Hamlet enters and launches an existential monologue.

> HAMLET To be, or not to be, that is the question —
> Whether 'tis nobler in the mind to suffer
> The slings and arrows of outrageous fortune,
> Or to take arms against a sea of troubles,
> And by opposing end them. (3.1.56–3.1.60)

Perlstein has proposed that Shakespeare based these lines on a passage in Aristotle's *Metaphysics*,

> Our present question is not whether it is possible for the same thing to be and not to be in regard to the locution, but whether it is possible in regard to the object. But if 'man' and 'not man' do not mean something different, it cannot be denied that 'not to be a man' will mean the same as 'to be a man,' and this will mean that 'to be a man' will just be 'not to be a man,' since they will be a single thing. (n.p.)

The opacity of this argument suggests that Hamlet's soliloquy is (among other things) a spoof of Aristotle.

Hamlet feels burdened by the circumstances that have suddenly enveloped him and his dilemma is not unlike that which is common in the lives of many—whether to fight or flee. He knows that suicide is an extreme antidote for oppression, yet he dreads what awaits him in the afterlife.

After Hamlet has had his say, he and Ophelia greet each other. She tries to get Hamlet to take back remembrances that she claims he gave to her. He denies giving her any, but she insists that he did, and that with his gifts came "words of so sweet breath composed / As made the things more rich" (3.1.98–3.1.99). It is unclear whether Ophelia's "words" refer to ones spoken or written (3.1.98n), and the ambiguity allows the eaves-dropping King to surmise the truth of the disinformation that Polonius has already fed him, particularly that Ophelia refers to written words and thus to the love letter he is supposed to have written to her. The lurking spymaster hopes that the concealed monarch will accept Hamlet's love letter as evidence in support of his misdiagnosis.

Despite Hamlet's denials, Ophelia unloads the alleged love tokens with the words, "There my Lord" (3.1.102). Her sudden show of assertiveness is out of character, suggesting that the time between 2.2.165 (when Claudius grants permission for Polonius to "loose" his daughter to Hamlet) and 3.1.55 (when the attempted entrapment begins) is sufficient for father and daughter to conspire to promote the idea that the love letter is really from Hamlet. However, the Prince sees through appearances and asks Ophelia whether she is honest and fair, whereas he himself admits to being merely "indifferent honest" (3.1.120) like a mingled yarn of good and ill. Both principals are blemished, as are the Sun and Moon with which they identify. "We are arrant knaves all," Hamlet says (3.1.125), and before he exits he leaves Ophelia with the advice that she repair to a nunnery to seek protection and care for her condition.

Ophelia frets over Hamlet's state of mind. She refers to him as, "The glass of fashion and the mould of form" (3.1.147). This evokes images of a molded lens and a glass mirror, which are elements of the perspective glass. The irony is that she calls her erstwhile beau, "Th'observed of all observers" (3.1.148). The lawful espials are observing Hamlet as she speaks, and Hamlet's real world counterpart is himself an observer—but of the heavens. Ophelia sustains the telescopic imagery with the plaint, "Oh woe is me / T'have seen what I have seen, see what I see" (3.1.154–3.1.155).

The surveillants emerge from behind the arras, and Claudius announces that Hamlet has little affection for Ophelia, which must mean he is not mad. Even so, the King fears that "There's something in his soul / O'er which his melancholy sits on brood" (3.1.158–3.1.159), meaning that he fears that Hamlet's melancholy will hatch misfortunes as a broody hen might hatch bad eggs—bad because they are the germs of the New Astronomy. Like Polonius, Claudius misdiagnoses Hamlet, thinking him melancholic, whereas in reality, Hamlet suffers from original thought and he is driven to his present state of angst because he is caught in a struggle between Good and Evil.

Polonius holds stubbornly to his ideas and prevails upon Claudius to retry the method of deceit, except this time the spymaster will arrange for Gertrude to interrogate Hamlet while he alone skulks behind an arras. With cunning aforethought, Polonius will not have the benefit of someone to verify his observations. Claudius agrees. "Madness in great ones must not unwatched go" (3.1.182), he says.

The Mousetrap

The thespians will stage *The Murder of Gonzago*, and Hamlet will adapt it to test the Ghost's honesty. Shakespeare places this episode in the middle of *Hamlet* because it is pivotal to understanding (Wilson 138); it emphasizes the centrality of epistemology and affirms the significance of evidence-based decision making. Hamlet gives the players last-minute instructions, again bringing Tycho's portrait to mind by speaking of hand sawing, "Nor do not saw the air too much with your hand" (3.2.3–3.2.4). In a reference to telescopy, he says the actors must "hold as 'twere the mirror up to nature," which refers to the mirror "which sets standards…by revealing things not as they seem, but as they really are" (3.2.18n).

After the players leave to dress for their parts, Hamlet tells Horatio of the upcoming performance. "One scene comes near the circumstance [of his father's death]" (3.2.66), he says, meaning one scene will be similar to the circumstances of Old Hamlet's murder, but it will not be exactly like it. He asks Horatio to observe Claudius to see whether he reacts guiltily, and if he does not, then Hamlet must conclude, "It is a damnèd ghost that we have seen" (3.2.72).

The players know of Hamlet's low opinion of dumb shows, yet they open their performance with one anyway. They owe allegiance to their geocentric employers and therefore ignore Hamlet's requests. The dumb show mimes the murder of a player King, after which a player Queen accepts the love of the murderer. Surely, Claudius could not fail to see that the unnamed murderer might represent him, but curiously, he does not bat an eye. Some think that he is preoccupied and does not see the mime. Others believe in the "second tooth" theory, according to which Claudius has the fortitude to withstand the imputations initially, only to cave in later.

The allegorical subtext provides an alternative reading. Beneath Claudius' human facade lies a metaphysical persona that is devoid of human feeling and incapable of emotion, and just as the subtext transcends the literal story line, so the King's otherworldly self is preeminent. As such, Claudius is proud of his achievements. The murder of Old Hamlet is part of the day's work, and he cannot react guiltily even to a precise rendition of his crime.

Ophelia senses that the dumb show foretells the plot of the play, and Hamlet confirms that it "means mischief" (3.2.122). The players have disregarded Hamlet's low opinion of dumb shows and have begun their performance with one, so Hamlet suspects that the players probably will not insert the lines he had requested. He expects that the players will continue with an oral equivalent of the mimed introduction in order to give Hamlet a second dose of geocentric medicine, and he knows that Claudius will not flinch at the telling any more than he did to the miming. His experiment is in danger of failing, but Hamlet knows what to do.

The Prologue begins the oral show and Player King and Player Queen pledge eternal love. Player King foretells his own death and prepares to offer advice on the Player Queen's next husband, but she interrupts him.

> PLAYER QUEEN In second husband let me be accurst:
> None wed the second but who killed the first…
> A second time I kill my husband dead
> When second husband kisses me in bed. (3.2.160–3.2.167)

The implication is that Gertrude is an accomplice to the death of Old Hamlet and may even be the murderer herself (Jenkins 3.2.175–3.2.180n). Player King expresses skepticism of Player Queen's continued fidelity if he should happen to die, "So think thou wilt no second husband wed, / But die thy thoughts when thy first lord is dead" (3.2.195–3.2.196), but she vows, "If once a widow, ever I be wife" (3.2.204).

Hamlet interrupts the proceedings and asks Gertrude how she likes the play, and she replies that she thinks the Player Queen "doth protest too much" (3.2.211). Hamlet mocks her, "Oh but *she'll* keep her word" (3.2.212; emphasis added), meaning that the Player Queen intends to remain faithful even if the real Queen Gertrude did not. This imputation catches the attention of Claudius, prompting him to ask Hamlet "Is there no offence in't?" (3.2.213). Claudius depends on Gertrude, perhaps even loves her, and here his human side emerges as he shows concern for her. Hamlet reassures Claudius, who inquires, "What do you call the play?" (3.2.215) and Hamlet replies "The Mousetrap" (3.2.216). Hamlet has renamed *The Murder of Gonzago* because after Hamlet has put his plan into action, the enactment will not follow the *Gonzago* script.

Claudius too has every reason to believe that the oral show will follow the gist of the dumb show, and thus that the Player King and Player Queen represent Old Hamlet and Gertrude (table 3.4). The allegory-within-the-allegory continues as a third player enters. Claudius confidently expects that this player represents himself preparing to kill the Player King and claim the throne, but Hamlet announces for all to hear, "This is one Lucianus, nephew to the king" (3.2.221). The only King with a nephew is the new King Claudius, whose nephew is Hamlet

TABLE 3.4. Imputed roles in the players' play.

Bard's *Hamlet*	Players' *Gonzago*	Hamlet's "Mousetrap"
Player King	Old Hamlet	Claudius
Player Queen	Gertrude	Gertrude
Murderer	Claudius	Hamlet

(table 3.4). Therefore, Hamlet has inculcated an impression of the Player King representing Claudius, and Lucianus representing his nephew, Prince Hamlet. Yet Claudius remains unperturbed by the mime and rhyme.

Hamlet chats with Ophelia, and at the opportune moment, he blurts out, "So you mistake your husbands. Begin murderer" (3.2.228). Lucianus pours poison in the ear of the sleeping King who supposedly is now the murderer's uncle—in other words Claudius himself. Moreover, just in case Claudius does not take the bait, Hamlet makes the point abundantly clear, "A poisons him i'th'garden for's estate. His name's Gonzago… You shall see anon how the murderer gets the love of Gonzago's wife" (3.2.237–3.2.239). Hamlet has sprung the trap. Claudius has already identified Player Queen with Gertrude, so he cannot avoid the conclusion that the sleeping figure is Claudius himself and the murderer is his nephew, young Hamlet!

Claudius loses composure. Ophelia announces, "The king rises" (3.2.240). Claudius calls for "some light," which marks the occasion when new insight is gained into the physical World.

The test has succeeded because Hamlet has bypassed Claudius' subtextual facade and targeted his emotional being. His coup de theatre has caught Claudius off guard and triggered a response that had the appearance of guilt. Before the entire court, Lucianus killed Claudius the same way that Claudius killed Old Hamlet, and Claudius is "simultaneously confronted with the image of his crime and the threat of its avenging" (Jenkins 3.2.248n).

Hamlet exults in the success of the experiment, which he attributes to "talk of the poisoning" (3.2.263). Hamlet has been the only one to talk of poisoning ("A poisons him i'th'garden for's estate"), which means that he circumvented the uncooperativeness of the players by becoming a player himself.

Guildenstern tells Hamlet that Claudius is marvelously distempered, and that the Queen wishes to speak to him. Rosencrantz threatens Hamlet with incarceration, and Hamlet responds that he lacks advancement because he has not been elevated to the status of King. Even so, the Prince is joyful and calls for music, which signals a return to the Pythagorean philosophy of the harmony of nature. Players enter with recorders, and Hamlet invites Guildenstern to play upon one of the pipes. "Govern these ventages with your fingers and thumb, give it breath with your mouth," he says (3.2.324–3.2.325). "I have not the skill," says Guildenstern, who refuses Hamlet's invitation four times (3.2.328). Hamlet rebukes him, "'Sblood, do you think I am easier to be played on than a pipe?" (3.2.334). The oath repeats the earlier one in scene 2.2 and refers again to actual and imagined losses of blood associated with Tycho's real and depicted anatomical deficiencies. Accompanying the oath on this occasion are mentions of "mouth," "ventages," "breath," "fingers," and "thumb," which revive images of his nasal disfigurement and swapped hands.

Hamlet asks Polonius and the courtiers to leave, and scene 3.2 ends with a soliloquy in which he vows to "speak daggers" to his mother, although of course in compliance with the Ghost's instructions, he will "use none" (3.2.357).

MISFORTUNE'S WHEEL

In scene 3.3, the courtiers listen as Claudius vents his ire. He tells them that he is dispatching Hamlet to England under their care because, "the terms of our estate may not endure / Hazard so near to us" (3.3.5–3.3.6). Guildenstern agrees.

> GUILDENSTERN Most holy and religious fear it is
> To keep those many many bodies safe
> That live and feed upon your majesty. (3.3.8–3.3.10)

Rosencrantz worries that the lives of many depend on the "weal" (3.3.14) or well-being of the King. "Weal" is a pun on "wheel," which is a term that Copernicus (26) uses to describe the majestic motions of planets and stars. Rosencrantz reinforces the conceit in a passage that warns of the collapse of the geocentered planetary and stellar spheres along with their rods, spindles, and gears.

> ROSENCRANTZ The cess of majesty
> Dies not alone…It is a massy wheel
> Fixed on the summit of the highest mount,
> To whose huge spokes ten thousand lesser things
> Are mortised and adjoined… (3.3.15–3.3.20)

Here, the number 10,000 refers to the number of stars in the sky visible to the naked eye (to the limit of magnitude $\approx$ 6.5) (Jones, *General* 302). It refers also to the ideal population of a city as proposed by Aristotle in his *Politics*, to which Hamlet had referred earlier in addressing Polonius (2.2.176–2.2.177). Thomas Digges would have been able to estimate the first number, which in turn leads to an understanding of the term "infinite space" (2.2.244).[10] Shakespeare writes ironically, as Aristotle in *Poetics* regards 10,000 as the archetypal large number.

Rosencrantz continues,

> ROSENCRANTZ …which when it falls,
> Each small annexment, petty consequence,
> Attends the boisterous ruin. Never alone
> Did the king sigh, but with a general groan. (3.3.20–3.3.23)

He thinks that the wheel of the Universe rotates about a Hub where the King resides and keeps the World turning. He believes that the monarch is like the mighty lawgiver who rules Heaven and Earth from atop Mount Olympus—that famous summit of classical mythology, colored blue because it blends into the scattered light of the daytime sky (5.1.221n).

When Shakespeare writes of the "majesty" of bounded geocentricism and a "boisterous ruin," he builds on Rheticus' ridicule of Ptolemy's

Almagest that Rheticus had included in his *Narratio Primus* ("First Narrative") on heliocentricism in 1539. *Almagest* derives from the Arabic *Al Majisti* ("The Greatest") which evolved into the Latin *almagestum*. Rheticus calls the *Almagest* a "majestic edifice," which plays on the Latin word *maiestas* meaning "greatness," "majesty," and "magnificence." However, he is not being complimentary because he compares geocentricism and the *Almagest* to a hut that a child might build out of mud and sand. Such an edifice has no lasting value and does not warrant serious attention (Danielson 177), the implication being that the *Almagest* is destined to become a "magnificent ruin." Rosencrantz fears that with the passage of time, the rays of the setting Sun will turn that mighty berg of the King's abode to gold and red, matching the color of the star Betelgeuse and signaling the time for the haughty despot to go west like the dying Sun.

If the Ptolemaic model were to collapse, the King's subjects could scarcely survive the shower of shards that would rain down upon them. The Tychonic model requires supporting structures with the magical properties of both strength and invisibility just as much as the Ptolemaic, although in Tycho's case there are many more of them, as each star in the encasing shell needs a separate support (Thoren 306). Shakespeare has both political and cosmological themes in mind as he satirizes the puffery of egocentrics.

Polonius leaves the scene to eavesdrop upon the Queen and her son, and Claudius shows his human side again by lamenting his errors. In the first 20 lines of his soliloquy, he refers to himself 17 times. Lines 3.3.54–3.3.56 are particularly self-centered.

> CLAUDIUS I am still possessed
> Of those effects for which I did the murder,
> My crown, mine own ambition, and my queen.

Claudius is egocentric and geocentric, a narcissistic monarch whose personality and ambition have left him vulnerable to manipulation by the spirits of the nether world. His human side is not devoid of conscience, however, for he betakes himself to a chapel to pray, and when Hamlet discovers him there, he decides not to kill him there lest his soul go to heaven.

Two Fathers

Polonius reaches Gertrude's chambers and hides behind an arras. Hamlet enters, speaks rudely to his mother and deliberately confuses his two "fathers." Hamlet's demeanor prompts Gertrude to call for help. Polonius exclaims, and Hamlet (mistaking him for a rat) draws his sword and kills him. When he discovers what he has done, he explains, "I took thee for thy better" (3.4.32). In Hamlet's mind, either Polonius ranks lower than a rat, or, if Hamlet mistook Polonius for Claudius who is Polonius' better, then Claudius ranks on a par with a rat. Claudius already has kinship with rodents because Hamlet snared him with "The Mousetrap," and not even Gertrude is immune to murid conceits because in the same scene, Hamlet refers to her as the King's "mouse" (3.4.184).

Hamlet berates his mother, and she asks what she has done to warrant an inclusion that "thunders in the index" (3.4.52). The prefatory nature of Hamlet's remarks (Jenkins 3.4.52n) becomes clear in Hamlet's reply as he refers to images of two brothers, Claudius and Old Hamlet.

> HAMLET Look here upon this picture, and on this,
> The counterfeit presentment of two brothers.
> See what a grace was seated on this brow;
> Hyperion's curls, the front of Jove himself,
> An eye like Mars, to threaten and command. (3.4.53–3.4.57)

Hamlet identifies his deceased father using the past tense ("what was seated"). Earlier, Shakespeare had developed certain qualities of Old Hamlet, including the fact that his devotion to Gertrude engendered comparison to the Sun god Hyperion (1.2.140), and that when emulating the god of war, Old Hamlet "smote the sledded Polacks" (1.1.63). The passage identifies Old Hamlet—not Claudius—with Jove, and it describes his appearance.

Old Hamlet has curly hair and an eye that is red like the color of Mars, the god of war. Hyperion is the progenitor of Helios and sometimes synonymous with it, and is like the Sun in color. "Curl" means a lock of hair or a ringlet and is "applied indifferently to a flat spiral…a (helix) or anything intermediate to or approaching these forms" (*OED*). The verb "to curl," means to bend round, wind, or twist into ringlets, and does not necessarily connote a helical form or even a shape that closes in on itself (as in *Henry IV part 2*, where Shakespeare describes the action of wind on ocean waves that curl their monstrous heads). The implication is that Old Hamlet's alias, Leonard Digges, had curly blonde or red hair with eye color to match.[11]

The passage suggests that the "front" of the planet Jupiter (3.4.56–3.4.57) has a feature like a red eye. In 1664–1665, Robert Hooke (1635–1703) and Giovanni Domenico Cassini (1625–1712) reported a Great Red Spot (the "GRS") on Jupiter.[12] Figure 3.5 shows Jupiter (left) as it appeared to a ground-based observer in 2002, compared to its appearance in 2000 (right) as recorded by NASA's Cassini spacecraft. The GRS is visible as an oval blemish in the southern hemisphere. The GRS varies considerably in color, and in its red or crimson phase (which Digges

FIGURE 3.5. On the left, Jupiter as it appeared to a ground-based observer in 2002. For comparison, the image on the right shows its appearance in 2000 as recorded by NASA's Cassini spacecraft. In both images, the oval blemish in the southern hemisphere is the Great Red Spot. In the Cassini image, the shadow of the Galilean moon Europa appears to the west of the Great Red Spot. Jupiter's equatorial radius is 71,400 kilometers (44,400 miles). For comparison, the Earth's equatorial radius is 6,400 kilometers (4,000 miles).

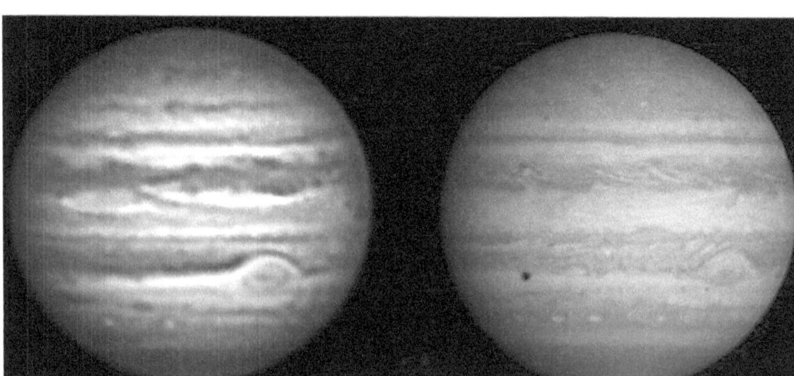

likely observed sometime in the sixteenth century) it would be strikingly visible.

In 1666 Cassini referred to the GRS as an "eye," which is an apt term because as Jupiter rotates on its axis the GRS disappears for more than 5 hours out of every 10, as if the Jovian face had a blinking "eye." The apparent disk of the planet Mars varies in size from 3.5″ to 26″, whereas the GRS on Jupiter's meridian has a size in the range of 3″ × 6″ to 5″ × 10″. Both ranges are well below the ordinary visual acuity of the human eye, requiring telescopic magnification in order to discern them. By such means, Donato Creti (1671–1749) depicted the GRS in an oil painting in 1711, but earlier in the sixteenth century, Shakespeare's astronomer would need a telescope to resolve ruddy Mars and the Jovian "eye."[13]

The ghost of Hamlet's real father appears and reprimands him for berating his mother and not getting on with his assignment. The spirit is invisible to Gertrude, perhaps because ghosts are fussy about who sees them. Some people have never seen one. Gertrude believes that her son is hallucinating, but Hamlet denies that he is mad and explains why he deserves his mother's blessing.

> HAMLET …heaven hath pleased it so,
> To punish me with this, and this with me,
> That I must be their scourge and minister. (3.4.174–3.4.176)

Hamlet announces that he is following the precepts of evidence-based inquiry. "I essentially am not in madness, / But mad in craft," he explains (3.4.188–3.4.189). This is true not just owing to methodology, but also to the skill of Leonard and Thomas Digges, who were tinkerers and craftsmen capable of producing a crafty device like a telescope (Gainer; Usher, "Elizabethan").

The Prince tells his mother, "I must to England, you know that?" (3.4.201), but the script is silent on how Hamlet came upon this news. Perhaps he is clairvoyant, as this is not the first time that he and Horatio have anticipated correctly events that were yet to unfold. In fact, after the Ghost has revealed details of Old Hamlet's murder, Hamlet mutters, "O my prophetic soul!" (1.5.40), which Jenkins suggests refers to "divination" of his uncle's true nature (1.5.41n). Despite the mandate to kill Claudius, Hamlet is accepting of the King's order because (as learned in scene 5.1) everyone in England is supposedly mad, and for the King to exile Hamlet to England is like throwing Brer Rabbit into the briar patch.

Responding to Hamlet's question, Gertrude replies "Alack / I had forgot" (3.4.201–3.4.202), which may mean that she had forgotten that he was to be dispatched to England, but if so the script does not explain how she came by the information. Perhaps she means that she had once known this and is excusing her poor memory, the rationale being that if Hamlet knows of his scheduled departure, then she should too. Another possibility is that to "forget" means to neglect to give attention to (Crystal and Crystal 184), which suggests that Gertrude had not even considered the possibility.

In 3.4.203–3.4.211 (which is not in F1) Hamlet supplies details of the mandate borne by his guards, and he vows to

> HAMLET ...delve one yard below their mines
> And blow them to the moon. Oh 'tis most sweet
> When in one line two crafts directly meet.

To "delve" is to dig, which refers to Hamlet's other self (table 3.1). The delver enjoys the prospect of the meeting of disparate crafts, one of which manufactures naked-eye sighting instruments, and the other telescopes. The context is Hamlet's pending sea voyage, and the pun anticipates meeting another craft, the pirate ship.

As for the corpse of Polonius, the Prince announces that he will "lug the guts" (3.4.213) into an adjacent room. Hamlet expresses no remorse for killing Polonius or for his attitude toward the courtiers. From actions in "The Mousetrap," it is clear that Hamlet's false father has two sides to his being, and it is reasonable to suppose that Hamlet does too. Hamlet's callous lack of regard for Polonius can be ascribed to Hamlet's other self, which is a subtextual persona indifferent to travails of real people and concerned solely with executing orders from on high in service to the New Philosophy.

Diet of Worms

Gertrude tells Claudius that Hamlet has killed Polonius, and Claudius concludes that Hamlet's liberty is full of threats to all. He announces his plan to ship Hamlet to England and plans to inform his wisest friends of the circumstances of the death of his chief minister. He dispatches the courtiers to bring Polonius' body to the chapel. They locate Hamlet and demand to know where he stowed the body, and Hamlet hints that he knows their agenda (4.2.11n). He insults Claudius whom he calls "a thing…of nothing" (4.2.25–4.2.27) and asks to see him.

Claudius communes with friends, and although he labels Hamlet dangerous, he conceals the real motive behind his actions (Jenkins 4.3SDn). Claudius demands to know where Polonius' body is, and the Prince declares that Polonius is at supper, not "where he eats, but where [he] is eaten" by a "certain convocation of politic worms" (4.3.19–4.3.20). This refers to the city of Worms, which in 1521 hosted an assembly or Diet before which the excommunicated Wittenberg priest, Martin Luther, appeared. Perhaps the passage is a dig at Luther, who denounced heliocentricism. Hamlet's disparagement of opponents of the New Astronomy continues when, in another reference to Tycho Brahe's disfigurement and in the presence of the courtiers who personify his World model, Hamlet tells Claudius that if he waits a month he "shall nose Polonius" (4.3.33–4.3.34) as he goes up the stairs into the lobby.

Hamlet learns officially that he is going to England, but he feigns ignorance.

> HAMLET For England?
> CLAUDIUS Ay, Hamlet.
> HAMLET Good.
> CLAUDIUS So it is if thou knew'st our purpose. (4.3.43–4.3.44)

Hamlet is unconcerned. "I see a cherub that sees them," he says, meaning that he knows that "heaven is watching" (4.3.45n). He knows that as part of the plan to establish the New Astronomy, events will occur for the better. After Hamlet exits, Claudius asserts that the ship bearing

Hamlet will set sail that night. This free radical is a menace to the body politic, and his guards, the courtiers, will carry letters ordering the English to execute the Prince immediately upon his arrival.

PATCH OF GROUND

In accordance with the outcome of the diplomatic mission to Norway, Fortinbras arrives with a captain of his army *en route* to Poland, and Fortinbras and the army depart forthwith. In F1, this episode takes up a mere eight lines, whereas in Q2 the scene is 58 lines longer.

The exscinded passage (4.4.9–4.4.66) helps understand the allegorical context. Hamlet and the courtiers enter, and the Captain, who remains behind, tells Hamlet that the Norwegian army under the command of Fortinbras is marching on a part of Poland. Hamlet asks whether the army is marching "against the main of Poland" or against some "frontier" (4.4.15–4.4.16), and the Captain replies,

> CAPTAIN We go to gain a little patch of ground
> That hath in it no profit but the name.
> To pay five ducats, five, I would not farm it... (4.4.18–4.4.20)

The patch of ground is associated with a "frontier" and is renowned and unsuited for agriculture. It has value only because of its name, yet remarkably its capture warrants a major sortie to "gain" it. Hamlet concludes, "the Polack never will defend it" (4.4.23), so it seems the invasion will be a walkover.

In *Historia Danica*, Saxo Grammaticus stated that graveyards are not arable, suggesting that the patch in question is a graveyard (Saxo Grammaticus 1: 217). This particular burial site must have great significance because of the "name" associated with it. The site contains the remains of a famous Pole, but in the present context, these are unlikely to be of a former King of Poland (cf. Jenkins 4.4.23n). It is more likely that the grave contains the remains of Copernicus, who (having died in 1543) cannot now defend himself in any tribunal of this world against charges of heliocentric heresy.

A hint at the nature of the burial site is that it is not big enough to hold 2,000 people. Research from 2005 to 2009 has identified remains in Frombork Cathedral, Poland, as those of Copernicus (Bogdanowicz et al.), and perhaps scholars can confirm that the number 2,000 is about right. It is odd that so prominent an intellectual and a church official in relatively high standing would have an unmarked grave, but perhaps Copernicus was already unpopular at the time of his death. Hamlet's comments support the idea.

> HAMLET This is th'impostume of much wealth and peace,
> That inward breaks, and shows no cause without
> Why the man dies. (4.4.27–4.4.29)

An "impostume" is an abscess or an inward swelling of corrupt matter. The slight to the memory of Copernicus might have been relieved if theocrats had publicized his achievements on a tombstone.

The courtiers are present and waiting to escort Hamlet to the bark, and they either have no idea what the conversation means or else choose to ignore it. Hamlet asks them to lead the way and he promises to follow shortly. When alone, he extols the wonders of humankind whose capacity to reason sets them apart from animals. This reemphasizes Hamlet's opinions from lines 2.2.286–290, in which he characterizes humans as noble in reason, infinite in faculties, angelic in action, and god-like in apprehension.

As Fortinbras forges ahead with his mission, Hamlet vows to screw his courage to the sticking-place. He foresees bloodshed. "Oh from this time forth, / My thoughts be bloody or be nothing worth" (4.4.65–4.4.66).

Reinforcements

The death of her father and the departure of Hamlet have stressed Ophelia, and her speech has become sometimes incoherent. In an exemplification of one of the fallacies of literary interpretation, a Gentleman remarks that those who listen to her "botch" her words to fit their own thoughts—in other words, they attempt to make sense of her words by appealing to their own preconceptions (4.5.10n). Claudius admits to having interred Polonius "hugger-mugger" (4.5.83)—that is, secretly. This is wise because Claudius wants to avoid drawing attention to the death of one who was a staunch Aristotelian and champion of pedantry.

Claudius announces that Laertes has secretly returned from France. On cue, a messenger arrives and announces that a mob is calling for Laertes to be King. "The rabble call him lord," he says (4.5.102). Earlier, it was revealed that Hamlet was once "loved of the distracted multitude" (4.3.4), but now the mob has switched allegiance. Laertes bursts into court and expresses anger at the death of his father, but Claudius denies involvement. Laertes demands to know why his father received an unceremonious burial. Claudius accepts the validity of the complaint and resolves to take action against the killer. "...where th'offense is, let the great axe fall," he says (4.5.213).

DEUS EX MACHINA

Sailors deliver a letter from Hamlet to Horatio that recounts his adventures during the sea voyage to England. The circumstances are odd. Pirates attacked the ship and captured Hamlet, leaving the rest to sail on to England, and there is no mention of plunder. Hamlet explains, "They have dealt with me like thieves of mercy, but they knew what they did: I am to do a good turn for them" (4.6.17–4.6.18). The "paradox of thieves showing mercy is wittily expressed by applying to thieves a phrase more commonly used of angels," whose mercy, however, "was calculated" (4.6.17–4.6.18n; Jenkins 4.6.19n).

When Hamlet later describes these events to Horatio, he calls the set to with the pirates a "sea-fight" (5.2.54). The script satirizes Aristotle's penchant for philosophical involution by alluding to his treatise, *On Interpretation*, which poses a weighty problem on inevitability by asking: is it inevitable that there will be a sea-fight tomorrow, or that there will not be a sea-fight tomorrow?

Conspiracy

Claudius explains to Laertes that Hamlet had regicide in mind, and Laertes asks why Claudius had not taken countermeasures. Claudius explains that he wished to please Gertrude. "She's so conjunctive to my life and soul," he explains (4.7.14). Gertrude's conjunction with Claudius represents her alignment with him, as when celestial bodies align. Laertes speaks of revenge, but Claudius counsels patience.

A messenger arrives with letters from Hamlet addressed to the royal couple. In it, Hamlet writes that he has returned to Denmark and desires to meet with Claudius. Laertes welcomes the news because he plans to avenge his father's death. Bereft of both courtiers and his chief minister, Claudius needs an ally, so he seizes the opportunity to ask Laertes, "Will you be ruled by me?" to which Laertes replies, "Ay my lord" (4.7.58). Laertes has made the fateful decision to join the cause of bounded geocentricism. Claudius plans to kill Hamlet in such a way that even his mother will think his death accidental, and Laertes requests that he be the means to accomplish this.

Claudius proclaims that, "Revenge should have no bounds" (4.7.127). Earlier, he had elevated his own opinion above empirical evidence by ascribing retrograde motion to an error of Nature, and now he adds blasphemy to his list of sins by implying that boundless revenge must necessarily extend beyond the limit of the natural World and into the sacred abode of the gods.

Claudius describes how a Norman had visited him two months earlier and had spoken highly of Laertes' swordsmanship. Laertes identifies the visitor, "Upon my life Lamord" (4.7.91). Because *la mort* is French for "the death," Laertes juxtaposes his own life with the death of someone other than himself, presumably Hamlet. France reclaimed Normandy in 1415, so it appears that in 1576, Lamord is indeed French. This is another French connection potentially inimical to the Hamlets.

Claudius asks whether Laertes would be willing to put words into action, and Laertes says that he does. Claudius describes how he will praise Laertes' swordsmanship and tempt Hamlet into a fight, for which he believes Laertes can easily select an "unbated" or sharpened rapier. This type is distinct from a foil, which is blunted to avoid serious injury (Jenkins 4.7.137n). Laertes adds that he will envenom the rapier with a poison so powerful that even the slightest wound will prove fatal, but just in case Hamlet escapes wounding, Claudius will provide a poisoned potion for the nonce, the hope being that Hamlet will partake of it to slake his thirst.

Gertrude announces that Ophelia has drowned, and Laertes reflects on her fate. "Too much of water hast thou, poor Ophelia," he says (4.7.185). This fulfills one of the omens of scene 1.1, implying that the other omen, a puncture wound to Hamlet, is imminent.

The final Act opens as two clowns prepare Ophelia's grave. The suspicion that she committed suicide prompts them to debate the death of James Hales (*c.*1500–1554) (1.2.132n, 5.1.9–5.1.10n), who landed in jail, having been a victim of political revenge, and (having failed to commit suicide there) succeeded in drowning himself in a stream soon after his release in 1554.

In a punning reference to Digges, Clown II calls Clown I "goodman delver," meaning "master digger" (5.1.12n). After discussing scripture and law, Clown I says that Adam was the first to bear arms. "The Scripture says Adam digged. Could he dig without arms?" he asks (5.1.31). The reference to Adam's profession occurs immediately after the jesters consider points of law, suggesting that "Adam digged" refers to a forebear, the judge Adomarus de Digge (*fl.*1310) (Kippis 5: 238; Philipott 60). The burlesque sets the stage for revelations to come.

Hamlet and Horatio enter and are still some distance away when Clown I sends Clown II on a mission. "Go, get thee to Yaughan, fetch me a stoup of liquor," he says (5.1.50). The name "Yaughan" is an eccentric spelling of "Johann" (5.1.50n), and its choice has not been satisfactorily explained (Hibbard, *Hamlet* 5.1.59n; Jenkins 5.1.60n). Other puzzles are why Clown II leaves the stage and never returns, and why Clown I is thereby denied refreshment. Perhaps Clown II's errand and the identification of Yaughan have less to do with refreshment and more to do with the approach of Hamlet and Horatio.

Three Skulls

Clown I unearths a skull (call it "Skull 1") and throws it up for all to see. A dozen lines later, Hamlet comments (5.1.75–5.1.76), seemingly out of context, "Here's fine revolution, and we had the trick to see't." Hamlet and at least one other person ("we") had the means to see the revolution. Hamlet's remark begins with the present tense and ends with the past tense, suggesting that something changed between the "trick" and the "revolution." The primary meaning of "trick" dates to the fifteenth century and means "a crafty or fraudulent device...an artifice to deceive or cheat," a clever device, or a "contrivance or invention" (*OED*). Shakespeare refers to the forerunner of the telescope, for which the term "trick" is appropriate because many believed it to be a fraudulent device like those that practitioners of natural magic use to deceive the gullible. The word can also mean a "characteristic quality" or "distinguishing trait," which also implies that Thomas Digges possessed a distinguishing trait. "Revolution" means as "alteration, change, mutation" (*OED*). Its astronomical meaning (the orbital motion of planets) was in use by 1390, but by 1450, the word came also to mean "great change or alteration in affairs or in some particular thing." When Copernicus made "revolution" essentially the entire title of *De Revolutionibus* in 1543, the possibility of a double meaning was already in place, and Shakespeare surely would have used it.

Shakespeare further clarifies the role of the Digges family when (in the passage 5.1.119–5.1.138) he reveals that Hamlet is 30 years old.[14] As learned later, Hamlet has just arranged for the English to kill Rosencrantz and Guildenstern, and he is about to finish off Claudius. Incidentally, Thomas Digges was also about thirty in 1576 when *A Perfit Description* signaled the demise of Tychonic and Ptolemaic astronomy. Age 30 also marks the end of innocence and of *adulescens* (Smith and Lockwood 22), so both Hamlet and Thomas Digges are just old enough to be taken seriously. In addition, if the advent of the New Star dates Act 1 of *Hamlet* to 1572, and Digges' *A Perfit Description* dates Act 5 to 1576, then another coincidence is that the five Acts of *Hamlet* equal the number of years from 1572 to 1576 inclusive. According to *Pantometria*, Leonard Digges died in 1571, so it seems on that authority that even from the grave, he is instrumental in overturning the Old Astronomy.

The chief gravedigger is a "Clown" because of the lowliness of his profession, and from the end of Act 1, the spirit Leonard Digges is an old mole, and therefore a low-down digger as well. Shakespeare celebrates the accomplishments of diggers of celestial data, but makes it clear that they too have feet of clay. Shakespeare is an equal-opportunity satirist because *he who digs is* as much of a clown as *he who is Digges*. Puns galore prompt Hamlet to decide, "We must speak by the card or equivocation will undo us" (5.1.115–5.1.116). Shakespeare must scatter an abundance of clues lest the cosmic implications of *Hamlet* pass unnoticed, yet he must disguise meaning sufficiently well that his cosmic exposé not imperil him—or England.

Clown I explains that Hamlet was sent to England where "the men are as mad as he" (5.1.130–5.1.131). The hyperbole would have elicited chuckles at the Inns of Court, some of whose members were probably sufficiently open-minded to warrant the epithet "mad." Hamlet muses over the profession of the person represented by Skull 1, and does so again when Clown I unearths a second skull ("Skull 2"). He supposes that Skull 1 belongs to a member of the class of politicians and nobles, and Skull 2 to someone in the legal profession. Clown I throws up a third skull ("Skull 3") and invites Hamlet to identify it, but he cannot.

An entourage bears Ophelia's body in a coffin and approaches the gravesite. Hamlet and Horatio retire and await developments. The priest announces that the manner of Ophelia's death is problematic and that the Church cannot honor her in the same way as "peace-departed souls" (5.1.205). Laertes has a few choice words for the priest, and leaps into the grave to embrace his sister one last time. Hamlet has overheard that the grave is for Ophelia, and he approaches the party. Laertes climbs out of the grave and grapples with him. Claudius orders them parted, and Hamlet explains how much he loved Ophelia. After Hamlet and Horatio depart, Claudius urges Laertes to be patient and stick to the plans that they had hatched the previous night.

Yaughan and Yorick

A stringent test of the allegorical hypothesis will be whether the script answers the questions:

> Y1. Does Yaughan represent anyone in particular?
> Y2. Why does Clown II not return to stage?
> Y3. What is significant about the stoup of liquor?
> Y4. Does Yorick represent anyone in particular?

Narrow the field to candidates who contributed significantly to the advancement of the New Astronomy. One can expect to make identifications with persons who contributed between 1543 (when Copernicus published *De Revolutionibus*) and 1576 (when Thomas Digges published *A Perfit Description*).

Yaughan may correspond to names like "John," "Johann," or "Johannes," and Yorick may correspond to names like "Jörg," "Georg," or "George" (5.1.50n1, 5.1.152n3). Consider two possibilities:

> P1. Yaughan represents John Field (*c.*1525–1587).
> P2. Yorick takes his name from Georg Joachim Rheticus.

Concerning P1, in 1556, John Field (Feild or Felde) authored *Ephemeris Anni 1557* ("Ephemeris for Year 1557"), and a year later he produced the *Ephemerides Trium Annorum* ("Ephemerides for Three Years," 1558–1560), which answers question Y1 because Yaughen does indeed represent someone in particular. Before testing P2, consider first the overlapping careers of Georg Rheticus and his colleague, Erasmus Reinhold (1511–1553), who was a professor of higher mathematics (including astronomy) at Wittenberg and senior to Rheticus (Danielson 104).

Rheticus and Reinhold

Georg Joachim Lauschen's mother was a woman of means, and his father (with the same first name) was a medical doctor and a humanist. The senior Georg fell under suspicion of sorcery and was convicted of theft. He lost his head in a public execution and with it his family's name, so his son eventually adopted the surname Rheticus derived from the region of his birth Rhaetia (Koyré, *Astronomical* 90n1). Melanchthon encouraged young Georg to study mathematics, and he secured a professorship for him in the field of mathematics at Wittenberg in 1536, the same year that Erasmus Reinhold joined the faculty.

In 1539 Georg Rheticus ventured to Frauenburg to visit Copernicus—a journey that Lutheran authorities did not object to despite the Catholic diocese at Frauenburg having just issued a proclamation against heretics. Science, however, exists in its own right regardless of the idiosyncrasies of its practioners, and Copernicus welcomed Georg graciously and continued to help and support him.

Copernicus' work on a theory of heliocentricism had laid fallow for 25 years or more, and as mentioned previously, Georg wrote *Narratio Primus*, which gave a preliminary account of heliocentricism and received wider coverage than Copernicus' *Commentariolis* (Hall 71). To be safe, Georg published the first edition of his book anonymously, but the second edition of 1541 bore his full name (Koyré, *Astronomical* 91n6).

Georg Rheticus returned to Wittenberg in 1540, and though his teaching assignment omitted mention of heliocentricism, he included the topic surreptitiously. In the summer of 1541, he resumed study under Copernicus, and this time on his return, the university awarded him an administrative position (Danielson 91). Nevertheless, he continued to introduce heliocentricism into his lectures.

While in Frauenburg, Georg Rheticus expressed interest in guiding Copernicus' manuscript through to press. He lined up a publisher in Nuremberg and entrusted the production to Osiander. Copernicus' magnum opus appeared in print at the time of his death in 1543. The irony is that despite Melanchthon's opposition to Copernican theory, it was owing to his good graces in promoting Georg's education and securing his appointment to Wittenberg that heliocentricism gained a foothold when it did. It is safe to say that without Rheticus, the heliocentric revolution would have occurred in a far different way. Georg Rheticus played a catalytic role like that of Edmund Halley (1656–1742), whose interest and industry promoted the publication of Newton's *Principia Mathematica* ("Mathematical Principles") in 1687.

In his lifetime, Georg Rheticus produced a map of Prussia, a biography of Copernicus, and a short treatise showing that heliocentricism does not contradict Holy Scripture (Koyré, *Astronomical* 90n1). He worked on calendars, and in 1550 produced an Ephemeris (or Almanac) in which he set forth the positions of planets according to Copernican theory (Berry 126; Danielson 137–138). He continued to work toward improving the quality of planetary predictions and developed what was to become the most important treatise on trigonometry of the sixteenth century.

Like Rheticus, Erasmus Reinhold calculated planetary ephemerides based on heliocentricism; however, he may have taken Osiander's bogus preface to *De Revolutionibus* too much to heart because he was never a convinced Copernican and merely adopted the heliocentric formalism for calculative purposes (Hall 92). In 1551, two years before his death, Reinhold published his tables under the auspices of a Prussian nobleman, and they became known as the Prussian or Prutenic Tables.

Fourteen Clues

When Hamlet fails to identify Skull 3, Clown I provides seven clues (5.1.146–5.1.153), here labeled C1 to C7. These do not enlighten the audience but are sufficient for Hamlet to identify the skull as belonging to Yorick. For Horatio's benefit, as well as the audience's, Hamlet adds seven more clues (5.1.156–5.1.162), here labeled H1 to H7, but they are just as puzzling. Clown I started the identification process, yet Hamlet does not direct his commentary to him but to Horatio. Perhaps the purpose of these clues is to educate Horatio, whose role it is to serve as the historian of the New Astronomy.

To test P2 from earlier, one must gather the clues. Clown I's clues are (5.1.146–5.1.153):

> C1. Skull 3 is of someone who died 23 years ago,
> C2. who was a whoreson,

C3. a mad fellow, and
C4. a mad rogue,
C5. who poured a flagon of Rhenish over his (Clown I's) head and
C6. was the "king's jester,"
C7. and whose name was Yorick.

Hamlet's clues are (5.1.156–5.1.162):

H1. that he too knew Yorick
H2. as a fellow of infinite jest and
H3. most excellent fancy who
H4. bore him on his back a thousand times.
H5. Hamlet's gorge rises at the abhorrence of it all.
H6. He kissed Yorick's lips many times.
H7. Yorick set the table on a roar with gibes, gambols, songs, and flashes of merriment.

C1 contains a temporal marker. Because Act 5 and the allegorical denouement occur in 1576, "Yorick" of C7 and H1 must have died in 1553, which is the year of Erasmus Reinhold's death. However, if this is relevant to the subtext, then there is a contradiction in the fact that the fourteen clues previously mentioned are supposed to test proposition P2 (that Yorick represents Georg Rheticus), yet the very first clue (C1) points to Erasmus Reinhold. It gets worse because clues C2–C6 lead to Reinhold as well. The disconnection is resolved only by solving the entire puzzle.

C2 claims that Yorick is a "whoreson." In Shakespeare's estimation, Erasmus Reinhold's appropriation of heliocentricism under false pretenses is sufficient to earn the epithet. C3 dubs the mystery man "mad," followed by "mad rogue" in C4. Shakespeare is unlikely to attribute these characteristics to Georg Rheticus because it was he who betook himself to Poland to study under Copernicus. Erasmus Reinhold, on the other hand, was never a convinced Copernican and subverted the enterprise of Georg Rheticus by appropriating heliocentricism for his own purposes.

C5 refers to Rhenish (Rhine) wine, whose only other reference in *Hamlet* occurs when Claudius consumes it copiously, which (judging from Hamlet's remarks) is sufficient for the liquor to go to his head. The events of scene 1.4 and Hamlet's dram-of-eale speech support the idea of liquor affecting someone innately, except that the geocentric Claudius actually imbibes the wine whereas Clown I is merely doused with it. Suppose that the wine instills spiritual significance into Clown I's being as water does in religious rites, or (in myth) as the apple of wisdom falling on Newton's head imparted the theory of gravity. Rhenish wine is from the *Rhein*, and a flagon is a flask used to *hold* wine. In keeping with the humorous tenor of the scene, "flagon of Rhenish" translates to *Rhein-hold*, which is reasonably a pun on "Reinhold." In this way, Yorick instills Erasmus Reinhold's heliocentric Tables into Clown I's mind, and the matching epithets of C2–C4 ("whoreson," "mad," "rogue") dub him an ersatz heliocentricist. Yorick pulled this dousing stunt 23 or more years ago (by C1, in or before 1553), and coincidentally, Reinhold's Prutenic Tables appeared in 1551.

C6 says that Yorick was a "king's" clown, but neither Georg Rheticus nor Erasmus Reinhold had ties to a monarch. A more likely interpretation is that "king" refers to Copernicus, to whom both were indebted; however, only Reinhold could be the "king's jester" because his deceit makes him a clown at the Copernican court. Shakespeare could refer also to the other chief component of the New Astronomy—the Diggesian infinite Universe of stars—as Hamlet once characterized himself as "king of infinite space." This possibility begs the question of whether Yorick has something to do with the Diggeses, which is addressed shortly.

C1–C6 concern Reinhold and indicate that Shakespeare did not hold him in very high regard, whereas C7 concerns Georg Rheticus, from whom it appears Yorick takes his name.

A full solution of the Yorick hypothesis (P2) must consider clues H1–H7, of which H1 connects to C7. From the remaining six of Hamlet's clues, the odd fact is revealed that Yorick is the only character in *Hamlet* that Hamlet speaks of with "unreserved tenderness" and who is a "thinly disguised father figure" (Eissler 84–85). *Hamlet* is the embodiment of the New Philosophy, which Thomas Digges established by hypothesis at the epoch on stage of 1576, facilitated by his father's invention; therefore, it seems reasonable to ascribe this display of affection to Hamlet kissing his father just as young Digges must surely have kissed his father. By H6, Hamlet would not have gone so far as to have kissed Yorick's lips when C1–C7 and H1 postulate that Yorick is a mix of Rheticus and Reinhold.

Perhaps, if Yorick represents the two real-life characters, Rheticus and Reinhold, he might also embody a third. Suppose that "Yorick" is a tripartite abstraction comprised of an amalgam of the first three post-Copernican mathematicians who made positive contributions to the rise of the New Astronomy: Rheticus, Reinhold, and Leonard Digges (table 3.5). The kissing then literally represents the affection of Thomas for his father and figuratively his love for the New Astronomy. The remaining items on Hamlet's list (H2–H5, H7) should deny or support the claim.

Concerning H5, Hamlet's gorge rises at the sight of the skulls that Clown I has cast up. In the fifteenth century, "gorge" meant the crop of a hawk or what a hawk has eaten, and by 1526, it meant to "cast up" or vomit gorge (*OED*). After 1586, to "cast up" also meant to throw up a heap of dirt with a spade, which is what the gravedigger does as he uncovers skulls, which in turn causes Hamlet's gorge to rise. The *OED* cites Hamlet's gag as an expression of extreme disgust. In these ways, both Georg Rheticus and Leonard Digges are associated with Skull 3, Georg via a hawk's gorge, and Leonard via the species leonard. This latter association has occurred twice before, and now H5 supports the possibility of Leonard being a third component to "Yorick" (table 3.5).

TABLE 3.5. Tabular properties of Yorick and Yaughan.

Date	Compiler	Ephemerides	Abstraction
1550	Georg Rheticus	Ephemeris	Yorick
1551	Erasmus Reinhold	Prutenic Tables	
1550 ± 1	Leonard Digges	Yes?	
1556	John Field	*Ephemeris anni 1557*	Yaughan

From H2 and H3, it is surmised that Yorick's third part, hypothetically Leonard Digges, was someone of "infinite jest" and "excellent fancy" (5.1.157–158). Concerning H4, Thomas Digges knew Yorick figuratively in the capacity of Leonard giving him piggyback rides, so Thomas was young at the time. Because he rode "a thousand times," an age of 3 to 5 seems reasonable. Thomas was born circa 1546, so Leonard was giving Thomas rides mainly during the triennium 1550 ± 1. This covers the times when Rheticus and Reinhold calculated their ephemerides (table 3.5), so it seems likely that H4 points to ephemerides as the common factor in tripartite Yorick. Perhaps, Leonard calculated ephemerides like the other two.

H7 refers specifically to Yorick's "table," which supports the suggestion because ephemerides are usually tabulated data. It does so in the context of Leonard being an affable fellow and a good father (H2–H4, H6) who—as part of Yorick—could well have entertained those seated at the dinner table (H7). From H2, Leonard can be recognized as a person of "infinite jest," which, in the scene's spirit of burlesque, recalls the imagery of infinite space. In addition, table 3.3 supports the possibility of Diggesian ephemerides because Shakespeare had data for the Moon and Venus which, given the volume of references to the Digges family, he most likely did not acquire from Wittenberg or anywhere else on the Continent. I conclude that, in the matter of planetary ephemerides, Leonard Digges was the equal of Rheticus and Reinhold and that his tables were coeval.

Postulate P1 further supports the present thesis, as in 1556 Yaughan (John Field) also became a computer of ephemerides (table 3.5). If Yaughan and two of the three parts of Yorick (Rheticus and Reinhold) have in common the calculation of ephemerides, it is likely that Yorick's third part—Leonard Digges—does too. The fact that Field is not part of the Yorick abstraction suggests that the putative tables of Leonard Digges existed prior to Field's (i.e., before 1557), and that unlike Field's tables, he calculated them from scratch, and not by adapting others' work.[15] This puts Leonard's tables on a par with Rheticus and Reinhold's, as in table 3.5.

The C and H lists have similar construction; the six clues C1–C6 pertain to Erasmus Reinhold and the six clues H2–H7 pertain to Leonard Digges. The central two items (C7 and H1) pertain to Georg Rheticus, who therefore knits the lists together and lends his name to the entire combination. This provides an answer to question Y4.

The remaining questions—Y2 and Y3—can now be answered. Yaughan's "stoup of liquor" in Y3 contrasts with Yorick's (i.e., Reinhold's) "flagon of Rhenish" in C5. If C5 refers to Reinhold's tables, then Y3 refers to John Field's tables, which were the first of several adaptations of Reinhold's tables for use at specific locations. Concerning Y2, Clown I thinks he needs refreshment from Yaughan, but Clown II never brings it because Clown I is already imbued with Reinhold's "flagon of Rhenish." He has no need for any from Yaughan because John Field's contribution is not basic and the show can go on without it.[16]

Divine Will

The final scene has Hamlet telling Horatio of his rescue at sea. He refers to divine will, "There's a divinity that shapes our ends, / Rough-hew them how we will," and Horatio replies, "That is most certain" (5.2.10–5.2.11). The remarks refer to "a higher power in control of us, directing us toward our destination, however much we have blundered in the past and impeded our own progress," so that "Hamlet feels the guiding hand of heaven in his own impulsive and unpremeditated actions, after the failure of his own willed efforts" (5.2.10–5.2.11n).

In Greek mythology, Apollo (as *Theos*) is a god of purity and light that resembles a law of nature as much as a god because the deity "accompanies the action on the divine plane [and] signifies that what happens below is the working of universal law" (Kitto 74–75). At the same time, Hamlet's rationalizations are like those that legions of saints and martyrs underwent as they came to grips with persecution and gave their lives in service to what they perceived as higher causes.

Hamlet describes how he sought out the commission that Claudius had given to Rosencrantz and Guildenstern, only to discover that it ordered the English to behead him (5.2.24), so he altered it by commanding the English to behead the two trusties instead. Horatio wonders how Hamlet sealed the new commission, and Hamlet replied, "even in that was heaven ordinant" (5.2.48), for he happened to have his father's signet in his purse, which was the model of that Danish seal. Hamlet must know that heaven has arranged this event, as in scene 1.5, the casuistic Ghost had informed him that he need feel no remorse at their deaths. "They are not near my conscience," he says (5.2.58). Hamlet's allegorical self is indifferent to their fate because the model that the courtiers personify deserves to die.

> HAMLET 'Tis dangerous when the baser nature comes
> Between the pass and fell incensèd points
> Of mighty opposites. (5.2.60–5.2.62)

Claudius had summoned the flunkies for moral support, and they are the first to go because at the turn of the century, the Tychonic model did not enjoy the same status as those mighty opposites, the time-honored Ptolemaic and the new Diggesian-Copernican World models. The passage brings to mind the ominous declaration of dews of blood and disasters in scene 1.1 and foreshadows a fight between opposites. Claudius will soon learn of the deaths of his lackeys, and Hamlet wonders whether the time has now arrived for him to "quit" Claudius "with this arm" (5.2.68). No sooner said, than a courtier named Osric enters in order to impart information from Claudius on the same topic.

Before Osric can relay his message, Hamlet criticizes Osric's garb in terms pertinent to Tycho's portraits (figures 3.3 and 3.4) and his hybrid geo-heliocentric model (figure 1.6). He tells Osric, "Put your bonnet to his right use, 'tis for the head" (5.2.90–5.2.91), which is precisely what artist de Gheyn did: he took the bonnet of the first engraving and placing it on the head in the second. Offering a reason for having cap in hand, Osric complains that it is very hot, which implies subtextually that he associates himself with the more southerly and warmer clime at Wittenberg, except that Hamlet corrects him, "No, believe me, 'tis very cold, the wind is northerly" (5.2.93). Tycho's World model is geocentric, and this cool influence emanates from Elsinore, which is NNW of Tycho's Hven. Osric is not there to pick a fight for himself and he concedes, "It is indifferent cold" (5.2.94). Both de Gheyn engravings reflect the state of the weather because they suggest that it is sufficiently chilly for Tycho to drape his greatcoat over his shoulders, yet too warm for him to don it fully. Hamlet adds to the confusion by saying, "But yet methinks it is very sultry and hot for *my* complexion" (5.2.95; emphasis added), which properly associates Hamlet with the southerly Wittenberg direction. These contradictions flummox Osric who humors the Prince by agreeing with him again.

Osric is able finally to announce the purpose of his visit, that if Hamlet engages Laertes in an ostensibly friendly sword fight, Claudius will lay a great wager on his head, the irony being that Hamlet's head is precisely what concerns him.

Hamlet knows that paranormal powers play a role in events, and he plays along, though not without satirizing his prospective opponent. He launches a sarcastic harangue (5.2.100–5.2.130) that until recently has made little sense (5.2.100–5.2.125n), but which has now been shown to be an excellent fit to the life and accomplishments of the mathematician and astronomer Thomas Harriot (table 3.1).[17] Of the items listed, several deal with his lack of published scholarship. Harriot's only work to appear in his lifetime was *A briefe and true report of the new found land of Virginia* of 1588, which he wrote after visiting the colony, and although he had promised a full account of his voyage, the larger discourse never appeared. His enduring reputation as a mathematician rests on the posthumous 1631 publication that virtually gave algebra its modern form, and his textbook on navigation was never published either (Shirley, *Harriot* 94–95).

Harriot's published report compares England's climate with that of Virginia, "the excellent temperature of the ayre there at all seasons, much warmer than in England, and never so violently hot" (Harriot 31), which fits Hamlet's spoof on Tycho's bonnet and overcoat. The parody leaves little doubt that in 1601, Shakespeare regarded Thomas Digges as England's leading mathematician, an opinion supported by the explorer

John Davis (*c.*1550–1605) who, just before Thomas' death, wrote that neither John Dee nor Thomas Harriot could compare to the "great arch-mastric" (Hotson, *Appoint* 115).

Stakes

Osric introduces the topic of the contest and announces the stakes that Laertes and Claudius have wagered.

> OSRIC The king sir hath wagered with him six Barbary horses, against the which he has impawned, as I take it, six French rapiers and poniards, with their assigns, as girdle, hangers, and so. Three of the carriages in faith are very dear to fancy, very responsive to the hilts, most delicate carriages, and of very liberal conceit. (5.2.134–5.2.138)

Poniards and assigns are daggers and accessories, a girdle is a sword belt, and hangers are straps that hold swords. Osric dismisses the rest of the accoutrements ("and so") as if they are not worth mentioning. His qualification "as I take it" suggests that he is repeating what Claudius told him and not that he has actually laid eyes on the items.

Hamlet needs a definition of terms. "What call you the carriages?" he asks, but before Osric can answer, Horatio interjects a puzzling and seemingly irrelevant comment, "I knew you must be edified by the margent ere you had done" (5.2.139–5.2.140). Its significance becomes clear when Osric defines a carriage as a hanger. The *OED* cites no other instance where "carriage" means "hanger," suggesting that the word is affected. Hamlet accepts Osric's use of the term by saying "I would it might be hangers till then" (5.2.143), indicating that in this high stakes drama, the three carriages are special.

Hamlet sums up the terms: "six Barbary horses against six French swords, their assigns, and three liberal-conceited carriages—that's the French bet against the Danish." (5.2.144–5.2.146)

The Danish bet must refer to the King's bet, so the "French" bet must be that of Laertes who, however, is Danish and not French, but by labeling his stakes as "French," Shakespeare reemphasizes that Laertes has become the latest champion of thirteenth-century Parisian scholasticism.

Osric spells out the terms of engagement: "in a dozen passes between yourself and him, he shall not exceed you three hits. He hath laid on twelve for nine," (5.2.147–5.2.149). Many assume that a pass and a hit are the same thing and that a bout or round ends after each hit, so that Laertes would win as soon as he makes eight hits and Hamlet would win with only five hits, for a handicap of three hits. If Laertes wins eight rounds, Hamlet may have won anywhere from none to four, but cannot have won five because the contest lasts only twelve rounds. Similarly, Hamlet wins with five hits because Laertes cannot then win eight. By these rules, a tie at six hits is impossible because Hamlet would already have won with five. Nevertheless, the terms lack precision and no explanation satisfies all critics (5.2.147–5.2.149n, 5.2.241n; Jenkins 5.2.162–5.2.164LN, 5.2.266n; Hibbard, *Hamlet* 5.2.128–5.2.130n, 5.2.216n), prompting reexamination here.

The King's ratio of "twelve for nine" comes right after the statement on the three-hit differential for a win, leading to the belief that it refers to bookie odds, but this is unlikely because the terms do not easily explain the digit "nine." The ratio 12/9 is the same as 4/3, but Shakespeare must have some reason not to divide out the common multiplier 3. Perhaps instead, as I have suggested in *HU*, "twelve for nine" has nothing to do with the handicap but refers to the relative value of the two stakes.

The King's stake is six Barbary horses and Laertes' stake is six French swords plus three liberal-conceited carriages. These items qua numbers are 6 and 9, and although they do not compare to the stated integers 12 and 9, the relative values 6 and 9 could transform to 12 and 9 by doubling the 6. This should occur only for good reason, which lies in Horatio's seemingly irrelevant statement formerly mentioned, "I knew you must be edified by the margent ere you had done."

"Margent" means the space between the edge of a page and the text, but in figurative use can mean the "edge" or "border" of anything. It can denote "an extremity or furthermost part of something" as in usage from about 1595, "eu'ry Margine of this earthy sphere." From the fifteenth century, a "margent" is the "ground immediately adjacent to a river or body of water…a shore," as in the brink of a sea. In exemplifying the use of "margent," the *OED* specifically mentions the sea past France. In fact, things French figure prominently in *Hamlet*: Laertes visits France twice; there are French rapiers, a "French" bet, and a Frenchman who esteems Claudius and speaks well of Laertes. The sea past France is the Mediterranean, and the land on the other margent of that sea is Africa, specifically the Barbary Coast, which is home to the very breed of horse in the King's wager. The prescient Horatio knew as much.

In round numbers, the distance from Elsinore to the Barbary Coast—whence come the horses—is a factor of two greater than the distance from Elsinore to Paris—whence come the "French" items. With the help of the ratio of the distances to the Barbary margent and Paris as measured from the political and cosmic center of Claudius' Universe, Shakespeare derives the factor of two that establishes the relative worth of his "Barbary" wager vis-à-vis the "French" wager. In short, the first number 6 concerns the actual number of the King's wagered items, but as far as their value is concerned, it seems that 6 receives a weight of 2 owing to the doubled distance, resulting in a relative worth of 12 to 9. This matter of stakes is a parody of Claudius' regicentricism.[18]

Claudius flatters Hamlet by leading him to believe that he has wagered a greater amount on his winning, but the stakes are part of the yarn that he spins to hide the fact that he has fixed the odds. Claudius expects Hamlet to die, if not by envenomed sword then by poisoned chalice, so that deceit and not skill will decide the outcome. The swindlers regard the stakes as irrelevant to their greater purpose because each stands to win by liquidating Hamlet. In reality, it is Hamlet's life that is at stake, and not horses, swords, and carriages.

Hamlet accepts the terms of the wager and tells Osric that he will try to win for Claudius but that if he does not he "will gain nothing but my shame and the odd hits," where "odd" means "extra" or "given over and above" (*OED*). If Hamlet loses, he will gain "odd hits," which are hits that score in his favor but which are "odd" because they are "given over and above" the hits that really count, which are the ones that Laertes inflicts.

Hamlet believes that he shall win at the odds because he has been in continual practice ever since Laertes went to Paris, but Hamlet's fitness concerns Horatio, who offers to present an excuse for him. The Prince will have none of it. "Not a whit, we defy augury," he says, adding, "[t]here is special providence in the fall of a sparrow. If it be now, 'tis not to come; if it be not to come, it will be now, yet it will come—the readiness is all" (5.2.192–5.2.195). The contest is supposedly not life threatening, yet Hamlet speaks fatalistically of his own death. He senses that a superior power is in charge, and he will not defy the paranormal script. He likens himself to a sparrow whose life is in the hands of a "special providence," and he resigns himself to whatever fate has in store for him. Edwards explains, "All occurrences show God's immediate concern and control, and [Hamlet] will therefore accept the circumstances which present themselves and not try to avoid them" (5.2.192–5.2.193n).

Carnage

The royal party enters and the stage is set for the contest. Claudius orders that Hamlet take Laertes' hand, and upon doing so, he offers Laertes a heartfelt apology (5.2.198–5.2.216). He explains that he is sorely distracted and that his "madness" is to blame for his wrongdoing. Hamlet's "madness" refers to his science, and he is apologizing in 1623 in F1 for his criticism of the science of Thomas Harriot in 1601 in Q2 (table 3.1). But Laertes is committed to revenge and will not back down.

Before the contest begins, Claudius orders cannoneers to fire their cannons either if Hamlet were to "give" the first or second hit, or if he should "quit" in answer of the third exchange (5.2.240–5.2.241). The customary meaning of "quit" in this context is the general sense of playing one's part (*OED*), as when Hamlet strives to win each bout. From the fifteenth century, "quit" also means to "give up" or "cease to be engaged in, or occupied with," often used in the sense of "to leave," "to leave the premises," or "to depart from a place or person," but the cannoneers do not know which meaning of "quit" applies. It appears that Claudius is supporting Hamlet, but in reality, he hopes that Laertes will wound Hamlet and that Hamlet will quit the premises feet first.

The contestants select their weapons, and Hamlet fails to notice that his opponent's sword is not blunted; or, if he does notice, he does not care because he is playing his part in the supernaturally scripted drama. The fight begins and Hamlet inflicts "a very palpable hit" (5.2.257), which knocks out the King's first secret hope because now Hamlet lives to fight a second round. Lest Laertes' swordsmanship fail him, Claudius tries his luck with his back-up plan by offering Hamlet the poisoned potion, but Hamlet declines to imbibe.

Hamlet makes the next hit as well and Claudius exclaims, "Our son shall win" (5.2.264). Hamlet is gaining on Laertes, and Claudius is saving the appearance of an honest bettor by mouthing words that others would construe as gleeful, but in reality, Claudius is alarmed because Hamlet has made the first two hits and from what Lamord told him, Laertes is highly skilled and should have ended the contest already. Gertrude takes the King's remark literally, and in anticipation of her son's win partakes of the viperous distilment, which sadly, Claudius is too slow to forestall.

Laertes wounds Hamlet with his envenomed rapier, but the cannoneers hold their fire because Hamlet is still standing and has not yet "quit." Claudius has no time to clarify his order because of the speed of events. The contestants grapple and accidentally swap swords, which Claudius notices. He becomes even more alarmed and orders the contestants separated, but Hamlet manages to wound Laertes with the toxic rapier.

The Queen falls and in her dying breath tells of the poisoned chalice. Laertes, who has not "quit" the scene either, lays blame upon Claudius. Despite being at death's door, Hamlet wounds Claudius with the tainted sword and forces him to drink from the poisoned cup. This is the metaphorical end of bounded geocentricism and literally the end of corruption at Elsinore.

Having exchanged rapiers, Hamlet and Laertes exchange forgiveness. Laertes gives up the ghost and Hamlet prepares to join the procession to the hereafter. He addresses Horatio.

> HAMLET Horatio, I am dead,
> Thou livest; report me and my cause aright
> To the unsatisfied. (5.2.317–5.2.319)

Horatio's response is puzzling, "Never believe it. / I am more an antique Roman than a Dane" (5.2.319–5.2.320). Aside from connotations of Horatio's future role as orator, Shakespeare may have named Horatio for the Latin poet Horace (Quintus Horatius Flaccus, 65 BC–8 BC), who exerted a strong influence on English literature (table 3.1).

Resolution of the allegorical plot begins with the arrival of Fortinbras from Poland, at which time the cannoneers finally have reason to fire their ordnance.

> OSRIC Young Fortinbras, with conquest come from Poland,
> To the ambassadors of England gives
> This warlike volley. (5.2.329–5.2.331)

This is the moment when Shakespeare connects the Copernican model from Poland to the Diggesian from England, and the cannons fire in celebration of the birth of the New Astronomy, and not its demise.

Past as Prologue

In his dying voice, Hamlet names Fortinbras as the next King. "I do prophesy the election lights / On Fortinbras," he declares (5.2.334–5.2.335). The choice of a French name—literally "Armstrong"—for the Norwegian Fortinbras, who is now the ruler of Denmark, restores France to the ranks of practitioners of the New Astronomy and dramatizes the role that post-Thomist Frenchmen played in hastening the ebb of scholasticism. By choosing the names Francisco and Barnardo in Act 1 for scholars with French connections who initiated enlightened anti-Thomist thinking, and by assigning the role of fostering and protecting the New

Astronomy to Fortinbras in Act 5, Shakespeare enfolds France in the new age. Under Fortinbras, free inquiry will enjoy the protection of the secular state, and in theory at least, future pioneers should no longer suffer bad dreams. The new regime will restore Pythagorean principles that obtained under Old Hamlet, and the New Astronomers will prosper in their newfound liberty. The death of the hero is a heavy price to pay, but *Hamlet* is a tragedy in appearance only because the cosmic subtext has a different cast (table 3.1).

The English ambassador announces the deaths of Rosencrantz and Guildenstern and presents Horatio with the first question of his new charge: who should assume responsibility for this outcome? Horatio declares that Hamlet "never gave commandment for their death" (5.2.353), which is a fine point best argued by lawyers at the Inns of Court.

Fortinbras accords the deceased hero full military honors. In his life, Hamlet has embodied the chief principles that constitute the ideal person. He has emotional and bodily needs and desires, but his capacity for language and ratiocination has set him apart from most other humans. The four pallbearers signify the natural order; that is, the true model of the World for which Hamlet gave his life. Words uttered by Fortinbras eulogize Thomas Digges in a manner appropriate to the military interests listed on his tombstone. Hamlet would have given a good account of himself had he been put on, in agreement with the opinion that Thomas Digges would have proved himself in combat had his mathematical, engineering, and other skills not steered him into other service. Fortinbras succeeds to the throne to lead the newly established form of the ideal political state. He shares heroic honors with Hamlet because as soon as he had perceived the disjointedness of the Danish frame, he wanted to attack it.

Hamlet's last words, "The rest is silence" (5.2.337) may mean that, henceforth, no beneficent deity will intervene to rescue humanity from political and philosophical folly. Hamlet dies when he has no further role to play, but his demise takes a back seat to enlightened governance and the advancement of the New Philosophy. Shakespeare leaves the play's final words to the one who transformed the political scene and held up a ray of hope for the future. Fortinbras orders the final tribute to the fallen hero, “Go bid the soldiers shoot,” (5.2.382).

In *Hamlet*, as throughout the Renaissance, science and ethics undergo a process of synthesis such as lies at the root of the “divine philosophy” of the Pythagoreans (Heninger 33). Following Pythagoras and Plato, Shakespeare establishes and dramatizes the covenant between a superior power and humankind to correct the flawed premises of the schoolmen and transform old perceptions of the World. The spiritual image is of a beneficent One, the Creator of Heaven and Earth and of all things visible and invisible, acting to free humankind from the bonds of ignorance. Political folly is not solely the province of the past, and *Hamlet* challenges the modern playgoer to consider the origins of science and its impact on worldview and hence on the mindset of the body politic.

CHAPTER 4

CYMBELINE

> They who in proof of any assertion rely simply on the weight of authority without adducing any argument in support of it, act very absurdly.
>
> —Vincenzio Galilei

In March of 1610, Galileo Galilei reported that he had used his newly constructed spyglass to discover four moons around the planet Jupiter. Shakespeare (*Cymbeline*) lauds this discovery by letting the deity Jupiter descend from the heavens at the behest of four ghosts, whose number equals that of the planet's newly discovered moons. *Cymbeline* has mystical and cosmic overtones, and this chapter shows that it contains a subtext that chronicles the course of astronomical discovery in 1610.

CONTEXT

It is useful first to examine the historical context of the play, which fuses the disparate worlds of Roman and Jacobean Britain, medieval Europe, and Renaissance Italy. Sources include the *Decameron* of Giovanni

Boccaccio (1313–1375), the tale *Frederyke of Jennen*, the anonymous play titled *The Rare Triumphs of Love and Fortune*, and *Chronicles* by Raphael Holinshed (?–*c*.1580)—a fanciful history of Britain (Warren 285). (In this Chapter, unattributed notes, lineation, and pagination refer to Warren's *Cymbeline*, and other editions of Shakespeare's plays are also identified by editors' names.)

In about 1,200 BC, after the sack of Troy, the Trojan Aeneas escaped and fled to Italy, where he married the daughter of Latinus and founded Rome. Brutus, a descendant of Aeneus and grandson of Silvius Posthumus, came to Albion and renamed it "Britain" by combining his own name with that of his wife, Innogen. He "divided Britain into three, christening Wales 'Cambria'" (37–38). Aeneas is therefore the legendary common ancestor of Romans and Britons. Legend also has it that Apollo founded old Troy, and in the eleventh century BC near Troynovant (New Troy) in Britain, there stood a Temple of Apollo (Hotson, *Hilliard* 107).

In 72 BC, King Lud (a descendent of Brutus and Innogen) rebuilt New Troy, which became known as "Lud's town" (5.4.482) or "London" (38). Lud's brother or son, Cassibelan, refused the demand of Julius Caesar (*c*.102–44 BC) for tribute because it seems Caesar had not bested the Britons in war. A younger son of Lud, Tenantius, succeeded Cassibelan. Tenantius' son, Cunobellus, grew up in Rome and ruled southeast England from 33 BC until his death 35 years later (Nosworthy xviii). Shakespeare names King Cymbeline for Cunobellus, and treats him as a contemporary of Augustus Caesar (63 BC–AD 14). Cunobellus was monarch at the nominal time of Christ's birth, which is the chief highlight of an otherwise uneventful reign. Cunobellus had two sons, Guiderius and Arviragus, for whom Shakespeare names Cymbeline's sons.

Overview

A brief overview of the play is helpful because *Cymbeline* is "possibly Shakespeare's most complex drama" (Maley 148). The literal story concerns the adventures of Posthumus, who is married to King Cymbeline's only daughter, Innogen. To open the play, Second Gentleman inquires about Posthumus' name and lineage. First Gentleman relates that Posthumus was the youngest of three children, that his mother had died in childbirth, and that his two brothers perished in combat. His father, Sicilius, was grief stricken and died. Sicilius had acquired the surname "Leonatus" for bravery in combat. He was not a member of the royalty, but Cymbeline adopted his orphaned child and gave him the patrilineal name "Posthumus Leonatus." The name "Posthumus" derives from Silvius Posthumus, the grandfather of Brutus (37), but the name is also apt because Posthumus was born "posthumously," after the death of his father (268).

Cymbeline had once unjustly accused a subject named Belarius of a transgression, and in retaliation, Belarius had kidnapped the two infant princes. Belarius pretended that his wife had died and left him with the infants, and for twenty years under assumed names, the three took shelter in a cave in Wales. With the disappearance of her brothers, Innogen became heir to the kingdom, but her marriage engenders her father's ire. So distraught is the King that he banishes Posthumus and threatens to imprison his daughter. The current Queen, who is Cymbeline's second wife, is secretly pleased with his treatment of Posthumus because her ambition is to arrange a marriage between Innogen and her highborn son, Cloten (who is the issue of her previous marriage). Cloten, whose name resembles that of an ancient Cornish King, Cloton (38), claims to be in love with Innogen and pursues her even after her marriage to Posthumus.

Before Posthumus departs for exile, he and Innogen exchange gifts as tokens of their love and fidelity. Her gift to him is a diamond ring, and his gift to her is a bracelet. In Rome, Posthumus meets up with his host, Philario, and four of his friends—one of whom, Giacomo, argues with Posthumus over Innogen's virtue. Giacomo (or Iachimo, the "little Iago") wagers that he will be able to seduce her, but Posthumus has supreme confidence in his wife and confidently accepts the wager.

Giacomo arrives in Britain but Innogen rebuffs him. Giacomo resorts to deceit in order to win the bet, and concocts a story about being concerned for the security of jewels that he and his friends have purchased as a gift for the Roman Emperor. Innogen falls for the ruse and offers to keep the jewels safe overnight in her bedchamber, whereupon Giacomo hides in the chest that supposedly contains the jewels and has it delivered to her room. During the night, he emerges and notes details of her scantily clad body and the trappings of her bedroom. Oddly enough, the interloper declines to assault her and is content merely to gawk—and to steal her bracelet.

On returning to Rome, Giacomo recounts the circumstantial evidence that he has gathered. He describes a mole that he spied beneath one of Innogen's breasts, which convinces Posthumus that Giacomo has compromised Innogen's virtue. Posthumus is so distressed that at the end of Act 2, he delivers a peevish rant bemoaning the shortcomings of the female gender, and then does not reappear until he opens Act 5.

Act 3 begins with the arrival at the British court of the Roman ambassador Caius Lucius, who demands payment of tribute exacted after Julius Caesar supposedly bested the British in battle, but the Queen and Cloten encourage Cymbeline not to comply because they feel that the British (aided by the weather) defeated the Romans. British defiance leads Lucius to declare war on behalf of Augustus Caesar. Meanwhile in Rome, Posthumus has become embittered over what he believes is Innogen's loss of chastity and plots revenge upon her. He writes to his manservant, Pisanio, telling him to deliver a letter to Innogen in which he tells her that he is at Milford Haven in Wales, and that she should go there. She prepares to do so, little knowing that Posthumus has also instructed Pisanio to kill her, but Pisanio suspects that someone has deceived Posthumus, and he reveals the plot to her instead.

Believing her reputation to be compromised and her marriage ruined, Innogen chooses not return to court. Pisanio recommends that she disguise herself as a boy and seek employment from Lucius, who is due to arrive at Milford Haven the next day. Innogen establishes her new identity and asserts fidelity to Posthumus by choosing "Fidele" as her new name. Pisanio intends to report Innogen's supposed death to Posthumus, which Innogen's absence from court would support; but in the meantime, seeking to help her, he gives her medicine, lest she suffer from ailments like gastro-intestinal upsets that afflict travelers. Pisanio had come by the potion when the Queen had asked her physician, Cornelius, to prepare a poisonous draft. She claimed that she planned to test it on animals, but the honorable doctor distrusted the Queen and prepared a harmless sedative instead.

Cymbeline discovers that Innogen is missing, and Cloten concludes that she has fled. Cloten coerces Pisanio into serving him instead of Posthumus, and he intimidates Pisanio into divulging the destinations of both Innogen and Posthumus. Cloten resolves to kill Posthumus and asks Pisanio to give him a suit of Posthumus' clothes, which he plans to wear in order to dupe Innogen into thinking he is her husband, and then ravish her.

On her way to Milford Haven, Innogen (alias Fidele) comes across the cave that is home to Cymbeline's sons and their abductor. The men take a liking to the young "lad" and invite "him" to stay. All are hungry and they eat. Innogen falls ill and the cave dwellers decide that "he" can remain behind when they go hunting. Innogen doses herself with the potion she had received from Pisanio and falls into a trancelike sleep.

Cloten arrives in Wales seeking Posthumus, but instead encounters the renegades. Guiderius strikes off Cloten's head and disposes of it. On returning to the cave, the brothers think that the slumbering Fidele is dead, and they recite obsequies. Belarius delivers Cloten's headless body and places it beside the inert Fidele. They leave, and in a grisly episode, Innogen awakens to find a decapitated corpse dressed in her husband's clothes lying beside her.

The Roman invasion of Wales proceeds. Lucius arrives and learns that troop reinforcements will soon arrive under the command of Giacomo. Lucius takes a liking to Fidele and offers employment, which "he" accepts. Meanwhile at the court of Cymbeline, the disappearance of Cloten causes the Queen to fall ill. Cymbeline suspects Pisanio of complicity in his daughter's disappearance and threatens torture, but Pisanio pleads ignorance. Pisanio, meanwhile, has falsely informed Posthumus of Innogen's death. Posthumus now feels that he has nothing to live for, and he repairs to Milford Haven resolved to fight until he dies.

Act 5 opens in Wales. Posthumus bears a cloth that Pisanio had given him, which he believes was bloodied when Pisanio supposedly killed Innogen. Posthumus acknowledges that Innogen was a far better person than he was, and he rues his order to kill her. He discards his Italian garb and dresses as a British peasant, vowing to fight to the death. He encounters Giacomo, sets aside his death wish, bests him in combat, and (in a surprising show of compassion) spares his life. The Romans have the upper hand in battle until Posthumus and the three cave dwellers join the fray. Aided by the terrain, they defeat the Romans and liberate the captive King.

Posthumus turns coat once again, leading the British to capture him and obligingly sentence him to death. He is asleep in prison, awaiting execution, when four specters appear that resemble the deceased members of his immediate family—his father, mother, and two brothers. They circle him "in imitation of planetary motions" (Simonds 291) and beseech the deity Jupiter to intercede on his behalf. As posited, this dance should imitate the motion of the Jovian moons.[1] The Olympian descends from the sky on the back of an eagle and orders the ghosts to place a tablet on Posthumus' breast. Jupiter is the apotheosis of joviality and is reputed to bring good luck, and sure enough, an order arrives to free Posthumus and bring him before the King.

At court, Innogen (as Fidele) does not recognize Posthumus (who is dressed as a Roman), but she recognizes the ring on Giacomo's finger as the one she gave her husband. Giacomo confesses that he won it from Posthumus by deceit. Fidele loses "his" disguise, Giacomo loses his ill-gotten gains, and Posthumus and Innogen are reunited. Cymbeline sentences Guiderius to death for killing Cloten, but forgives him once he discovers that Guiderius is his son.

With the restoration to court of all of Cymbeline's natural children, the royal lineage is all but assured. A Roman soothsayer interprets the inscription on the tablet and announces the advent of a prosperous and peaceful age. Even though the British have triumphed militarily over Rome, as a token of good will, they pledge to pay the tribute anyway. Britons and Romans celebrate their common ancestry, and the détente moves Cymbeline to declare, "Pardon's the word to all" (5.4.423). His final words pronounce a happy conclusion, "Never was a war did cease, / Ere bloody hands were washed, with such a peace" (5.4.485–5.4.486). *Cymbeline* has the distinction of being the last play in F1, which suggests that the collected works speak to the prospect of peace and prosperity.

FAILINGS

Efforts to understand *Cymbeline* have engendered more disagreement than consensus. Some question why Shakespeare fashioned the play the way he did, and even why he wrote it at all. Commenting in 1756, Samuel Johnson (1709–1784) felt it senseless to remark upon the "folly of the fiction, the absurdity of the conduct, the confusion of names, the impossibility of the events," "unresisting imbecility" and faults "too gross for aggravation" (Nosworthy xl). Defects include the prolonged absence of Posthumus for two Acts, his clumsy return to the stage, his "dumb-show fight" with Giacomo, the "jingling pageant of his deceased relatives" (Granville-Barker 111), and its overall incomprehensibility (10). Other brickbats concern plot proliferation, anachronisms, incongruities, improbabilities, convoluted writing, and multifarious meanings.

The several subplots of *Cymbeline* make classification difficult, although the category "romance" is often used (15). Apologists proclaim that it is "experimental," but the burden of alleged defects invites the opinion that critics are missing something important and that "the dramatist knew pretty well what he was doing" (Nosworthy lxxviii). Nosworthy sees *Cymbeline* as "a comprehensive piece of impressionism" that "expresses something which Shakespeare never quite achieves elsewhere" and which "must be reckoned among his supreme utterances" (lxxviii). Because it is incumbent upon readers to exhaust interpretive possibilities before blaming the poet, the present hope is that a fresh approach might help to "remove...difficulties" (Rowse, *Cymbeline* 18).

JAMES I

It is helpful to consider initially the extent to which Shakespeare uses political allegory to celebrate the first years of the reign of the first Stuart King, James VI of Scotland and I of Britain, known also as Jacobus Pacificus. His ascension to the throne in 1603 was a first step toward restoring unity following the divisions created by the Germanic invasions of the fifth and sixth centuries. The discovery of Cymbeline's lost sons supposedly symbolizes the flourishing of the Stuart line and the ensuing age of national and international peace. Milford Haven was once the site of an invasion by King James' great grandfather, who would later be crowned Henry VII (1457–1509), thereby "by extension" defining James' right to the British throne. These aspects characterize the hopefulness attendant to the ascendancy of James and the fact that in 1610, peace in Europe actually did break out, briefly (67n1).

Early in his reign, James aspired to ideals of humility, forgiveness, grace, civility, and generosity, which in *Cymbeline* are appropriate to the Augustan period and the time of Christ's birth. Nosworthy sees *Cymbeline* as "a partial comprehension" of peace procured in part via the play's most dramatic moment, the Jupiter incursion, which could represent an irate monarch swooping down on a parliament stubbornly resisting his vision of a *Pax Britannica* (lxxxiii).

However, the suggestion that the two deceased sons of Sicilius Leonatus "mirror" the two lost sons of Cymbeline (Holland, *Cymbeline* xliii) discounts the gender difference of the third child. In addition, Shakespeare might refrain from comparing the families of commoners to those of royalty, for no matter how esteemed the family of Sicilius Leonatus was, it was certainly not royal. Instead, the bard may have selected Cymbeline's children to equal (both in number and gender) those of James' offspring that were alive in 1610, namely Henry Frederick Stuart (1594–1612), Elizabeth Stuart (1596–1662), and Charles I (1600–1649), the future King of Britain.

Despite transparent references to the reign of James, the play "stubbornly refuses to make sense at the level of Stuart interpretation," and some prefer the confusion of a "textual palimpsest" to "the spectre" of Shakespeare celebrating a Stuart political cause (Marcus 136–137). As a result, *Cymbeline*'s Jacobean subtext is more of a "grace note" (63) than a melody.

The New Source

If one is not to concede the play's alleged defects, efforts at interpretation must appeal to "some extra-dramatic fact unspecified in the text" (Thorne 177). I suggest this is a hitherto unrecognized source for the play, Galileo's *Sidereus Nuncius*, whose dedicatory letter of March 12, 1610 is consistent with the nominal time of the completion of *Cymbeline*, which is generally accepted to be no later than autumn of 1610 (67). Even before that date, news of Galileo's discoveries had reached Augsburg in Bavaria by letter from Padua (*Sidereus* 90), and had possibly trickled into England as well. The news travel time from Italy to England during the warmest months of the year was reasonably about a month (*Sidereus* 87), leaving ample time for an able playwright either to write *Cymbeline* from scratch or to revise the draft of an existing play. By early October of 1610, Kepler had verified the existence of the Galilean moons (Drake, *Work* 163; *Sidereus* 105), so by year's end, the dramatist in question—if he harbored doubts at all—would have felt free to incorporate that information into a play.

Sidereus Nuncius is unique because historians agree that it is the first work explicitly devoted to astronomical telescopic results. This may explain the proliferation of documents in *Cymbeline*, which is one of "many unaccountable oddities" that "would seem to point to the working out of some deeper preoccupation" in Shakespeare's own writing (Thorne 179).

From the preceding chapters 2 and 3, one wonders whether one such preoccupation involves astronomical telescopy. In *Cymbeline*, the idea of a heavenly apparition bearing a message accords with the book's title, "*Sidereus Nuncius*," which means "The Starry Messenger" (*Sidereus* x–xi), making the Jovian deity either a sidereal messenger or an agent delivering a sidereal message. Either possibility has a precedent in *Hamlet*, wherein the Ghost acts as an agent for a higher power intent upon liberating humans from the prisons of their minds.

Cymbeline has many tropes involving seeing, reading, writing, and narrating, which further draw attention to Galileo's work, especially his letter from January 7, 1610. On stage, at least eight written communications appear, and the tablet itself has another. Posthumus' worth is esteemed from a "report" (1.1.55) and a "catalogue" of endowments (1.4.5). He communicates by written "notes" (1.1.172), and Innogen knows what kind of man he is because his virtue "may be truly read" (1.1.53). Giacomo feels that Posthumus has an overrated reputation because his marriage to the King's daughter "words him…a great deal from the matter" (1.4.14–1.4.15). "Covenants" and "articles" are "set down by lawful counsel" (1.4.138, 1.4.150, 1.4.159). Giacomo logs his data, saying, "I will write all down" (2.2.24). The ghosts dispose of a "tablet" (5.3.203) or "book" (5.3.227) that has an inscription (5.3.232–5.3.238). Cymbeline's order is to "publish" the peace (5.4.479), and of the plays in the Canon, *Cymbeline* has the greatest incidence of the word "note." The allusions seem to honor Galileo's opus and involve the play's most puzzling document—the Jovian tablet.

TABLET, BOOK, LABEL

Although Jupiter delivers the tablet in the final Act, it is discussed next not only because of its overt theatricality, but also because it helps establish the paranormal context and because of its relevance to what I have proposed is a new literary source, Galileo's *Sidereus Nuncius*.

Posthumus' fortunes decline steadily as the play progresses. He loses his bet with Giacomo, and as he lies in a British cell awaiting execution, he dreams of the four specters of his immediate family interceding on his behalf. The one resembling his father, Sicilius Leonatus, scolds the god Jupiter and accuses the mighty deity of neglect. "Thou shouldst have… shielded him / From this earth-vexing smart," he says (5.3.135–5.3.136). Posthumus "deserved the praise o'th'world" (5.3.144–5.3.145), but instead Jupiter allowed Giacomo to taint his nobler heart and make a mockery of him. The phantom siblings chime in and point out that Posthumus has an obligation to uphold the "Leonatus" title that Tenantius bestowed on Sicilius, but Jupiter tells the ghosts to mind their own business and reprimands them for losing confidence. The gods are not about to let Posthumus die.

Jupiter appears astride an eagle, the Bird of Jove. Eschewing rhyming fourteeners in favor of iambic pentameter, the deity issues the command to the ghosts, "This tablet lay upon his breast, wherein / Our pleasure his full fortune doth confine" (5.3.203–5.3.204), which means, "our pleasure confines his unrestricted good fortune within the strict limits of this tablet" (Nosworthy 5.4.110n). Jupiter departs for his "palace crystalline" (5.3.207), the ghosts vanish, and Posthumus knows upon waking that he has just seen his dead relations. He says that he has a "golden chance," but knows not why (5.3.226). The opportunity relates to his next concern, the oracular legacy resting on his bosom, and in astonishment, he exclaims "A book?" He recognizes it from its cover because without even opening it, he describes its properties.

> POSTHUMUS O rare one,
> Be not, as is our fangled world, a garment
> Nobler than that it covers. Let thy effects
> So follow to be most unlike our courtiers,
> As good as promise. (5.3.227–5.3.229)

Puzzles proliferate. The "tablet" that has metamorphosed into a "book" replete with covers, is now also "rare." The property of rarity is no surprise because Posthumus recognizes it as *Sidereus Nuncius*, the world's first monograph devoted to reports of telescopic data. The words "rare" and "eagle" each occur in *Cymbeline* with the highest incidence of any play in the Canon. The tablet is rarer still, considering its mode of delivery.

Posthumus' perplexity persists into the next scene, at which point the god-given gratuity undergoes yet another taxonomic shift to a "label."

> POSTHUMUS This label on my bosom, whose containing
> Is so from sense in hardness that I can
> Make no collection of it. (5.4.431–5.4.433)

He finds the inscription incomprehensible ("can make no collection of it," 5.4.432n), so he requests help from a soothsayer, who interprets it. Whereupon, Posthumus has nothing further to say in the play! Somehow, the resolution of the riddles posed by the Jovian bequest—terminology, ambiguity, cover, contents, rarity, and promise—follows from the soothsayer's explanation and antecedent revelations, and the seer's *explication de texte* silences the one whom First Gentleman calls "our theme" (1.1.39). This is perplexing, as are the descriptions themselves.

The "tablet" (5.3.203) is a "book" (5.3.227), and it is a "label" (5.4.431) that Posthumus fears has a cover better than its contents ("a garment / Nobler than that it covers," 5.3.228–5.3.229). The word "tablet" occurs only once in the Canon. It can mean a "small smooth inflexible or stiff sheet or leaf for writing upon" which could comprise "one of a pair or set…fastened together" (*OED*). Because only two stone tablets bore the testimony that Moses carried down Mount Sinai, it seems reasonable that Jupiter could use only one tablet to deliver a message that is only about five dozen words long.

The "most primitive" meaning of "book" is a "writing tablet" (*OED*), so at a stretch and with recourse to language dating to very early times, the terms "tablet" and "book" could be synonymous. Also in its original sense, a "label" is "a small strip of paper," so on these flimsy grounds all three definitions could share a shade of synonymy, but it is more likely that Shakespeare uses three different words because he has three different meanings in mind.

In post-cuneiform literature and in "all ages and countries," a "book" can be a treatise of "separate leaves [joined] together" (*OED*). This implies that it contains words sufficiently numerous to fill many sheets and so comprise a "volume" or a "literary composition." Such a book might be printed, such as in 1600 when *As You Like It* was described in Register Stationers' Company as a "booke," but such a production would hardly resemble a "tablet" containing a scant 60 words.

In the fifteenth and later centuries, "book" could be an item "in which we may read and find instruction or lessons," and that might contain "allusive reference to various real or reputed books, records, etc."

(*OED*). *Sidereus Nuncius* fits this definition because it contains "instructions" (e.g., on how to measure telescopic magnification), gives "lessons" (on the nature of celestial objects), and has "records" (of celestial appearances and events).

A "label" can mean "a supplementary note" or "comment" (*OED*). Galileo's tract is supplementary because, from present understanding of earlier plays in the Canon, it contains only a single item that is new. In Galileo's words, this is the discovery of "four planets flying around the star of Jupiter at unequal intervals and periods with wonderful swiftness" (*Sidereus* 26). Shakespeare had reported Galileo's other results earlier, and has no need to make a fuss about them again.

In this context, it follows that the distinction between "tablet," "book," and "label" arises because "tablet" refers to an entity from supernatural space, whereas the other two are existents in the natural World. This distinction is made at the instant that Posthumus' dream ends, and he reengages the reality of his pending execution. The Jovian bequest itself then magically parts company with the ethereal spirits of the supernormal world and enters the real world of a jail cell. The implication is that *Sidereus Nuncius* is a work of supplemental value, which divine powers bring to the Briton's attention.

Suppose further that Posthumus signifies the store of knowledge about the Universe as it had accumulated up to the start of 1610. This comprises the Diggesian empirical evidence, which (from the discussion of *Hamlet* in chapter 3) proves heliocentricism and postulates an infinite Universe. This explains why Innogen knows that Posthumus' virtue "may be truly read"—in Shakespeare's plays.

The dance of the ghosts and Jupiter's spectacular coup de theatre require Posthumus to add the Galilean moons to the inventory. The imagery makes sense because in 1610, Thomas Digges had been dead for 15 years, and by then (like Posthumus at the court of Cymbeline), he had already received high praise and was unlikely to receive any more posthumously. In fact, one might as well postulate that Posthumus' spirit is that of Thomas Digges, who by 1610 had graduated to a heavenly post and needed educating on new discoveries. However, in

March 1610, there is but one new discovery—the moons of Jupiter—which is why Shakespeare makes such a splash of the Jupiter intervention, as a tribute to Galileo. This is not unusual, as Galileo reported that he "contrived" his spyglass because he had been "inspired by divine grace" (*Sidereus* 36).

Nevertheless, before Posthumus can fulfill his role, he must first survive the death sentence imposed by the King. A messenger arrives with a reprieve, tempting an argument from First Cause that because the event occurred after an intervention, it occurred because of it. Jupiter on a conveyance is literally the *deus ex machina* that rescues Posthumus.

Therefore, a reasonable proposition is that from the start, a transcendent power had arranged for Cymbeline to raise Posthumus, and thusly create the circumstances for the adoptee to fall in love with Innogen and undergo the travails that the play documents. Later, during his stint on stage, the deity Jupiter explains, "Our Jovial star reigned at his birth, and in / Our temple was he married" (5.3.199–5.3.200). The spirit likely arranged for the clergy to abet Innogen's marriage, and because Cymbeline occupies an inferior position in the pyramid of power, he has no chance to affect the agenda. All the British King can do is react, and rather poorly at that, but in the end the deftness with which he resolves the chaos speaks to his competence and the steadfastness of supernatural governance in aid of good causes.

BIRDS

The Bird of Jove establishes the primacy of eagles in *Cymbeline*. In a peculiar pair of lines with eleven or twelve syllables (1.1.140n), Innogen compares Posthumus to an eagle, which hints at his parallel primacy and reinforces his role as the play's theme. With Posthumus as a sharp-eyed eagle (1.4.11), Innogen invites reference to the capabilities of a leonard's visual acuity and thereby to Leonard's perspective glass. She herself is likened to the legendary "Arabian bird" (1.6.17), the Phoenix,

Table 4.1. Bird hierarchies in *Cymbeline*.

SPECIES	DIET	PART
eagle	live prey	Posthumus, Augustus
puttock (kite)	small reptiles, insects, carrion, refuse	Cloten
crow	scavenger	Belarius
jay	eggs, seeds, nuts	seductress
raven	rodents, insects, grain, carrion, refuse	Giacomo

which is unique and supposedly about the size of an eagle; though as Fidele, she is simply a "bird" (4.2.198).

More generally, *Cymbeline* has a high incidence of ornithic identifications. Table 4.1 lists species taxonomically from two orders of birds in "pecking order" according to their "breeding" (i.e., by the supposed nobility of their diet). The Phoenix is omitted because it exists only in mythology, which hints that Innogen has a subtextual role.

As a Prince, Cloten should soar with falcons and tercels (Berners and Gryndall), but he is instead a puttock (1.1.141) resembling a cock and a capon (2.1.21). Belarius is a crow (3.3.12), but Shakespeare takes care to mention that his two princely wards are "unfledged" (3.3.27) because they have not yet flown the coop. Posthumus' imagined seductress is a jay (3.4.49), and Giacomo is a raven (2.2.49) whose diet resembles that of Cloten's puttock. The Romans invade Britain like eagles, but flee like chickens (5.3.41–5.3.42), and the soothsayer foresees the Roman eagle of empire winging its way to Britain.

Posthumus' Properties

Cymbeline concerns the marriage, separation, and reunion of Posthumus and Innogen, which justifies classifying its genre as "romance;" but before the play can be understood on a deeper level, one must examine the characteristics of these principals.

Consider Posthumus first. In words reminiscent of Hamlet's makeup ("th'exterior" and "the inward man," *Hamlet* 2.2.6), First Gentleman praises him.

> FIRST GENTLEMAN I do not think
> So fair an outward and such stuff within
> Endows a man but he. (1.1.22–1.1.24)

Like Hamlet, Posthumus has shortcomings, and First Gentleman praises him reservedly.

> FIRST GENTLEMAN I do extend him, sir, within himself,
> Crush him together rather than unfold
> His measure duly. (1.1.25–1.1.27)

He lauds Posthumus only "as far as he deserves it" (1.1.25n) and will "minimize his virtues rather than do them justice" (1.1.26–1.1.27n), which suggests that before the play opens, Posthumus' signification has received all the praise it will ever get. Again, this is an odd state for a romantic hero, the play's "theme," and *Cymbeline*'s "most interesting riddle" (Marcus 141).

First Gentleman's reservations occur in the context of torture: "extend him" as on a rack, and "crush him," as in death by pressing (1.1.26–1.1.27n). The theme persists. Cymbeline threatens Pisanio with a "sharp torture" (4.3.12) and Giacomo with a "bitter torture" (5.4.133), to which Giacomo responds, "Thou'lt torture me to leave unspoken that / Which to be spoke would torture thee" (5.4.139–5.4.140). Belarius fears "death / Drawn on by torture" (4.4.13–4.4.14), Posthumus speaks of "torturers ingenious" (5.4.215), and in fact, the word "torture" occurs more often in *Cymbeline* than any other of Shakespeare's plays. Torture abounded during the reign of Elizabeth, with the period from 1580 to 1603 being the heyday of its practice in England. Under Burghley's reign, *Regnum Cecilianum*, unkind treatment was not limited to mere physical pain but included psychological strategies designed to induce terror (Breight 70–76). Perhaps Shakespeare also refers to the torture and execution of Giordano Bruno, who suffered his last 8 years under a delusion of reprieve. Shakespeare is nothing if not prescient, and with *Cymbeline* in 1610, he anticipates the harms and fears that Galileo's publishing of *Sidereus Nuncius* will engender.

Second Gentleman asks, "What's his name and birth?" (1.1.27) and First Gentleman replies that he "cannot delve him to the root," where "delve" means "dig." In *Hamlet*, one Clown calls the other "goodman delver," or "master digger," and the puns persist in *Cymbeline*. In addition, the spelling "posthumus" arises from a mistaken association with "humus" (meaning "earth") (*OED*), wherein delvers too attain a state of eternal repose.

Posthumus developed a fine reputation while still a fledgling at court. To the mature, he was "A glass that feated them, and to the graver / A child that guided dotards" (1.1.49–1.1.50), all the while setting a standard for youngsters to follow. "Dotards" are pedants. They hold up the New Astronomy because they uphold the Old Astronomy. Posthumus showed them how to behave, for he was "the reflection of what they aspired to but could not attain" (Nosworthy 1.1.49n). The imagery reflects the role of the mirror in perspective glasses, by means of which mature scholars might be "feated" or brought to a higher state of refinement or grace.

Innogen's properties complement those of Posthumus and originate in part in folklore, as adapted from the ninth tale of the second day of Boccaccio's *Decameron*. There, Bernabò boasts that he is blessed with a wife who is "endowed with all qualities of an ideal woman" and who is the world's "most chaste and honest woman" (27). When asked whether the Emperor had granted him the privilege of a "perfect wife," Bernabò replies that it was God who made the concession to him, and that God is "a little more powerful" (27). A villain wagers that he can compromise her virtue, and he bribes a servant to convey him in a chest to the lady's bedroom. He emerges at night and spies under her left breast "a mole…surrounded by a few strands of fine, golden hair" (31). The intruder declines to take advantage of the sleeping beauty and instead takes items from her room. On returning to the chest and after its removal, he hastens to present his trophies as evidence that he has won the wager.

Shakespeare follows this tale closely. Innogen is a "gift of the gods" (1.4.81) that they created "chaffless" (1.6.177–1.6.178) and "divine" (2.1.54), like the heavens in the Old Astronomy. In her married state, she is "chaste" (1.4.57, 2.4.165) and "more goddess-like than wife-like" (3.2.8). Posthumus admits that Innogen denied him his connubial rights, so the couple practiced what may be termed the "larger virtue" of "married chastity" (Nosworthy xlvii). Further properties give the impression that Shakespeare was obsessed by threatened female chastity, but he is rather more preoccupied with the New Astronomy.

The wager subplot threads its way through from start to finish and contributes fundamentally to the play's subtext. As it reaches its conclusion in the final scene (5.4), it joins with the reconciliation of Posthumus and Innogen to become the two chief concerns of the subtext and of this chapter.

ORLEANS

The wager subplot begins in Rome, where Philario and four guests await the arrival of the exiled Briton. One of the guests is an Italian, Giacomo, and another is an unnamed Frenchman. The remaining two are a Dutchman and his Spanish *paisano*, both also unnamed. Giacomo recounts that he had seen Posthumus in Britain when he was "of a crescent note, expected to prove so worthy as since he has been allowed the name of" (1.4.2–1.4.3). The "crescent note" signifies that the exile's "star [was] rising, like the increasing, crescent-shaped moon" (1.4.2n). Still, Giacomo believes that Posthumus' reputation is less than stellar. "I could then have looked on him without the help of admiration, though the catalogue of his endowments had been tabled by his side and I to peruse him by items," he says (1.4.3–1.4.6). Giacomo is "mocking the literary habit of itemizing a person's qualities" as if they were a shopping list (1.4.6n), like the items that identify Harriot (see chapter 3). Nevertheless, the language is equally evocative of record keeping, which Giacomo later undertakes.

As noted, Philario feels the need to clarify Giacomo's remarks. "You speak of him when he was less furnished than now he is with that which makes him both without and within," he says (1.4.7–1.4.9). Posthumus possesses inner and exterior qualities corresponding to the two facets of the New Astronomy, and in light of the role of Galileo's *Sidereus Nuncius*, one expects his "endowments" to increase like that of a waxing crescent Moon. To "furnish" can also mean to "equip," which helps explain the next comment by Frenchman, who picks up where Giacomo left off. "I have seen him in France," he says. "We had very many there could behold the sun with as firm eyes as he" (1.4.10–1.4.11). The ability to stare directly at the Sun is a mythical prerogative of the eagle (1.4.11n), a bird to which Innogen has already compared Posthumus and which I propose refers to keen vision and the optical equipment of the Diggeses.

Philario and his guests greet Posthumus, and Frenchman recalls that they had met in Orleans where Posthumus had touted Innogen's virtues. Frenchman recounts how someone present had taken exception to the Briton's bluster, and the consequent discord might have been resolved at the point of a sword had Frenchman not intervened.[2] Frenchman says that Posthumus had supposed Innogen's virtue to be out of this world, which Posthumus had not denied because in those days, he held Innogen in just such high regard. Frenchman thinks that the matter in Orleans was "slight and trivial" (1.4.39–1.4.40), but Posthumus demurs, "[U]pon my mended judgement—if I offend not to say it is mended—my quarrel was not altogether slight" (1.4.44–1.4.45).

Giacomo asks for details, whereupon Frenchman retrieves from the files of his pluperfect memory the properties that Posthumus had attributed to Innogen. He had believed her to be fair, virtuous, wise, chaste, constant, qualified, and less attemptable than the rarest French lady. "Fair" implies "unblemished," and "virtue" is a property of gods and goddesses whom mortals may adore but not make love to. Her mind is as rare as the Phoenix is unique, and being chaste and less attemptable would challenge any Lothario. "Constancy" is an attribute like that of the stars in the Old Astronomy, and "qualified" means having good or fine qualities (1.4.57n).

Giacomo scoffs, "[Such a] lady is not now living" (1.4.59), but Posthumus reiterates his position, that she "holds her virtue still" (1.4.61).

Although Posthumus has altered ("mended") his judgment, he does not divulge the aspects that have changed, nor is it known who or what caused him to change his mind. These gaps contribute to the complexity of *Cymbeline* because Giacomo and the audience are in the dark as to what Posthumus truly believes. Here as elsewhere, Shakespeare conflates physical and metaphysical properties. Prior to Orleans, Posthumus thought that Innogen epitomized the heavenly perfection of the Old Astronomy, but now he has altered his judgment in ways not specified. His revised judgment refers to celestial perceptions that update the new Worldview, whereas in the real world he still believes that she will remain virtuous and faithful.

Thus by willful intent of the dramaturge, Giacomo believes that Posthumus ranks Innogen ahead of the earth-angel Artemis, twin sister of Apollo, whom the Romans knew as the chaste moon goddess Diana. Posthumus had admitted that he is her "adorer," but not her lover ("friend") (1.4.64–1.4.65), and Giacomo later recalls that Posthumus had spoken of her "as Dian had hot dreams / And she alone were cold" (5.4.180–5.4.181), implying that Innogen was more chaste even than the epitome of chastity. Shakespeare refers to Diana four more times, yet all five references are in an unchaste context (1.6.133n, 2.3.67, 2.4.82, 2.4.159, 5.4.180). Toward the end of the play, the soothsayer disabuses Posthumus of his remaining idealism by calling Innogen a "woman" ("mulier") (5.4.449), implying that she has down-to-earth qualities after all. To understand her "symbolic meaning," therefore, her "erratic tissue of inconsistencies" (Nosworthy lxi) must be explained. As mentioned, this and the wager subplot are the chief topics of this chapter.

WAGER

Giacomo senses opportunity and the badinage turns rancorous. He compares Posthumus' "unparagoned mistress" to the diamond she gave to Posthumus, and Posthumus replies that Innogen is "not a thing for sale" but is "the gift of the gods," which "by their graces" he will keep (1.4.76–1.4.83). Giacomo brags that he can "get ground" of fair Innogen

(1.4.100) and is prepared to wager half of his estate against Posthumus' diamond. His stake amounts to ten thousand ducats, or roughly ten times the annual income of a well-to-do Italian nobleman (1.4.122n). Posthumus confidently accepts the wager, adding that if Giacomo fails, he will have to answer to his sword. In Rome as in Orleans, his rodomontade of spousal virtue has led to a threat of combat.

Giacomo is multiply motivated. His target is a married Princess, and his goal is both to experience the satisfaction of sexual conquest and to own the token of fidelity that the braggart husband flaunts upon his finger.

Suppose that, after Orleans, Posthumus knows of cosmic advances accumulated up to the time of death of Thomas Digges in 1595 and that Giacomo's test of Innogen's virtue amounts to Galileo testing the verities of the Old Astronomy in 1610. Giacomo in his role as Galileo may soon add to them as he attempts to gainsay what he has been led to believe is Innogen's perfection. This is the crux of the wager subtext.

Moroseness afflicts Posthumus as it does Hamlet. Innogen describes her husband's gloom.

> INNOGEN When he was here
> He did incline to sadness, and oft-times
> Not knowing why. (1.6.62–1.6.64)

Perhaps here as in *Hamlet*, moroseness afflicts earthlings fingered by metaphysical powers for special duty in aid of the Good.

Posthumus is human except that "a kind of honour sets him [apart]" (1.6.170) and makes him "rare" (1.6.175). He is a blend of worldly flesh and heavenly spirit. Shakespeare assigns Giacomo the task of inquiring into the nature of the heavens and bringing results to the attention of Posthumus, whereupon the potential cuckold will receive the news literally in his role of a lovelorn husband, and figuratively as the scribe of new astronomical discoveries. If Giacomo fails to sully Innogen, her status would remain unaltered, Posthumus would continue to revere her as before, and there would be nothing to add to his store of knowledge. However, if Giacomo succeeds, he could potentially compromise one or more of her qualities, whereupon (subtextually) Posthumus would need to add the new information to his database. In the sensate world, Posthumus suffers the pangs of married celibacy, to which Giacomo threatens to add the agony of betrayal, and in his other role, he would be duty bound to record his grief as evidence of advances in the New Astronomy.

THE COMMITTEE

Returning to the topic of Philario's welcoming committee, recall that his guests are a Frenchman, a Dutchman, and a Spaniard, plus the Italian Giacomo (who is the only guest named). What is odd is that the Dutchman and the Spaniard have nothing to say during the entire episode. A short digression explains why.

Having assembled a spyglass from two lenses, the Flemish-Dutch optician Lippersey tried to patent it in 1608 but failed. He presented it to Count Maurice of Nassau (1567–1625), Prince of Orange of the Dutch Republic (Reeves 3; Drake, *Work* 137; Pannekoek 227), who showed it to diplomats who had gathered in The Hague for a peace conference. These included Ambrogio Spinola (1569–1630), who was an Italian serving Spain and who commanded the forces occupying the Spanish Netherlands (Reeves 4–5). News of the device spread quickly, and soon copies were in the hands of "the most important European authorities," these being chiefly Spanish and French political leaders and the Pope in Rome (*International* 2; Reeves 4). The Dutch had the spyglass from the outset, and thus all four members of Philario's welcoming committee hail from countries that are the leading players in the saga of Continental telescopy. Of these, the Dutch and the Spanish failed to use Lippersey's invention to study the heavens, and therefore both contribute nothing but silence to the subtext, whereas Giacomo and the Frenchman play more active roles.

In July of 1609, Galileo learned of the Dutch device from a noble Frenchman, Jacques Badovere (*c.*1575–*c.*1630) (Drake, *Discoveries* 29), and immediately set about duplicating it. Like Badovere, Shakespeare's Frenchman is a good communicator, but like Dutchman and Spaniard, he goes unnamed because neither Badovere nor any other Frenchman saw fit to use the Hollander's device to study celestial bodies. By contrast, Shakespeare assigns a name (Giacomo) to the Italian guest because an Italian (Galileo) did follow up on the Dutch design. Shakespeare associates this guest with Galileo, but why name him "Giacomo"?

Jacques Badovere also went by the name Giacomo (Drake, "Galilei" 78; Drake, *Work* 439). He belonged to a branch of the Badoer family from Venice that had migrated to France for the same reasons of faith that pitted the Prince of Orange against Spain. Badovere had studied under Galileo at Padua in 1597–1598, and by alerting Galileo to the merits of the newfangled device, he assured himself a talkative (if anonymous) role in the Canon. Into this religious and political mélange steps the naive idealist Posthumus. Giacomo is ignorant of Posthumus' true position regarding Innogen just as Galileo in 1610 knew nothing of prior telescopic results on the nature of the heavens.

New Madness

The next stage of the wager story finds Giacomo landing in Britain where, escorted by Posthumus' servant, Pisanio, he meets Innogen. In accordance with his state of ignorance, Giacomo calls her "fairest" (1.6.31), meaning that he regards her as unblemished. He delivers a letter from Posthumus, which she reads aloud. The letter praises the letter bearer, and Innogen promises to aid him in whatever way she can. One might expect that he would engage his target in small talk conducive to intercourse, but instead, he blathers about seeming irrelevancies.

He asks, "What, are men mad?" (1.6.32) whereupon he speaks of gatherers of celestial and terrestrial data.

> GIACOMO Hath nature given them eyes
> To see this vaulted arch and the rich crop
> Of sea and land, which can distinguish 'twixt
> The fiery orbs above and the twinned stones
> Upon th'unnumbered beach, and can we not
> Partition make with spectacles so precious
> 'Twixt fair and foul? (1.6.32–1.6.38)

As posited, Giacomo plays the part of a pretelescopic Galileo who subscribes to heliocentricism and (for want of information to the contrary) must also subscribe to old perceptions of the perfection of individual objects in the heavens. The imagery speaks to "madness" like that which afflicts Prince Hamlet, as the issue in *Cymbeline* also concerns cosmic images and their significance to the New Astronomy.

In accordance with Aristotelian doctrine, one presumes that there are discernable differences between "fiery orbs above" and "twinned stones" below, the former (in accordance with Aristotelian doctrine) being celestial and "fair," the latter terrestrial and "foul." By stating that grains of sand are "unnumbered," Giacomo alludes to the calculation made by Archimedes of the number of grains of sand that would fill the Universe (Pannekoek 121). Giacomo asks whether "precious spectacles" can distinguish them. This refers to Galileo's first telescope that he assembled from commercially available spectacle lenses. A connection between "eye" and "spectacles" (1.6.37n) also occurs in *Henry VI part 2*. In Galileo's case, it recalls the report (made prior to October of 1610) by another former student, John Wedderburn (1583–1651), that Galileo had rearranged the lenses of his spyglass and devised a microscope (Drake, *Work* 163). He used this to examine insects at close range, and may also have distinguished and counted grains of sand.

Innogen asks what so engenders Giacomo's admiration (i.e., causes his amazement) (1.6.38n), and Giacomo continues his circumlocution.

> GIACOMO It cannot be i'th' eye—for apes and monkeys,
> 'Twixt two such shes, would chatter this way and
> Contemn with mows the other; nor i'th' judgement,
> For idiots in this case of favour would
> Be wisely definite; nor i'th' appetite:
> Sluttery, to such neat excellence opposed,
> Should make desire vomit emptiness,
> Not so allured to feed. (1.6.39–1.6.46)

This speech widens "the focus" from "sight" to "judgment" and "appetite," and is renowned for "impressionism" and "convoluted complexity"

(1.6.39–1.6.46n). The pronoun "it" in the first line quoted refers to that which engenders Giacomo's admiration, which Giacomo (Galileo) says is not "i'th' eye" of the beholder because optical images exist independent of the observer.

Giacomo's tone is depreciatory. He affirms that pedants who deny the reality of optical images are like "apes and monkeys" that chatter approvingly even as they remain ignorant of their source ("Contemn with mows") (1.6.40–1.6.41n). Like the dogs of Heraclitus that bark at everything they see, they look no further than appearances. They cannot accept a celestial "she" who is "foul," so by default they favor a "she" who is "fair" because that is how the heavens are supposed to be. The nattering simians are "gross creatures" (1.6.39n), but they cannot have it both ways, marveling at images and yet judging them false, for even idiots can judge a beauty contest ("this case of favour," 1.6.42n) and could decide wisely ("be wisely definite," 1.6.43n). The passages satirize those who believe that lensed images are misleading yet have no qualms about using spectacles to improve their vision.

Innogen can make no sense of this, so she asks Giacomo, "What is the matter?" (1.6.47). But the visitor persists, interlacing his doublespeak with images of debauchery.

> GIACOMO The cloyèd will,
> That satiate yet unsatisfied desire, that tub
> Both filled and running, ravening first the lamb,
> Longs after for the garbage. (1.6.48–1.6.51)

The criterion of sexual desire is unsuited to decide between "fair" and "foul" because it is like a leaking tub of water. Even when sated, it soon becomes unsatisfied. The lustful, having taken advantage of an innocent, yearn still for the slut ("the garbage," 1.6.51n). The word "ravening" again associates Giacomo with a raven, whose diet includes garbage (table 4.1).

Innogen asks, "What, dear sir, / Thus raps you? Are you well?" (1.6.51–1.6.52) Giacomo denies that he is ill, but realizes that if he is to seduce Innogen, he must be alone in her company. He sends Pisanio on a mission, but by now Pisanio has heard enough to become suspicious of the Roman visitor.

The sexual innuendos prompt Innogen to inquire after Posthumus. Madness joins deceit as Giacomo presses home his advantage. He paints Posthumus as a licentious reveler who is undeserving of her. He pretends to pity Innogen for having an unfaithful husband, and hints that others are enjoying Posthumus sexually. Giacomo elaborates on the perils of intercourse with prostitutes and suggests that she exact revenge on her husband by committing adultery with him, but Innogen is unimpressed and declares that she will report Giacomo to her father the King. Giacomo quickly backtracks, begs her pardon, and pretends that he was merely testing her.

Having failed to seduce her, Giacomo resorts to flattery. He compliments the absent Posthumus by calling him "a holy witch," meaning that he is a gender-nonspecific enchanter who practices magic (1.6.166n) and "enchants societies into him" (1.6.167), or wins converts to his cause. The lines refer to telescopy and its supposed capacity to mislead the unwary. He belabors his admiration of Posthumus.

> GIACOMO He sits 'mongst men like a descended god.
> He hath a kind of honour sets him off
> More than a mortal seeming. (1.6.169–1.6.171)

If Giacomo is not being facetious, Posthumus has properties "like" those of "a descended god," which implies a dual physical-metaphysical nature.

CHEST

As if it were an afterthought, Giacomo requests that Innogen keep safe a chest of jewels that he says he and his companions have purchased as a gift for the Roman Emperor. "Willingly," she says (1.6.193). The sexual imagery persists as she acknowledges that Giacomo has an "interest" in the jewels (1.6.195), and she stakes her honor on their safekeeping.

Giacomo does indeed have an interest in jewels, but not those that are the figments of his deceit; rather, he desires that rare gem, Innogen, and the diamond she gave to Posthumus. He explains how he will get the chest to her.

> GIACOMO They are in a trunk
> Attended by my men. I will make bold
> To send them to you. (1.6.196–1.6.198)

Are the gems safer in Innogen's bedroom than under the guard of Giacomo's own men? Innogen has rare intelligence and has encountered Giacomo's duplicity already, so her generosity challenges credulity. Improbabilities suggest either a poor dramatist or one who has gods pulling strings on behalf of the subtext.

Giacomo, in the guise of Galileo, will use "a trunk" to achieve his goal. A trunk is a container that might contain personal effects, and could refer to the human body exclusive of head and limbs (*OED*) as in 4.2.354–4.2.355 when Lucius refers to Cloten's headless body, "Soft ho, what trunk is here / Without his top?"

However, "trunk" is also another term for a perspective glass, as in 1610 when an observer wrote concerning the Pleiades that, "With one of our ordinary Trunks I have told eleven stars in the Pleiades" (*OED*). On the Continent, Galileo's two-lens device was known as a *perspicillum* in Latin and an *occhiale* in Italian, and only in April 1611 was the term *telescopium* coined (Drake, *Discoveries* 27n3; Drake, *Sidereus* 26n2, 112). Therefore, when writing *Cymbeline* in late 1610, Shakespeare would have felt free to use the contemporary English term, "trunk."

When the god Jupiter commands the ghosts to place a tablet on Posthumus' "breast," the allusion is to both Posthumus' (bodily) trunk and Galileo's (telescopic) trunk. Jupiter chooses the breast of the one who needs educating on the new device, and where the tablet is placed, Posthumus could hardly fail to see it upon regaining consciousness in the real world. This peculiar location tells Posthumus that a trunk is at issue. Giacomo's trunk is both a dramatic subterfuge and the instrument for observing the fair "shes" of heaven. The term "heavenly body" in connection with Giacomo's glasses has a double meaning too trite for explanation, other than to note that the *OED* mentions it as a term pertaining to objects in the natural heaven. "Heavenly" means "divine" or "having excellence [or] beauty that belongs to heaven," so with the help of a clever ruse, stratagem, or trick, the wily Giacomo sets out to destroy the reputation of incorruptibility that he thinks Innogen enjoys, just as with the aid of a trunk, Galileo set out to inquire into the supposed perfection of the heavenly bodies.

WEATHER

That evening, Giacomo conceals himself in the trunk and has it delivered to Innogen's bedroom. When she is asleep, he opens it and steps out. The bedroom scene is said to have a "strangely oblique, dream-like quality" (Thorne 182), which is an apt description, given Innogen's slumbering state and the play's transcendent purpose. Giacomo can see what he is doing because, luckily, Innogen had asked her maid in waiting to keep a candle lit in her room. Evidently, she had had no fear of zephyrs extinguishing the flame.

The interloper provides a running commentary on his thoughts and actions. The sound of crickets greets him immediately upon his emerging from the trunk and his first remark is that "The crickets sing" (2.2.11). "Crickets" might refer to ladies-in-waiting, as in *The Winter's Tale* (Snyder and Curren-Aquino 2.1.31n), but it is past midnight and it is unlikely that they are chattering. The reference is probably to the Old World insect *Acheta domesticus*, which chirps in the quiet of the night and which (after 1325) became known as the common house cricket (*OED*). Innogen might even have kept specimens as pets. To "sing" could mean to anticipate trouble (*The Merry Wives of Windsor* 3.2.33), which fits the context.

House crickets thrive at temperatures between 26–32°C (79–90°F), which is also a comfortable range for sleeping without nightclothes, as at the peak of summer's heat. Yet the window of Innogen's bedroom is shut and the curtains drawn (Nosworthy 2.2.21–2.2.23n), which is not what one expects in the summertime. To assuage his appetite for fame and riches, one would confidently expect that Giacomo will hie for Rome as soon as he completes his business in Britain, yet shortly before he arrives in Rome, Posthumus remarks, "Quake in the present winter's state, and wish / That warmer days would come" (2.4.5–2.4.6). How can Romans shiver from the winter's cold when a few weeks earlier in a chillier, more northerly clime, crickets chirp and Innogen slumbers naked?

Suppose that it is wintertime and that Innogen and the crickets are beneficiaries of the surge of architectural innovation that occurred in the Augustan era. In the interval from 54 BC (after Julius Caesar had left Britain for the second time) to AD 43 (when the Roman occupation of Britain began in earnest), contact between Rome and Britain grew through trade and the expediency of tribal chiefs sending their heirs to Rome to be educated. Reciprocally, Romans spread their architectural and engineering expertise throughout their empire and into Britain.

One innovation was the *hypocaust* ("heated from below"), which heated public baths and villas. It consisted of a space below a raised floor into which flowed hot exhaust from a furnace. Ducts ended at flues that vented the fumes, thereby maintaining the purity of the inside air. In villas, for example, the reticulation resulted in a temperature gradient across rooms, with those closest to the furnace being warmest. In the Augustan era, well-to-do citizens heated their homes centrally, and in view of Cymbeline's intimate relationship with Rome, it is likely that his palace is state of the art. Plausibly, proximity to the furnace and room temperature would correlate with social status, so a Princess would qualify for one of the warmest. Lacking a thermostat, Innogen's bedroom might well have been comfortable for her to sleep in as long as she was sans pajamas, whereas with an open window, her room would be chilly and liable to fill with smoke. In a closed chamber, radiant heat rising from the floor would cause only a slight movement of air, as when, pathetic fallacy aside, "The flame o'th'taper / Bows toward her" (2.2.19–2.2.20).

How do crickets enter the story? In southern Britain, feral crickets can survive outdoors in the egg stage, but with the onset of colder weather, mature specimens may swarm indoors seeking warmth. By crawling, jumping, or flying, they can enter windows one or even two stories high. They become active at night and are attracted to light. The insects derive nourishment by consuming cloth and related material that bedrooms have in plenty. Adults live two to three months or longer, and could still be chirping toward the end of November, and of course, members from families of year-round residents could chirp at any time, winter included.

Thus, Giacomo—homeward bound for a few weeks—arrives in Rome in the early winter, and the scenario leads to a confirmatory dating of astronomical events and the emergence of the play in late December of 1610.

GIACOMO'S DESIGN

Continuing the wager story, Giacomo emerges from the trunk and hints at his purpose using a passage reminiscent of *Lucrece*,

> GIACOMO Our Tarquin thus
> Did softly press the rushes ere he wakened
> The chastity he wounded. (2.2.12–2.2.14)

"Rushes" is from line 138 of *Lucrece*, and "Our Tarquin" refers to the Roman King notorious for his rape of Lucrece, where "our" establishes commonality of nationality with the ostensible goals of Tarquin and Giacomo. Shakespeare introduces Tarquin at the outset of the trespass in order to arouse concern that Giacomo will rape Innogen, just as Tarquin raped Lucrece (2.2.12n), but those familiar with the *Decameron* know that the attack will not occur (28–29). This suits Shakespeare's intent exactly.

Giacomo calls the sleeping beauty "Cytherea" and describes her as a "fresh lily, / And whiter than the sheets!" (2.2.14–2.2.16). The lily is the traditional image of purity and chastity (2.2.15n), signifying Innogen's virgin state. The comparison of her complexion to white bed linen recalls the passage from *Venus and Adonis*, "Who sees his true love in her naked bed, / Teaching the sheets a whiter hue than white" (397–398).

Venus is "in her naked bed" (i.e., naked in her bed), and it is likely that Innogen is naked in hers too. Later in the scene when Giacomo removes Innogen's gold bracelet, he removes the gild from the lily. Giacomo's comparison of immaculate Innogen to heavenly Cytherea (Venus) indicates that Galileo was unaware of the phases of Venus at the time corresponding to Giacomo's intrusion. This agrees with the earlier conclusion that the play was written in late December 1610, and in fact, it was only on New Year's Day of 1611 that Galileo revealed the existence of Cytherean phases.

Giacomo once had designs on Innogen's chastity, but now—inexplicably—he gets cold feet and chooses merely to observe her and her environs. Shakespeare follows Boccaccio's *Decameron*, wherein the trespasser does not take the treasure of the maiden's honor, but Giacomo's reason is supposedly his "design," in accordance with which he vows "To note the chamber" and "write all down" (2.2.23–2.2.24). He no longer has designs on the maiden. He aims to gather data to convince skeptics that he was in her bedchamber and leave the rest to their imaginations.

In the subtext, however, "design" means something different from machination. A "design" is a plan, scheme, or conception of an idea effectuated by action (*OED*). In the semi-titular heading of *Sidereus Nuncius*, Galileo states that he made observations using "a New Spyglass," which in the first paragraph he calls an "instrument" (*Sidereus* 35, 26n3). In the seventh paragraph, he credits "a certain Dutchman" with making it ten months before he devised his own. Galileo's "design" is the telescopic concept that he effectuated by the action of building the instrument, aiming it on the sky, and recording his observations, just as Giacomo's aim and "design" is to use a preexisting trunk to note the chamber and write down what he sees.

The Chamber

The intruder refers to the flame of the candle that "would underpeep" the lids of Innogen's eyes to "see th'enclosèd lights, now canopied / Under these windows" (2.2.21–2.2.22). The punctuation is uncertain and the lines are "complex and ambiguous," but Nosworthy (2.2.21–2.2.23n) concludes that the language is suggestive of a "secondary strand of cosmic imagery."

"Windows" can refer to Innogen's eyes, by which she may observe her surroundings, and "canopied" means curtained or covered, like the sky enclosing the observer. Suppose that the walls and ceiling of the bedchamber are like the sky, and that Giacomo's ensuing actions simulate those of an observer collecting information about objects within the celestial chamber. To Giacomo, Innogen is the figurative icon of heavenly perfection, just as in pretelescopic times, the Sun, Moon, and five known planets were supposedly perfect. Any flaws in Innogen's makeup would correspond to flaws in the old visualizations of the seven wanderers, which Posthumus (in his posited role of recorder of celestial data) would ultimately need to know about.

Difficulties with parsing and punctuation persist into Giacomo's next descriptions, here identified by line number.

GIACOMO Such and such pictures,
there the window, such 2.2.25
Th'adornment of her bed, the arras, figures, 2.2.26
Why such and such; and the contents o'th'story. 2.2.27
Ah, but some natural notes about her body 2.2.28
Above ten thousand meaner movables 2.2.29
Would testify t'enrich mine inventory. 2.2.30

Lines 2.2.25 and 2.2.27 contain five instances of the word "such," comprising two "such and such" pairs, plus a solitary "such." Sandwiched between them are unidentified "pictures," "the window" of the chamber, the bed's "adornment," an "arras," and "figures." The remainder of 2.2.27 switches the subject to the contents of the "story," which could mean the associated narrative of the arras, the bedroom (Nosworthy 2.2.27n), or even the wager story itself. No adornment of the bed can compare to Innogen, about whose body Giacomo spies "some natural notes," and ten thousand "meaner movables" to complete his "inventory."

RESULTS

The test of the present hypothesis is to see whether these references, taken as a whole, correspond to what Galileo observed and announced in *Sidereus Nuncius*. Let "P*n*" denote his book's *n*th paragraph. In P2–P5 (*Sidereus*), Galileo summarizes what he considered his chief results:

> P2. "add to the countless multitude of fixed stars…others never seen before"
> P3. "the Moon…rough and uneven"
> P4. "end…the debate about the…Milky Way," "nebulous [stars] very different [from fixed stars]"
> P5. "four wandering stars…around [Jupiter]"

"Fixed stars" are what is known in modern times simply as "stars," (see *G* and *H*, chapter 1). These are those points of light in the sky whose positions are fixed relative to one another as the entire ensemble appears to rotate westward with respect to the horizon. Let "telescopic stars" refer to fixed stars visible only through a telescope. "Nebulous" stars are fuzzy patches that are also "fixed" in the same sense as the "fixed stars."

TABLE 4.2. Correspondence between Galileo's chief results in *Sidereus Nuncius* and Giacomo's descriptions in *Cymbeline*.

Galileo's results	Paragraph	Giacomo's descriptions
G1. Telescopic stars	P2	2.2.25
G2. Nebulous stars	P4	2.2.27
G3. Milky Way	P4	2.2.25, 2.2.28, 2.2.29, 2.2.30
G4. Maculate Moon	P3	2.2.26, 2.2.28
G5. Jovian moons	P5	2.2.26, 2.2.27

For ease of reference, let the five items in P2–P5 have labels G1 to G5, as in table 4.2. Their correspondence to the *Cymbeline* script is as follows:

G1. "Such and such pictures" (2.2.25) refers to two star charts that Galileo sketched and published (*Sidereus* 60–61). One is of the asterism of belt and sword in the Orion constellation in which he discovered more than 500 telescopic stars; another is the Pleiades where he found more than 40.

G2. The second "such and such" (2.2.27) refers to Galileo's descriptions of two nebulous stars that his spyglass revealed as congeries of faint stars too close together for the naked eye to distinguish (*Sidereus* 63). He counted 21 stars in the image at the head of Orion and more than 40 in Praesepe.

G3. All four instances of "such" in G1 and G2 pertain to telescopic stars, and the fifth (2.2.25) supposedly does too. It follows hard upon "there the window," which refers to an old interpretation by Geoffrey Chaucer (*c.*1340–1400) that the Milky Way is a region of relative transparency—a window—through which one sees many more stars than elsewhere in the sky. Correspondingly, Giacomo enriches his "inventory" (2.2.30) of telescopic stars by looking at the Milky Way and claiming in P4 to have "put an end to the debate about the Galaxy or Milky Way" (*Sidereus* 36). Giacomo's observation of "above ten thousand meaner movables" (2.2.29) supposedly refers to 10,000 "pieces of furniture" or "items of property" (2.2.29n; Nosworthy 2.2.30n), but it is hard to understand why there are so many of them or why they would lie "above." Likewise, if they were crickets, the infestation would warrant explanation. Although Giacomo's avian analogue is a peckish raven whose diet includes insects (table 4.1), and 10,000 crickets would certainly assuage that aspect of his appetite, his other self would need only a few to examine through his spyglass turned microscope. Instead, following the analogous use in *Hamlet*, let the 10,000 "meaner movables" lying "above" be the naked-eye stars. They lie on what Giacomo describes as the roof of the chamber fretted with "golden cherubins" (2.4.88), where "cherub" means an angel, or more commonly a celestial being or heavenly beauty. This parallels Hamlet's "brave o'erhanging firmament" and "majestical roof fretted with golden fire" (*Hamlet* 2.2.283–2.2.285) and accords with the Neoplatonic perception that a hierarchy of angels make their home in the stars. The word "movables" refers paradoxically to "fixed" stars, which as previously explained are fixed relative to one another as they appear to move in unison from east to west. They are plausibly "meaner," meaning "smaller," because less bright stars seem smaller. They are "meaner" too because they are "less important" (2.2.29n) in the sense that in the prevalent hierarchical pyramid of Worldly significance they have a lesser status. "[N]atural notes about her body" (2.2.28) refers *inter alia* to these fixed stars and the music emitted by the Ancient Planets, which by ancient tenet emit notes naturally. The word "note" has many meanings (*OED*). It can mean a symbol used in writing or printing, like the asterisk-like symbol that Galileo used to denote a naturally occurring fixed star or a moon of Jupiter (figure 1.10). It can also mean a brief written observation or record that might "serve as a basis for a more complete statement or for future action." "About" can refer to a "circuitous course," especially "orbiting objects, wheels...in a strictly rotational movement around," as in the diurnal motion of the Firmament (*OED*). The description sustains the celestial imagery because Galileo wrote that he could not count telescopic stars owing to their enormous number, and that "lack of time" caused him to "put off this assault until another occasion" (*Sidereus* 59).

G4. From 2.2.26, the bedroom contains a number of "figures," one of which is a sculpture or carving of "Chaste Dian bathing" (2.4.81–2.4.82; Nosworthy 2.2.26n, 2.4.82). In all probability, the bathing beauty is naked, and because it has been noted that Shakespeare consistently refers to the moon goddess Diana in an unchaste context in *Cymbeline*, one can presume he implies that the Moon's surface is marred. This corresponds to Galileo's description of the Moon as not "smooth and polished." Giacomo's observation of "some natural notes about her body" (2.2.28) can also refer to a distinctive feature, a distinguishing mark, a stigma, or a mark of disgrace (*OED*). The notes occur "about" her body, which can mean "on," "attached," or "connected" to it (*OED*). The phrase foretells the existence of the mole that Giacomo will later discover on her breast (2.2.37), but subtextually it means that Innogen in her epitomic role of heavenly perfection is marred.

G5. This item concerns Jupiter's moons, which Shakespeare regards as the pièce-de-résistance of *Sidereus Nuncius* and whose introduction he defers to scene 2.4 and the final Act.

Bracelet and Mole

In *Decameron*, Ambrogiuolo filches a number of items belonging to Bernabò's quintessential wife, but in *Cymbeline*, Giacomo takes only one item—Innogen's bracelet. Then he steals another glance at her and spots a mole under her breast. The mole in *Decameron* is "surrounded by a few strands of fine, golden hair" (28) whereas in *Cymbeline*, it has five parts.

> GIACOMO On her left breast
> A mole cinque-spotted, like the crimson drops
> I'th'bottom of a cowslip. (2.2.37–2.2.39)

A "mole" is a "spot or blemish on the human skin" (*OED*), and a "cowslip" is a kind of yellow primrose with a red spot on each of five petals ("cinque-spotted").

Shakespeare departs from *Decameron* for good reason. Innogen is blemished, implying that as a representative of the Old Astronomy, the heavens are not perfect, and Giacomo asserts that the imperfection has five parts, which happens to equal the number of items G1 to G5.

It is not surprising that at this point Shakespeare ends Giacomo's (i.e., Galileo's) data collection because G1 to G5 account for essentially everything Galileo observed. "No more. To what end?" (2.2.42), Giacomo asks himself. He has recorded information on the bedroom in writing, but he declines to treat data on the mole in the same way because, he says, its appearance is riveted and screwed to his memory. Curiously, the mole's pentamerousness—its "fiveness"—receives no further mention in the play, and later Giacomo reports merely that he observed "a mole" (2.4.135). Withholding information on its compound structure is a dramaturgical ploy to keep Posthumus ignorant of G5 (the Jovian moons) until the final Act, when the four circling specters and the starry messenger bring the moons to his attention.

Close Up

When Giacomo commenced observing, he referred to Tarquin's rape of Lucrece (2.2.12n), and as he prepares to cease observing he notices that Innogen has been reading about the classical rape of Philomel by Tereus (2.2.45, 2.2.46n). His entire observing run is sandwiched between descriptions of violations of a "heavenly angel" (2.2.50), which buoys his hope that he has gathered enough evidence to impute a similar feat to himself. His voyeurism equates to Galileo using a spyglass to violate the privacy of the gods, an act that he claims is the first ever ("their newness, unheard of through the ages;" *Sidereus* 35).

In 1610 it was not hard for an insightful student of human nature to foresee that Galileo's ventures would land him in a peck of trouble, and so it is not surprising that Giacomo, on stepping back into the trunk, admits, "I lodge in fear" (2.2.49). Early on, he had had no misgivings, but now the craven raven realizes his predicament. He knows that Diana (to whom he repeatedly refers) once targeted an entire "city of unjust men" (Graves, section 22*e*), and would likely make quick work of him.

The intruder shuts the trunk in anticipation of the "dawning" (2.2.48), which is the time when celestial observers close their logs and shut down their instruments. Giacomo will leave the next day and return to Rome to claim his prize.

Timing

The timing of events in scene 2.2 supports the conjectures. As the scene opens, attendants bring the trunk holding Giacomo into Innogen's bedroom followed by a bed supporting Innogen. It is not known how much time elapses between these events, but soon thereafter, the audience learns that it is almost midnight and that Innogen, who has been reading, has become drowsy. Innogen concludes that her reading has lasted three hours, so she must have started reading at about 21:00 hours (9 p.m.). She asks her lady-in-waiting, Helen, to wake her at four o'clock in the morning. Innogen falls asleep and Giacomo emerges from the trunk. At the stroke of three, Giacomo repairs to his hiding place and attendants remove the trunk, the bed, and their occupants (2.2.51SD).

Galileo first mentioned the Jovian moons in a letter of January 7, 1610. Before midnight local time on January 6, which supposedly is the time on stage when scene 2.2 begins, Giacomo is inside the trunk and unable to observe. He emerges after midnight, as Galileo observes Jupiter 45° to 50° or less above the southwest horizon. This occurs early in the morning of January 7, and Galileo would have completed his letter later that day.

In the early morning of January 7, 1610, the first conjunction of the Moon and Jupiter occurred. At that time, these were the only Ancient Planets visible, the other five being virtually opposite in the sky below the northern horizon. Jupiter was eye-catching because it was close to Opposition (see figure 1.5), at which time the planet and its moons were virtually at their brightest. The Moon was also close to Opposition (i.e., nearly Full), with an illuminated fraction of 96%. In addition, the constellation Virgo peeked over the ENE horizon at 22:00 hours, and at midnight when Giacomo exits the trunk, Virgo is rising, again pointing to Innogen awakening and rising in a virgin state.

Crux Redux

In Rome, Posthumus and Philario await Giacomo's return. Posthumus seems confident that Innogen's "honour / Will remain hers" (2.4.2–2.4.3), but when Giacomo arrives and affirms that Innogen is, "One of the fair'st that I have looked upon" (2.4.32), Posthumus agrees hesitantly, which hints "at a greater anxiety" about her than he had previously admitted

(2.4.34n). This is known to be the case because he has "mended" his judgment (1.4.45), and in his subconscious role as a memory bank of heavenly imperfections, he might well wonder whether Giacomo's data will add to his list and further disabuse him of his beloved's erstwhile perfection.

Giacomo arrives with letters from Innogen to Posthumus, who reads them and concludes, "All is well, yet" (2.4.39). However, Giacomo announces that he has won the wager and Posthumus asks for proof. Giacomo presents his data, which fall into three categories: records of her surroundings, possession of her bracelet, and knowledge of her mole. Take these in turn.

Giacomo describes the artwork in Innogen's bedroom.

> GIACOMO The chimney
> Is south of the chamber, and the chimney-piece
> Chaste Dian bathing. Never saw I figures
> So likely to report themselves; the cutter
> Was as another nature, dumb, outwent her,
> Motion and breath left out. (2.4.80–2.4.85)

To Giacomo, these "figures…report themselves," by which the lecher analogizes naked females like Dian to some as yet unidentified celestial figures that rivet one's gaze. The figures "outwent" Dian, possibly because they attracted more attention than her analogue, the Moon. To "outgo" can mean to attract more attention than another referent, or to outdistance or go faster than (*OED*). The comparison is between certain celestial figures and the Moon, which, if one compares comparables, implies that at least some of the figures are moons.

Jovian moons "report themselves" because they attract attention, as when Galileo saw them arranged exactly along a straight line parallel to the ecliptic, which is the apparent path described by the Sun across the sky. This indicated that the moons lay in that plane containing the Earth's orbit, which is close to the planes of the other planets as well. To Galileo, the moons were much brighter than fixed stars of the same apparent size, and not twinkling like bona fide stars (*Sidereus* 66–68).

The equivalence of the figures and moons finds support in the fact that the Moon's godly equivalent, Dian, achieved the exalted state of perennial virginity only with the help of her father, Jove (Ovid 1: 671), from which another reference to the planet Jupiter can be attributed. Galileo detected no blemish on the Jovian attendants and reported that they were "smooth," "exactly circular," and "covered with light" (*Sidereus* 58, 62). The Galilean moons have orbital periods of 1.8, 3.6, 7.2, and 16.7 days—compared to the Moon's orbital period of 27.3 days—which explains why they "outwent" the Earth's Moon because their motion back and forth is faster than the Moon's. This sets them apart from the Earth's Moon, which according to G4 is blemished and shows phases.

Shakespeare's description that the Jovian moons are of "another nature" is apt, even though they are still moons in the sense of being satellites of another body and appear as round spheres as Aristotelians suppose all heavenly figures to be. The cutter artist was "like a second nature, though a dumb one, who surpassed nature except that he could not provide movement or breathing" (2.4.84–2.4.85n). In a thoroughgoing portrayal of celestial phenomena, an ideal cutter would have imparted motion to the figures as if they were alive and breathing (as deities are supposed to be), but instead omitted their "motion and breath" as if they were lifeless or frozen in time. In the art versus nature debate of the time, an artist might mimic nature, but could not instill life.

Posthumus protests that a description of Innogen's bedchamber does not prove Giacomo's claim others might have imparted this information to him. Posthumus denies, or does not understand, that Giacomo's description pertains to Jovian moons, which means he must learn of them later. This, as has been described, occurs in Act 5 via the four specters and the descent of Jupiter.

Therefore, for the time being, Posthumus gets to keep his ring. Giacomo proceeds to the next category. He produces the bracelet and, in anticipation of winning the bet, he says that it "must be married / To that your diamond" (2.4.97–2.4.98). Innogen's bracelet and Posthumus' diamond are their emblems of mutual love, and Giacomo will be pleased to own them both. Posthumus is so surprised to see the manacle of love that he exclaims "Jove!" (2.4.98). His is not an idle oath, for Jupiter takes heed and will rescue him in Act 5.

Posthumus declares that the bracelet had been his gift to Innogen, whereupon the quick-witted Roman justifies his possession of it, saying that she parted with it voluntarily. Posthumus concludes that he has lost the bet, so he hands over his ring. Philario intervenes, pointing out that mere possession of the bracelet does not prove the case. Posthumus realizes that the bracelet might have been lost or stolen, so he demands its return, and Giacomo complies, albeit not without protest. With an oath and a half truth, Giacomo defends his possession of her bracelet. "By Jupiter, I had it from her arm" (2.4.121), he says.

Each bettor has now invoked Jupiter. Giacomo's oath is solemn (2.4.121–2.4.122n), and it sways the Briton's opinion. Over Philario's objections, Posthumus concedes that his beloved "hath been colted" (2.4.133). He hands the ring back to Giacomo, who with wanton malice introduces evidence from the third category. The scavenging raven crows that under Innogen's breast lies "a mole," and he asks Posthumus scathingly, "You do remember this stain upon her?" (2.4.138–2.4.139) The crux concerning Innogen's nature resurfaces with Posthumus' reply,

> POSTHUMUS Ay, and it doth confirm
> Another stain as big as hell can hold,
> Were there no more but it. (2.4.140–2.4.142)

A "stain" is "a blot" or a "moral disgrace" (2.4.139–2.4.140n), and Posthumus reveals that he knows of both a "stain" and a "mole." Posthumus has been educated since Orleans, and he knows of both blots, but Giacomo knows only of the mole.

Giacomo offers to elaborate on the nature of the mole, but the formerly brash Briton, thinking that all he will hear are details of the Roman's conquest, declines. "Spare your arithmetic, never count the turns," he says (2.4.142). "Arithmetic" refers to the science of numbers and the art of computing, which from the thirteenth century were closely associated with the celestial arts (*OED*). "Turn" can mean a sexual act (2.4.142n), but in the subtext, it means "change," an "act of turning," or a "revolution"

(*OED*). Ironically, Posthumus refers to the orbital revolutions of the Galilean moons ("periods around a certain star," *Sidereus* 36). Cognoscenti who stayed abreast of developments would appreciate the irony because by 1610 (via *Hamlet*), Posthumus' cosmic data bank already contains G1 to G4, and lacks only G5 on Jupiter's moons. Here, in Act 2, this translates to Thomas Digges knowing of G1 to G4 but not G5, and the astute can foresee that Posthumus' Diggesian spirit eventually must learn of G5.

In addition, a "turn" can mean a "piece of work," an "event," "occurrence," or a "change for better or for worse" (*OED*). This last possibility allows for each of the numbered turns G1 to G5 to refer to occasions of discovery. They are for the better because they concern pieces of work up to the time of *Sidereus Nuncius*, and they are for the worse because Posthumus thinks that Giacomo counted the acts of his sexual conquest.

Posthumus' has had enough and will hear no more. In Act 2 therefore, his psyche never learns of the Galilean moons because he has denied their existence and does not care to know more.

In his distress, Posthumus has acknowledged that the mole confirms a "stain" that is "as big as hell can hold." If Giacomo had spotted this outré blemish, he would have made a show of it. Posthumus must have learned of it after the events of Orleans, but by then he was already in love with Innogen and love, they say, is blind.

What could the stain be? In another instance of interpretation leading to prediction, one might expect to understand the "stain" as something that Galileo did *not* observe.

The net result of the wager subplot, as it has developed so far, is twofold: that Posthumus has opted not to receive knowledge of the fifth part of the mole G5, and that Giacomo is ignorant of the stain. Meanwhile, Posthumus' ring has changed hands a few times and is now in Giacomo's possession, along with Innogen's bracelet.

PENTADS

In the multiplex of allusion, Shakespeare sets the number of mole parts at 5 because there are 5 chief Galilean observations G1–G5, of which the 5th leads to a new 5-member class of planetary moons (the Moon and Jupiter's four) whose nature becomes apparent in the fifth Act. The number of male Leonati dead and alive is 5, there are 5 spectral visitations (Jupiter's and the four spirits), and 5 real-life species of bird (table 4.1), and the following discussion concerns a 5-fold event. In the last scene, the color of Innogen's mole and the "crimson" color of the cowslip's drops identify Innogen's brother, Guiderius, who has "Upon his neck a mole, a sanguine star" (5.4.365). It resembles Innogen's mole, but it is star-like, which implies that it too has five parts, like a pentacle—a 5-pointed star and a symbol of magic, the natural variety of which Galileo is supposedly practicing.

Ire

Posthumus is convinced that he is a cuckold and he vents his ire in the presence of Philario and Giacomo.

> POSTHUMUS O that I had her here to tear her limb-meal!
> I will go there and do't, i'th'court, before
> Her father, I'll do something. (2.4.147–2.4.149)

He stalks off, so Philario is unable to aid his defense, and he tells Giacomo, "You have won," (2.4.150). The two depart to keep an eye on the Briton, but Posthumus gives them the slip and promptly reenters to resume his churlish invective.

Posthumus wonders why men must share the world with women. "Is there no way for men to be, but women / Must be half-workers? We are all bastards," he says (1.4.153–1.4.154). He knows that the two genders must work together and thinks that the shared load makes us all "bastards," like hybrids or androgynes.

He shifts attention to his parents, starting with his father.

> POSTHUMUS And that most venerable man which I
> Did call my father was I know not where
> When I was stamped. Some coiner with his tools
> Made me a counterfeit… (2.4.155–2.4.158)

He believes he is illegitimate like a counterfeit coin (2.4.157n, 2.4.158n), but the play's *dramatis personae* and stage directions (5.3.123SD) list Sicilius as father to Posthumus. "Counterfeit" can mean "imposter," so perhaps Posthumus feels illegitimate because he is a royal adoptee. To the audience, he is a mix of natural and supernatural properties with one foot in the real world and the other in the next. He worries that a supernatural "coiner" put him into this state, and he curses his misfortune. He lays blame upon his parents, but in reality, he should credit incorporeal influence.

He cries for vengeance because Innogen denied him his rightful pleasure.

> POSTHUMUS O vengeance, vengeance!
> Me of my lawful pleasure she restrained,
> And prayed me oft forbearance, did it with
> A pudency so rosy the sweet view on't
> Might well have warmed old Saturn, that I thought her
> As chaste as unsunned snow. (2.4.160–2.4.165)

Innogen's "pudency" was "modest" (*OED*), and modesty carried to extremes connotes prudery—unless for a noble purpose. Innogen imposed her proscription against sexual intercourse so sweetly that it "might well have warmed old Saturn." The passage imputes to Saturn the properties of coldness ("warmed") and time ("old"). In mythology, Saturn is the god of Time, and in astronomy, he is the yellowish planet that takes the longest time to traverse the Zodiac (29½ years). As such, to geocentered ancients, its orbit was large compared to the Sun's, making it the farthest Ancient Planet from both Earth and Sun. According to the New Astronomy (to which Posthumus subscribes), planets shine by virtue of the sunlight they receive, and regardless of World model, Saturn receives the least solar energy and is thereby comparatively "unsunned."

COLD

The excerpt quoted earlier is difficult to parse (Nosworthy 2.4.163–2.4.164n). Perhaps in keeping with the overall problem of ambiguity and impressionism, Shakespeare deliberately expresses something subtly, in keeping with what I believe is his entire approach to cosmic telescopic revelation.

A possible paraphrase is that Innogen withheld her favors so that Posthumus "thought her / As chaste as unsunned snow." Suppose that this means that Posthumus thought Innogen was frigid. She denied his connubial rights coyly—so coyly in fact that she "*Might* well have warmed old Saturn" (2.4.164; emphasis added). Either she did not actually do so, which would leave Saturn in its frigid state, or else she could not do so because she remained as cold as or colder than Saturn, and Shakespeare knew that heat does not flow naturally from cold to warmer objects. If Innogen shares with Saturn the property of frigidity, this property likely masquerades as the virtue of chastity.

In his rancorous delirium, Posthumus rebounds from frosty Saturn to hellish heat. "O all the devils!" he exclaims (2.4.165). Similarly, when Giacomo shut his trunk in Innogen's bedroom, he warned, "hell is here" (2.2.50). Posthumus shifts his perception of Innogen from perfection to wantonness, from a heavenly to a hellish state, and he lets his imagination run wild as he imagines how Giacomo reached his goal.

> POSTHUMUS This yellow Giacomo in an hour—was't not?—
> Or less—at first? Perchance he spoke not, but
> Like a full-acorned boar, a German one,
> Cried 'O!' and mounted; found no opposition
> But what he looked for should oppose, and she
> Should from encounter guard. (2.4.166–2.4.171)

He calls Giacomo "yellow," which is not unlike the skin color of Saturn and which may refer to the Italian's complexion (2.4.166n). The fact that Saturnia is the ancient name for Italy (Bulfinch 281) sharpens the imagery. Posthumus imagines the traducer's seduction of Innogen occurred over a time that shrinks from an hour or less to right away. He imagines that vulgar copulation occurred without verbal intercourse, and he thinks that Giacomo encountered no hymenal resistance.

Posthumus uses the female gender as a scapegoat for his own shortcomings and those that beset the world. He blames women for lies, flattery, deception, ambition, covetousness, change of prides, disdainfulness, longing, slander, mutability, and all faults "that man can name" and that "hell knows" (2.4.174–2.4.179). To this litany, he adds that even in matters of vice, "They are not constant, but are changing still" (2.4.182), which implies that vice would be more acceptable if it were constant.

The supposed inconstancy of women is a gender bias having roots partly in *Rosalynde* by Thomas Lodge (*c.*1558–1625), which declares that "if man had growne from man, as Adam did from the earth, men had never been troubled with inconstancie" (Nosworthy 2.4.153–2.4.154n). This rant against fickleness is an outgrowth of the wager subplot, and its relevance will emerge in the final Act.

The end of the soliloquy is "incoherent" and "ludicrous" (Nosworthy lxiv) as Posthumus threatens to pen broadsides. He concludes that women should wallow in their own sinfulness (2.4.184–2.4.187), yet the audience knows that Innogen has remained loyal to her husband. Can this marriage be saved? The wager subplot has not yet played out, but for now, Shakespeare uses Posthumus' uncharacteristic ravings as a hint for playgoers to expect the unexpected.

DENOUEMENT

Posthumus' angry and defiant mood ends Act 2, yet his mood on opening Act 5 is penitent and purposeful. As told in the précis of the play at the top of this chapter, he joins the Romans, and once in Britain he turns coat again and disguises himself as a British peasant. In scene 5.2, he encounters Giacomo, vanquishes him, and disarms him. There is no dialogue during the combat, and the action alone bespeaks the important message of the fray, that Posthumus spares Giacomo's life. Posthumus turns coat yet again, the British capture and sentence him to death, and after the god Jupiter alights in a breathtaking violation of physical space, Posthumus receives a reprieve and a summons to appear before the King. In the final scene (5.4), Shakespeare resolves the threads of all the literal subplots, including those of the cosmic allegory, in which the wager subplot and the oddities of the romance between Innogen and Posthumus reach a conclusion.

The denouement begins as the King knights the Welsh cave dwellers. The physician, Cornelius, reports the machinations and announces the death of the Queen, and Posthumus, Giacomo, Lucius, and other Romans arrive under military escort. The party includes Fidele (alias Innogen), who recognizes the ring that Giacomo brazenly flaunts. With the King's help, she inquires how Giacomo acquired it. The King prompts Giacomo twice and Giacomo, after some self-serving dialogue by which Shakespeare importunes the audience to consider him as less than an out-and-out scoundrel, the Italian starts his confession, "Your daughter's chastity—there it begins," (5.4.179). He tries to lay some of the blame on Posthumus; justifiably so because the story of Galileo's discoveries would never have played out if Posthumus (i.e., Thomas Digges) had published his data to begin with.

Giacomo relates how he met up with Innogen and acquired simular proof of adultery sufficient to win the forfeit. He explains that he returned to Rome armed with her bracelet and notes about her bedroom, as well as knowledge of a mark upon her bosom, all of which he used to mislead Posthumus and win the wager. Imagine how Posthumus feels now! Innogen was true to their marriage after all, and thanks to his own bragging and the duplicity of Giacomo, Posthumus rues his order for Pisanio to kill Innogen. His remorse is heart wrenching. "O Innogen!" he cries, "My queen, my life, my wife, O Innogen, / Innogen, Innogen!" (5.4.225–5.4.227)

Posthumus confesses to arranging for the murder of Innogen, which comes as news to Fidele (who is Innogen in disguise). Fidele runs to embrace him but before she can do so, Posthumus strikes "him" down. Pisanio reveals her identity and Cymbeline is in a state of disbelief that his daughter is back at court. "Does the world go round?" he asks (5.4.232).

It takes 37 lines from the time that Innogen rushed toward Posthumus for her to succeed in embracing him (5.4.263SD), and it appears as though her affection is unrequited. Surely, the reunion of the loving young couple would be an occasion for joy, yet Posthumus stands idly by as Shakespeare continues to tidy up loose ends. After the misfortune that the couple has endured, all that Posthumus can bring himself to say is, "Hang there like fruit, my soul, / Till the tree die" (5.4.263–5.4.264). This is famously regarded as poetry at its best (Nosworthy 5.5.263–5.5.264n), but note that there are no more displays of affection or any words of love portending a state of marital bliss. After the embrace, Posthumus does not apologize for inadvertently striking her, nor does he beg forgiveness for attempting to kill her. When Posthumus says "my soul," he might not have Innogen in mind at all, but his own soul, which might mean his other, figurative, self. Innogen "throws her eye / on him…with joy" whereas Posthumus merely "anchors" his eyes on her (5.4.394–5.4.397). Such boorishness is comparable to his eruption at the end of Act 2, and again is out of keeping with what seems to be his better nature. The romance of Innogen and Posthumus, already beset by strange attestations of married celibacy and heavenly purity, has taken on the appearance of estrangement.

Temporarily forgotten in the riot of revelations is the fact that Posthumus must establish his role as a British fighter. Luckily, he had spared Giacomo's life, and he now turns to him to validate his valor, "Speak Giacomo, I had you down, and might / Have made you finish" (5.4.411–5.4.413). Giacomo kneels before Posthumus and begs him to end his life, but not right away because he wants to return his ill-gotten gains.

> GIACOMO …but your ring first,
> And here the bracelet of the truest princess
> That ever swore her faith. (5.4.416–5.4.418)

Posthumus refrains from exacting mortal revenge. "The power that I have on you is to spare you, / The malice towards you to forgive you" he says (5.4.419–5.4.420), adding in a reprimand that elicits chuckles from jaded auditors, "Live, / And deal with others better" (5.4.420–5.4.421). As if for emphasis, this is the second time that Posthumus has spared Giacomo's life. Shakespeare must not knock off Giacomo, not just because Posthumus serves ideals that are antithctical to mundane jealousy or revenge, but also because Galileo will likely continue to do science after 1610 and therefore must not be harmed.

Incongruities persist. Giacomo returns the ring to Posthumus, but there are no stage directions to indicate that Giacomo returns Innogen's bracelet to her, nor, if he does give it to Posthumus, that Posthumus passes it on to her. It is as if Shakespeare lets the bracelet fade into obscurity. In addition, both males deny the Princess the deference that her station demands, and neither she nor her father protest. Oddly enough, she has nothing to say for the rest of the play, not even to demand attention or proclaim her happiness.

On top of that, Giacomo neither forfeits his stake of 10,000 ducats nor offers to do so. Perhaps he plans to settle his debt as soon as the banks open, or, because Posthumus has forgiven him, he may think that he forgives his debt as well. More likely, Shakespeare lets Giacomo keep his golden ducats to symbolize that Galileo can "keep" his golden stars because effectively these minor collections contribute little to the New Astronomy. As ducats, "the moiety" of Giacomo's estate represents a sizable sum of 10,000 ducats, but in *Lucrece* and *The Winter's Tale*, Shakespeare also uses "moiety" to mean "a small portion" (*OED*). As stars, the moiety refers variously to Galileo's lack of quantification of the number of visible stars (about 10,000 as previously mentioned) and his gross underestimate of the total number of telescopic stars.[3]

The relationship of the couple after their reunion is anticlimactic, which calls into question the conviction that *Cymbeline* is classifiably a romance. Instead, perhaps the goal is to devise a plot resolution that equates the conjunction of the star-struck lovers with an additional and hitherto unrecognized subtextual theme. In other words, the wager subplot coupled with the odd romance has left us hanging without satisfactory explanations of the difficulties mentioned, and one is obliged to press on in order to see whether there is more to *Cymbeline* than so far revealed.

INSCRIPTION

The inscription on Jupiter's tablet, far from being "absurdly infantile" (Bevington 1436), "dramatically redundant," and a "relic of an earlier play" (Nosworthy 5.5.436–5.5.453n), lends support to the posited subtext, and in the process, helps to unlock the riddle of the Innogen and Posthumus' relationship. It also further ameliorates the "sovereign difficulty of *Cymbeline*," which is the absence of "hard facts" (Nosworthy xi).

Cymbeline contains two recitations of the inscription—a repetition unprecedented in the Canon. Posthumus is first to read it.

> POSTHUMUS Whenas a lion's whelp shall, to himself unknown, without seeking find, and be embraced by a piece of tender air; and when from a stately cedar shall be lopped branches, which being dead many years, shall after revive, be jointed to the old stock, and freshly grow, then shall Posthumus end his miseries, Britain be fortunate and flourish in peace and plenty. (5.3.232–5.3.238)

Superficially, the passage means that when the son of Sicilius Leonatus finds something unknowingly and marries Innogen, and after Cymbeline's sons return to court, then Posthumus' travails will end, and Britain will flourish in peace and plenty. This does not seem particularly earth-shattering, but the simplistic meaning is not the only one, as it causes Posthumus to think either that he must still be dreaming or that he is turning mad (5.3.240n).

The second reading occurs in the next and final scene. At the King's request, a soothsayer reads the message and then addresses the riddle of the name Posthumus Leonatus.

> SOOTHSAYER Thou, Leonatus, art the lion's whelp,
> The fit and apt construction of thy name
> Being leo-natus doth import so much. (5.4.444–5.4.446)

The seer identifies Posthumus as the lion's whelp, but does so via a Latin construction—*leo-natus*—whose meaning he values as "so much." This hints that there is more to "Leonatus" than is immediately apparent.

King Tenantius conferred the name "Leonatus" on Posthumus' father Sicilius, who had "served with glory and admired success" and so had gained "the sur-addition 'Leonatus'" (1.1.33). In the *OED*, "suraddition" has only one meaning—an "additional name or title." For further help, the *OED* refers readers to "sur-" and "addition." Following the trail, the prefix "sur-" implies an overburden or excess, as in "sur-stretching" which was used in 1560 for example to express the vast extent of the sky. "Addition" can mean "something annexed to a man's name to distinguish him," an honorary title, as used by Shakespeare in *Troilus and Cressida*, "I came to kill thee, cousin, and bear hence / A great addition earnèd in thy death" (4.5.141–4.5.142).

In essence, "addition" means, "surname," and "sur-addition," means an "excess" of "addition," or so to speak, an "excess of surname," as if the surname Leonatus had added meaning. When Tenantius conferred the sur-addition "Leonatus" on Sicilius, he must certainly have regarded his soldier as brave. The Old German etymons *leon* (lion) and *hardu* (bold or hardy) lead to the German name Leonhard. A fifth-century Frankish saint was called Leonard, and the name survived in Gaul, whence the Normans introduced it into England. By about 1200, "Leonard" occurred there as a Christian name and in the Middle Ages also as a surname, along with the variant "Lennard" (Withycombe 193–194). However, the incidence of these names in England was low (Dunkling 164–165). Overall, the rarity of "Leonard" in the early modern era decreases the pool of possible identifications of Leonatus Sicilius and increases the probability that the referent is Leonard Digges.

There are other aspects to the name "Leonard." From the first millennium, the "heart" is considered the seat of mental and intellectual acuity and of individual courage (*OED*), and from the sixteenth and seventeenth centuries, "hart" is an alternate spelling of "heart" (*OED*). Both can mean "hard" (i.e., firm), steadfast, unyielding, or hardened (*OED*). Just as pairing "leon" and "hard" leads to Leonhard and Leonard, so pairing "leon" to "heart" or "hart" leads to "lion-hearted," which aptly describes Sicilius (Nosworthy 1.1.33n). This interpretation is the more appealing one because Leonard traces his ancestry to the time of Richard I (1157–1199), known as Richard the Lion-Hearted (Kippis 238; Philipott 60).

In addition, the Latin *natus* in *leo-natus* can mean poetically "a son," and in Posthumus' case, the seer interprets "leo-natus" to mean literally "lion-born" or "lion's whelp" (5.4.444). "Lion's whelp" means "a valorous youth sprung from a valorous race" (*OED*), which in Posthumus' case is indeed a "fit and apt construction" (5.4.445) because he helped to defeat the Romans and rescue the King. As a sharp-eyed eagle (1.1.140, 1.4.11n), Posthumus has properties of both a brave lion and a bird of prey. During the dance of the spectral relations, this bravery did not escape the notice of the oldest sibling who remarks that Posthumus performed deeds of valor for the King just as his father had done for Tenantius. "Like hardiment Posthumus hath / To Cymbeline performed," he proclaims (5.3.169–5.3.170), where "hardiment" recalls the root "hard," meaning valorous. It follows that "leo-natus" is both the "brave lion" or the "lion's whelp," and one may now appreciate why *leo-natus* "doth import *so much*" (5.4.446; emphasis added) because *either* Leonard *or* Thomas Digges can be the referent.

Father-son pairings are as prominent in *Cymbeline* as in *Hamlet*. Jupiter is the offspring of Saturn, Posthumus is the son of Sicilius, and Thomas is the son of Leonard. The god Saturn established itself in Rome and raised the ethical standards of the natives, rewarding them with peace, prosperity, and abundant harvests. The golden age of Rome resembled the so-called Golden Age of Saturn, which thirst for blood and lust for gold brought to an end (Bulfinch 280). In the play, Shakespeare imputes golden ages to the reigns of Cymbeline and James I, and the inference seems reasonable that during the reign of James, the Digges father and son have a role to play.

The tablet's inscription also mentions "Posthumus," which derives from Late Latin, meaning generally "anything appearing after the death of its originator" (*OED*). The challenge is to identify the "originator" and the definitional "anything." An animate meaning was in use from 1591 and meant a child born after the death of its father. The originator is the father, and the "thing" in "anything" is the child, but these cannot be Leonard and Thomas because Thomas was about 25 or 26 years when he reported the death of his father.

In 1608, however, the definition broadened to include the inanimate option, including "action, reputation, etc., occurring, arising, or continuing after death," where the death in question need not necessarily refer to the one performing the action or acquiring the reputation, as in a posthumous confession made by a perpetrator after the other conspirators are dead (*OED*). With reference to the ambiguity concerning "leo-natus" mentioned previously, the death could be that of the brave lion Leonard Digges, or it could refer to Thomas himself.

The "etc." in the definition allows latitude in interpretation, and inasmuch as "reputation" requires memory, the definition could cover observations that occur, arise, and continue after death. A straightforward reading is that Posthumus Leonatus is associated with data appearing after the death of either Leonard or Thomas Digges. As posited, Posthumus represents the accumulation of celestial telescopic knowledge up to the time of writing of *Cymbeline*, and in the interests of the coming age of scientific and political enlightenment, Shakespeare saves the life of Posthumus because "the Artificer of all things" (Copernicus 8) has work for him to do. This explains why, despite vowing to fight until he is killed, Posthumus survives.

Like Posthumus, Thomas Digges carried on the family reputation for bravery, if not actually in battle then in pursuit of science at a time when original thought carried risks even worse than those borne by soldiers. This explains Posthumus' concern.

> POSTHUMUS Let me make men know
> More valour in me than my habits show.
> Gods, put the strength o'th'Leonati in me. (5.1.29–5.1.31)

Subtextually, this association draws attention to the need for history to acknowledge the role that Thomas Digges played in the development of the New Astronomy and the need for valor in the face of adversity.

As if to remove all doubt, at the start of Act 3, Shakespeare connects Posthumus directly to astronomy. Pisanio informs Innogen that he has a letter to her from him.

> INNOGEN That is my lord, Leonatus!
> O learn'd indeed were that astronomer
> That knew the stars as I his characters—
> He'd lay the future open. (3.2.26–3.2.29)

There is no obvious reason why Innogen would associate "Leonatus" with "astronomer," for none of the Leonati had an avowed interest in the celestial arts. Although a possible meaning for "astronomer" is "astrologer" (3.2.27n) and although in the early modern age, the two were virtually synonymous, Shakespeare disdained astrology *per se*, as his disavowal in Sonnet 14, lines 1–3 makes clear (Vendler 104):

> Not from the stars do I my judgment pluck,
> And yet methinks I have astronomy —
> But not to tell of good or evil luck.

Note that Innogen uses the past tense when she says that her astronomer "knew the stars." This is appropriate because the cover of the Digges almanac of 1576 (depicted by Best, Maene, and Usher 39) is replete with zodiacal signs that could well be the "characters" that Innogen's astronomer knew (3.2.28n). As argued, Leonard and Thomas Digges "knew the stars" both empirically and theoretically, whereas Innogen's knowledge probably extended merely to astrological signs ("I [knew] his characters"), as expected of a nonspecialist. It is reasonable to suppose that "that astronomer" refers to Thomas, and that (as posited) his subtextual representative on stage, Posthumus, has a store of information residing in memory comprised of contemporary celestial facts.

Immediately after addressing the topic of Posthumus, Soothsayer advises Cymbeline that "we" (i.e., Romans and Britons) should regard Innogen as "mollis aer" and "mulier."

> SOOTHSAYER The piece of tender air thy virtuous daughter,
> Which we call 'mollis aer'; and 'mollis aer'
> We term it 'mulier'… (5.4.447–5.4.449)

As befits the pending alliance between Rome and Britain, the definition of Innogen, like "leo-natus," is a riddle posed in Latin. *Aer* describes the lower atmosphere as opposed to *aether*, which refers to the pure air of the uppermost reaches of creation, or simply to heaven itself. The *aer* of the lower atmosphere is responsible for weather and climate, and *mollis* means mild, calm, or tender, so *mollis aer* likens Innogen to "tender air," as stated. However, *mollis* also refers to character, in which case the sense used by Cicero (106–43 BC) and Vergil (70–19 BC) means "changeable" (Smith and Lockwood 442–443). In accordance with Aristotelian physics, the *aether* is immutable, whereas *aer* is changeable because it undergoes temperature fluctuations and is not always calm. Through erroneous etymology, *mulier* derives from *mollis* and means "woman" or "wife" (as distinct from "maiden"), and Shakespeare acknowledges this confusion by writing, "'mollis aer' / We term it 'mulier'." Thus, the Soothsayer instructs the King that his daughter continues to be virtuous, but that she is nevertheless still a woman and a wife.

The soothsayer shifts his attention to Posthumus and repeats his advice by continuing with the 'mulier' theme.

> SOOTHSAYER …which 'mulier' I divine
> Is this most constant wife, who even now,
> Answering the letter of the oracle,
> Unknown to you, unsought, were clipped about
> With this most tender air. (5.4.449–5.4.453)

The soothsayer tells Posthumus that Innogen is his "most constant wife" (5.4.450), a "woman" ("mulier"), and not a maiden. Unspoken is the command that Posthumus reestablish his relationship to Innogen, and that they spend the commodity that her frigidity has safeguarded so far.

Soothsayer says that Posthumus has been "clipped about / With this most tender air," where as noted, "tender air" refers to Innogen 5.4.450n). The oldest and most common meaning of "clip" dates back to the first millennium AD, and means to "embrace" or "hug" (*OED*), which eventually she does when their identities are at long last disclosed ("She embraces him," 5.4.263SD). "Clip" can also mean "to check" or to "cripple action," as used by Marlowe in 1588 in the context of aviculture ("to clip the wings of"), and "to curtail" or "diminish," as used two years later by Shakespeare. Innogen clipped Posthumus' wings because she prevented the consummation of their marriage, and in the Jupiter descent, Shakespeare clips Thomas Digges' wings by celebrating the discovery of the Galilean moons before his lidded eyes.

The seer foresees that after the events described in the tablet's inscription have transpired, "then shall Posthumus end his miseries" (5.4.441–5.4.442). Posthumus understands fully the meaning of the events, yet remarkably, he remains silent for the last 50 lines of the play. His silence is golden. In imitation of Giacomo taking notes in Innogen's bedchamber, one supposes that he stands silently by, scribbling away in his own pseudo-metaphysical notebook for the benefit of future generations. This sets priorities in the subtext and explains his neglect of Innogen.

In answering the letter of the oracle, Innogen has fulfilled the terms laid upon her (5.4.451n). She remains virtuous, undefiled, and celibate, and continues in this state "even now" (5.4.450) at the end of the play. The seer predicts that she will be a "constant wife," a woman, *mulier*, and no longer a maiden. Her frigidity has served the gods well, and now she will warm up to Posthumus. The two are what Marcus calls "unknowing precursors of a new era" (149–150). In the political arena, this is the prospect of international peace and British parity with Rome and her empire, and in the allegorical context, it is the internationality of the New Astronomy. The King is a study in understatement as he concludes, "This hath some seeming" (5.4.453). By the play's end, all literal loose ends are accounted for, but in the subtext, a few remain. One is the identification of Innogen's epidermal "stain."

Roman Ears

In one of *Cymbeline*'s sources, *Frederyke of Jennen*, the gifted ring is a "rynge with a point of diamond" (Nosworthy xxiv). Suppose, therefore, that Innogen's gift of the diamond ring represents what nowadays is known popularly as the "jewel of the solar system" (Lynch 99), the "diamond" Saturn set in a golden ring (figures 1.11[c], 1.12). Innogen represents the state of heavenly perfection, but after the events in Orleans, Posthumus must learn of her "defects" in his capacity as the keeper of cosmic data. Shakespeare casts these as blemishes, which comprise the mole that Giacomo later discovered but whose detailed structure he did not divulge, and a "stain" that is "as big as hell can hold" (2.4.140), which only Posthumus knows about.

Let this "stain" represent the ringed planet Saturn.[4] The hyperbole "as big as hell can hold" is fitting because in a superstitious age rife with spirits and demons, a devotee of the Old Astronomy seeing a ringed planet for the first time would be dumbfounded and prone to ascribe the image to the work of the devil. Even in modern times, magicians' cloaks often depict a planetary ball and ring.

The fact that Shakespeare keeps Giacomo ignorant of this "stain" fulfills the prediction discussed previously—that it refers to something Galileo did *not* observe. As related in quasi-secret communications beginning on July 30, 1610, Galileo thought (erroneously) that Saturn was an orb with two ears attached (figure 1.11a, 1.11b), which *Cymbeline* implies is a datum not worth communicating to Posthumus. Galileo did not revise his opinion of Saturn's image until 1612, when—to his astonishment—he saw that the ears had disappeared.[5]

Giacomo is the first to refer to Innogen's gift as a "ring" ("Your ring may be stolen," 1.4.85–1.4.86), but the script never actually calls it a "diamond ring." The omission attracts attention, supporting the conjecture that the conjunction "diamond + ring" is reasonably analogous to "Saturn + ring." Giacomo is keen to arrange for the bracelet to "be married" to the "diamond" (2.4.97–2.4.98), but Giacomo's desires are as inappropriate as is his union "diamond + bracelet," because the diamond is already spoken for, or in other words, is attached to a ring.

This does not end the matter, however. In Giacomo's confession, he refers to the ring as a "bond of chastity" that is "quite cracked" (5.4.207). A "crack" is a flaw or a defect, as in *The Winter's Tale* when Camillo tells Leontes, "I cannot / Believe this crack to be in my dread mistress" (1.2.318–1.2.319). In Giacomo's capacity as a raven, it is appropriate for him to use the word "cracked," as the Latin word for raven is *corvus*, which derives from *crepare* (to crack), which can also mean to break (5.4.207n) or split asunder, as in *King Lear* 3.2.1.

There is no evidence that the bracelet suffered any physical damage in the course of its travels, so it makes sense to at least entertain the possibility that this metaphorically cracked state denotes a division in the ring. This may refer to the Cassini Division (figures 1.12 and 1.13).[6] In good observing weather, the gap is detectable from Earth even with a small telescope, provided it has excellent optics. If Shakespeare knew of this gap (as suggested here), its detection places a stringent limit on the resolution of Digges' telescope, which cannot be attained with the prototype as presently envisioned and constructed (figure 1.9). An ensuing discussion indicates how this might have been achieved.

VARIABILITY

During his fulminations at the end of Act 2, Posthumus accuses women of the vice of mutability. In the last scene 5.4, the soothsayer establishes that she has been constant in her wifely role and hints that she has been constant in her sexual coolness too, but he associates Innogen with *aer*, so that in other respects, she is as changeable as atmospheric air. When Shakespeare likens her to the Phoenix, he propagates the myth that arose *circa* 1400 that the Phoenix constantly renews itself by rising from the ashes of its former existence. Innogen does that twice, when she disappears from court and returns, and when she returns seemingly from the dead in Wales. Paradoxically, she is constant in one way and not in another, a condition that part G5 of the mole fulfills because the Galilean moons change position even as their orbits are constant. Saturn and the "stain" also fulfill this condition through the appearance and disappearance of its rings. Does Shakespeare confirm this?

In scene 1.1, Innogen hands the diamond ring to Posthumus (the vector I → P in the third column of table 4.3); in 2.4, Posthumus hands it to Giacomo (P → G), then reclaims it (G → P), only to give it to Giacomo again (P → G); and finally, in 5.4, Giacomo returns it to Posthumus (G → P). Column 1 of table 4.3 labels these five instances as subcycles I–V, with Galileo's data of 1610 (and 1612) identifying the end of the fifth. Column 2 gives the associated scenes. Column 4 gives the approximate times of ring disappearance and the duration of each subcycle, and column 5 gives the two viewing aspects "+" and "–" of which a consecution of any two comprises a complete cycle. The number of (diamond) ring exchanges corresponds to five (Saturn) ring changes and is yet another pentad. This would have the property of symmetry because a pentad would pertain to the fifth component (G5) of the "mole" and the five components of the Jupiter system (planet plus 4 moons) (G4, G5 of table 4.2), and to a property of the "stain" of Saturn (table 4.3).

In 1610 subcycle V was well on its way to completion. Working back from then, the end of subcycle IV occurs in about 1597.3, which is after the death of Thomas Digges, who therefore could not have contributed data to that aspectual change. There is no record of Thomas Harriot having done so, as his telescopic ventures began only in the seventeenth

Table 4.3. Inferred correspondence between diamond ring exchanges and Saturn ring changes in *Cymbeline*.

Subcycle	Scene	Vector	Duration	Aspect
I	1.1	I → P	1538.2–1554.1	+
II	2.4	P → G	1554.1–1567.7	–
III	2.4	G → P	1567.7–1583.6	+
IV	2.4	P → G	1583.6–1597.3	–
V	5.4	G → P	1597.3–1612.8	+

century, and even then, his depiction of the Moon is scarcely better than an artist could draw by the naked eye (*Sidereus* 9).

At the start of subcycle II, Leonard Digges was about 29 years old and his son about 8, which would indicate the Leonard was as much an observer as his son. However, to make the thesis of periodicity plausible, Leonard Digges, who was born about 1521, must have commenced observing in his 20s or early 30s in subcycle I, which would provide at least two full cycles of data to justify the claim. In writing for future ages, Shakespeare needed as long a time series as possible to ensure for the benefit of future auditors that his interpretation was correct and that the ring changes would repeat.

CROWNING GLORY

The present interpretation retains the surreal quality which critics have decried but which is a feature common to each of the plays analyzed in the present volume. This agrees with the opinion that Shakespeare wrote *Cymbeline* with reference to his earlier works. By placing contemporary astronomical progress into context, Shakespeare relates a "peculiar, and perhaps decisive turning point" in his career that has significance "beyond any last curtain or final Exeunt" (Nosworthy xi, lxxxiii).

The prospect of an enlightened political age under James I and the precedent set by Galileo's scientific disclosures mean that *Cymbeline* ends with the prospect of a new worldview built on rational discourse, atonement, and forgiveness. Through ratiocination and love of truth, Shakespeare attains the loftiest rung of Plato's ladder of knowledge of the intelligible world. From a broad base of mathematics and empirical knowledge, he scales the pyramid of World understanding and attains the highest Platonic Form, which is the Idea of the Good whence springs moral purpose and all things bright and beautiful.

CHAPTER 5

THE MERCHANT OF VENICE

Errors, like straws, upon the surface flow:
He who would search for pearls must dive below.
—John Dryden

In Venice, Bassanio wishes to finance his romantic pursuit of the fair damsel Portia, who lives in Belmont. His good friend, Antonio—the merchant of the title—is short on cash, so he borrows money from Shylock. The moneylender terms the agreement a "merry bond," but Antonio stands as surety, and in the event of a default, he will have to forfeit a pound of his flesh. In Belmont, Bassanio meets the condition laid down by Portia's father, namely that to win her hand, a suitor must read the inscription on each of three chests of gold, silver, and lead, and then select the one he thinks best suits his vision of the future. In celebration, Portia gives Bassanio a ring. Shakespeare (*The Merchant of Venice*) crafts a secondary romance between Bassanio's friend Graziano and Portia's companion Nerissa, which springs from this success, and Nerissa gives her beau a ring too. In the literal storyline, the loan defaults, and Portia helps rescue Antonio from certain death, but the subtext is concerned mainly with understanding both the Portia-Bassanio and Nerissa-Graziano romances, and the "doubling of the ring business" (Halio 16). (In this Chapter, unattributed notes, lineation, and pagination refer to Halio's edition of *The Merchant of Venice*, and other editions of Shakespeare's plays are also identified by editors' names.)

ANTONIO

The play opens with Antonio telling his friends, Salarino and Solanio, that he does not know why he is so sad. He wonders how his melancholia has become so severe that he scarce knows who he is (1.1.7n), but he concludes, "I am to learn" (1.1.5). In the course of the first seven lines, Shakespeare establishes the framework of the subtext of the play—that Antonio suffers from melancholy and that he will learn something over the course of the play, as will the audience. In line 1.1.50, Solanio reinforces the theme when he speaks of Janus, the Roman god whose two faces point in opposite directions, as if looking to the future while still acknowledging the past.

Salarino mentions a ship named Andrew in line 1.1.27, referring to the San Andrés that the English renamed after capturing it in Cadiz harbor. News of this event reached England on July 30, 1596, which sets an early limit on the time of writing of the play. *The Merchant of Venice* (herein *MV*) is a play for which the time of composition and the date on stage are the same (Sohmer, *Wiser* 38), and Brown (xxvi–xxvii) cites evidence that sets a late limit of July 22, 1598. A more recent estimate (27–29) puts the date simply as "1596–1597," for which the early limit as set by the Andrew citation is July 30, 1596. The interval from this date to the end of 1597 gives a range of 1597.3 ± 0.7. The fact that Shakespeare establishes this temporal constraint so early in the play, right after Antonio's certain pronouncement "I am to learn," signifies a relation between the time of the action and the merchant's ignorance.

Salarino and Solanio suggest that Antonio is sad because he is either worried about the safety of his merchant ships, or he is in love, but

Antonio dismisses both suggestions. Antonio's dear friend Bassanio arrives, accompanied by Graziano and Lorenzo, and they too remark on Antonio's physical affect. "You look not well…you are marvellously changed," says Graziano (1.1.73–1.1.76). "Marvellously" occurs usually in a positive sense to mean exceedingly, extremely, wonderfully, or to an astonishing or extraordinary degree, and at least until the sixteenth century the word could have had supernatural connotations (*OED*).

Antonio accepts his lot in life. To him, the world is a stage "where every man must play a part," which for him happens to be "a sad one" (1.1.78–1.1.79). How is it that he is a chronic sufferer, yet at the same time is "marvellously changed"? Antonio is the axis about which the play's imbroglio revolves, and the agency of change, it seems, is supernatural and not easily detected in appearances. When the play opens, the agency has already begun to effect processes and outcomes.

Graziano's name resembles that of a canonically comic character (1.1.79n), and as such, he sees fit to joke around. "Let me play the fool," he says (1.1.79) and asks, "Why should a man…creep into the jaundice / By being peevish?" (1.1.83–1.1.86) Emotional upsets like grief were supposed to induce an excess of yellow bile, causing jaundice (1.1.85–1.1.86n), and the context suggests that Graziano relates Antonio's chronic condition with yellowness for comic effect. In mythology however, melancholy is a saturnine temperament associated with and arising from the slow forward advance of the yellowish planet Saturn relative to the stars.

Graziano warns Antonio not to be silently melancholic in order to appear wise (1.1.101–1.1.102n), for such persons are in reality often ignorant. Antonio's ignorance will soon abate. Graziano promises to expound on this "another time" (1.1.100). Lorenzo thinks he must be such a "dumb wise" man because "Graziano never lets me speak" (1.1.107), but Graziano responds that if Lorenzo keeps him company "but two years more" (1.1.108), he won't recognize the sound of his own voice. This can mean that some event occurring two years hence will dumbfound Lorenzo.

To pin down this date, let 1597.3 ± 0.7 be the date on stage as discussed previously. Having been alerted to the possibility that planet Saturn is at issue, a reference to table 4.3 shows that 1597.3 happens to be the median epoch in whose associated interval (± 1 month or so) the rings of Saturn are effectively telescopically invisible. Graziano foresees that something startling will manifest itself 2 years later, which sets that interval at 1599.3 ± 0.7. This occurs soon after the beginning of subcycle V of table 4.3, during which time the band around Saturn would become telescopically visible. Graziano plays a pivotal role in the subtext of *MV* and may well know as much, intending to surprise Lorenzo at that time.[1]

The first scene of Act 1 ends with the penurious Bassanio seeking money from Antonio in order to pursue fair Portia in Belmont. Antonio has invested his cash in his merchant ships, which he expects to return in two months, so in the meantime, he encourages Bassanio to seek a loan in Venice for which he (Antonio) will stand as surety.

PORTIA

In scene 1.2, Portia tells her companion, Nerissa, that she has grown weary of "this great world" (1.2.1–1.2.2). Her weariness is unlike Antonio's sadness, as hers stems from *ennui* (1.2.1n). No doubt this is a consequence of her unwed state, which in turn results from the stricture placed on potential suitors by her well-intentioned and now deceased father, who according to Nerissa was "ever virtuous" and "holy" (1.2.27). The challenge to credulity is whether any person of this world merits such praise. One criterion would be to see the consequences of a test that he devised, by which a suitor to his daughter must read notes attached to each of three caskets and select one that he thinks would best foretell their future state. One of the caskets is made of gold, another of silver, and the third of lead, and if a suitor were to make the wrong choice, he would not qualify to romance his daughter. If Portia's father was holy and forever virtuous, perhaps he embodies an element of the divine, in which case one must wonder whether his test serves more than a mundane goal.

Portia describes six suitors who had traveled to Belmont but failed the test. She describes each in unflattering and stereotypical terms—a commonality that prepares for the stark characterization of Shylock. She dotes on their "very absence" and wishes them "a fair departure" (1.2.107). Nerissa recalls a Venetian scholar and soldier that Portia met in her father's time. Portia remembers his name—Bassanio—and tries not to appear eager to meet him again (1.2.112n).

A Servingman enters and says, "The four strangers seek…to take their leave, and there is a forerunner come from a fifth…" (1.2.120–1.2.122). The definite article in "The four strangers" implies that only four suitors are at issue, and that Portia knows about them and their condition. Portia has just enumerated six suitors, so which two needed no permission to leave? The script refers to "four" and as well to a "fifth" suitor (1.2.120, 1.2.122, 1.2.124, 1.2.125), so a compositional or transcriptional error is implausible (1.2.120n; Brown 1.2.117n). This raises the possibility that this blatant discrepancy is not accidental. I posit that Shakespeare creates it to draw attention to the numbers involved.

Shylock

The next scene 1.3 opens in Venice as Bassanio approaches the moneylender, Shylock, for a loan of 3,000 ducats for which "Antonio shall be bound" (1.3.4–1.3.5). Shylock asks to meet Antonio, who enters, and each expresses alienation from the other's ethnicity, profession, culture, and religion. Vituperation persists through the play. Shakespeare did not blot out these lines and if (supposedly) he writes with noble intentions, one wonders what they might be.

Like Antonio, Shylock is tight on cash, but he agrees to raise the required amount with help from a friend. The duration of the loan is three months, and the cost of forfeit is a pound of flesh cut from whatever part of Antonio that Shylock pleases. Because Antonio expects his argosies to return in two months, he persuades Bassanio to accept the terms of what Shylock calls a "merry bond" (1.3.170). They agree to meet at the notary's office to seal the deal.

THE GOBBOS

Act 2 opens in Belmont with Portia greeting the fifth wooer (by Servingman's count), the Prince of Morocco. The dullness of dialogue in scene 2.1 belies the surprise in store in the next. This opens with the entrance of a clown, Lancelot Gobbo, who works for Shylock. Lancelot is dissatisfied with his position and desires to leave it. He retains a semblance of loyalty to his "master," Shylock (2.2.2), yet a "most courageous fiend" (2.2.9) bade him leave Shylock's employ. Lancelot is conflicted because he regards the fiend as "the devil himself," yet at the same time, he regards Shylock as "the very devil" incarnate (2.2.22–2.2.25). In the end, the fiend wins out because it "gives more friendly counsel" (2.2.28).

Lancelot's father, Old Gobbo, enters carrying a basket containing a present for Shylock. Old Gobbo is virtually blind ("sand-blind, high gravel-blind," 2.2.33–2.2.34), and does not recognize his son. He asks Lancelot for directions to Shylock's home, but Lancelot makes merry of his father's disability and gives him nonsensical directions. He then tells him that his son is dead. The father is distraught, but Lancelot relents. Kneeling, he begs his father's forgiveness and identifies himself as the old man's son. "I am sure Margery your wife is my mother," he says, and Old Gobbo confirms, "Her name is Margery, indeed" (2.2.84–2.2.86).

In *Henry VI part 2*, Margery Jordan is a "cunning witch" who (with a conjurer named Roger Bolingbroke) can raise a spirit from the "depth of underground," but who is sentenced to "be burnt to ashes" (*Henry VI, part 2*, 1.2.75–1.2.79, 2.3.7). Margery Jordan represents Margery Jourdemayne (?–1441), the witch of Eye (Ebury), who ignored warnings to abandon her alleged practice of political necromancy. Her scholarly friends abandoned her, and she perished in a flaming pyre. A collection of political poems of *circa* 1559 reveals that Margery wrought wonders through the land, and that "Both feendes and fayries her charmyng would obay / And dead corpsis from grave she could uprere" (Freeman 344). The facts that fiends obey Margery and that a fiend goads Lancelot make it plausible that Shakespeare is referring to the same Margery in *MV* that he immortalized in the *Henry VI* trilogy. As for Margery's command of fairies, John Middleton Murry and Harley Granville-Barker regard *MV* as a fairy tale. Yet how does one reconcile the fact that Margery died in 1441, while, as previously suggested, the date on stage is 1597?

Gobbo is Italian for a hunchback, although "Shakespeare may not have intended a family of hunchbacks…even if Old Gobbo was one" (2.2.0n). Alternatively, from an Italian dictionary of 1598, a "gobbo" is a kind of falcon (Brown xxii). Lancelot Gobbo's appearance on stage in the final scene (5.1) supports the latter choice. He calls, "Sola, sola! Wo ha, ho!" where "Wo ha, ho!" is the cry of a falconer (5.1.39n2). Previous chapters associate Leonard Digges with falconry, so it is reasonable to associate Old Gobbo with him, and his son Lancelot Gobbo with Leonard's son Thomas.

However, Thomas reported his father's death in 1571, and he himself died in 1595, so by the date on stage of 1597.3 ± 0.7, both Diggeses had departed the visible world. How then can Shakespeare represent them as alive and well on stage? Margery died over a century and a half earlier than the stage date, a martyr to her cause of necromancy, but perhaps from her subsequent position of honor in the phantom world of fiends and fairies she would continue to wield her influence. She once had the power to animate corpses and raise them from the grave, and presumably she still has these powers and uses them on the Diggeses. In a word, the Gobbos (also known as the Diggeses) are *zombies*.

In real life, the given name of Thomas Digges' mother was Sarah Bridget (*DNB*), whereas Lancelot's mother is Margery. However, Margery gave "birth" to Lancelot by liberating him from his tomb like a mother who births a baby from her womb. In her heyday in the 1430s, Margery hobnobbed with men of learning, one of whom was an astronomer who correctly predicted an eclipse (G. L. Kittredge, as cited by Freeman 355), thus it is not surprising that she might revivify the foremost astronomers of the late sixteenth century. Having done so, what should be their dispositions? Shakespeare will have his audience believe that when *MV* opens, the younger zombie is in the employ of Shylock, while the father, "an honest exceeding poor man" (2.2.48), ekes out an existence with the help of his fill-horse, Dobbin.

Lines 2.5.45–2.5.50 demonstrate that there is no love lost between Shylock and Lancelot. Shylock calls Lancelot a dullard who is "a huge feeder" and a drain on his resources. Tellingly, Shylock says that Lancelot "sleeps by day." In olden days, it was the lot of most observational astronomers to work at night, therefore the identification of Lancelot with Thomas Digges seems reasonable. A corollary is that Shylock is the victim of forces beyond the pale of natural experience, and his displeasure with Lancelot and his other shortcomings are therefore eminently forgivable.

Lancelot knows of Bassanio and wishes he could work for him. On cue, Bassanio enters accompanied by Leonardo. Bassanio hires Lancelot, and Old Gobbo embraces the deal. Thus, the benevolent fiends of the preternatural world have wrested the Gobbos from the grasp of Shylock and placed them under the purview of Bassanio.

LEONARDO

The Gobbos depart and Bassanio tells Leonardo, "I pray thee, good Leonardo, think on this" (2.2.162). Although Bassanio mentions equipage for his pending trip to Belmont (2.2.163n, 2.2.166n), Bassanio's "this" could refer to the Gobbos' transfer of allegiance. Either way, it is unclear why Leonardo needs to pay attention to "this." Further, when Graziano enters, Leonardo directs him to where Bassanio walks. What purpose does this serve? Leonardo utters only eleven words in two sentences, and although he is undoubtedly present on stage in later scenes, he never utters another word. The episode does little if anything to advance the literal story, but perhaps the subtext benefits.

Leonardo's name differs from Leonard's by the minimal amount of one letter, "o," which could pass as a little zero and signify "no difference." One supposes that Leonard(o) plays a role in future events, the first occasion being, as was just mentioned, gratuitously to direct Graziano to Bassanio and thus to establish a subtextually significant connection between them. If so, one can expect that the meaning of the subtext will depend on that connection. There is a snag, however. Old Gobbo and

Leonardo are both on stage from 2.2.107 to 2.2.160, but they cannot both represent Leonard Digges. Because Old Gobbo is Leonard's revivified corpse, Leonard(o) could be the body's spirit. If so, the deceased inventor of the world's first telescope is doubly present on stage; his body is the zombie and his spirit is the ghost. It seems that Bassanio's contingent is a phantasmagoria of people, zombies, and a spook, cavorting in a conflation of natural and supernatural space. Alternatively, if such a menagerie is too much to swallow, perhaps *MV* is simply a fairytale.

Graziano meets Bassanio and has an urgent request. Concerning which, he tells Bassanio, "You must not deny me. I must go with you to Belmont" (2.2.171). As if taken aback, Bassanio replies, "Well then, you must" (2.2.172), but Graziano does not explain the urgency. Before departing, Graziano says that he "must to Lorenzo and the rest" (2.2.197). Leonardo is not involved in this interaction, so it can be concluded that "Lorenzo and the rest" do not rise to the level of Bassanio's and Graziano's roles and the putative connection between them.

Jessica

Shylock's daughter, Jessica, has learned of Lancelot's plans. She bids him a fond farewell, and Lancelot reciprocates tearfully. In the meantime, Graziano reaches Lorenzo, followed by Lancelot who bears a letter from Jessica to Lorenzo. Graziano bids Lancelot to return and tell Jessica that he "will not fail her" (2.4.20). He divulges that Jessica will take her father's gold and jewels and she and Lorenzo will elope. She plans to disguise herself as a page and be Lorenzo's torchbearer in a masque planned for that evening. At his home, Shylock tells Jessica that he is dining out and asks her to guard the house, but after he departs, she states the facts, "I have a father, you a daughter, lost" (2.5.56).

Graziano and the masquers arrive and Lorenzo joins them. Jessica appears at a portal of the house disguised as a boy. "Here, catch this casket" (2.6.33), she tells her beau. Lorenzo catches this casket in Venice, whereas Bassanio is yet to choose one in Belmont. Lorenzo bids Jessica to descend and join him, as she is to be his torchbearer at the masque.

Jessica obliges but not without first securing the doors of the house and gilding herself with more of her father's ducats. A shift in the wind causes a change in the time to set sail, the masque is cancelled, and Bassanio and Graziano prepare to depart for Belmont.

Salarino and Solanio confirm that Bassanio and Graziano are under sail, and Lorenzo is not. Shylock discovers his daughter missing along with valuables and a diamond costing two thousand ducats. His anguish is palpable. Usurer or not, a man's a man for all that. Shylock vows revenge on Antonio should the merchant default on the loan and rejoices at news that one of Antonio's ships has foundered. In exercising his right to have a pound of Antonio's flesh, he senses an increased probability "to have the heart of him" (3.1.120), meaning he intends to kill him.

SUITORS

I have noted a disparity in the enumeration of suitors, with Servingman referring to four suitors and Portia describing six, and that by Servingman's count, the next suitor, the Prince of Morocco, is the fifth. He arrives and prepares to choose one of the three caskets. He asks how he will know if he has chosen correctly, and Portia answers that the correct casket is the one that contains her picture. Morocco selects the gold casket and discovers a scroll with doggerel beginning with "All that glisters is not gold" and ending "Fare ye well; your suit is cold" (2.7.65–2.7.73). Morocco departs, and a "sixth" suitor, the Prince of Arragon, arrives to try his luck. After a long discourse, he selects the silver casket, and on opening it, finds his pedantic prattle justly rewarded by a message that concludes, "So be gone: you are sped" (2.9.71). He too departs.

A messenger arrives with the news of yet another suitor who, by Servingman's count, is the seventh. Portia and Nerissa hope that the new suitor is Bassanio, and to their delight, their hope is fulfilled. Portia urges Bassanio to tarry before choosing a casket, lest she would lose his "company" (3.2.3) if he were to choose the wrong one. She says that something tells her that she must not lose Bassanio, but that that something "is not love" (3.2.4). Apparently, Portia has more than mere matrimony in mind. Bassanio, on the other hand, is more direct. He declares his love for her and wishes to make his selection without ado.

The aura of riches that the gold and silver caskets radiate does not impress Bassanio and he selects the one made of lead. He opens it and discovers Portia's picture along with a scroll that congratulates him on choosing without regard to appearances. Portia and Bassanio are overjoyed. She names him her lord and gives him her house and servants, albeit conditionally.

> PORTIA I give them with this ring,
> Which when you part from, lose, or give away,
> Let it presage the ruin of your love
> And be my vantage to exclaim on you. (3.2.171–3.2.174)

If Portia is to be taken seriously, and if Bassanio were to relinquish his ring, she would precipitate a major change in their relationship. Bassanio is carried away by the momentum of the moment and goes so far as to declare that parting with the ring would be tantamount to death.

> BASSANIO ...when this ring
> Parts from this finger, then parts life from hence.
> O, then be bold to say Bassanio's dead. (3.2.183–3.2.185)

As discussed earlier, Graziano associates Antonio's chronic melancholy with yellowness and that Bassanio is suitor number 7. It happens that Saturn is yellowish and the seventh Ancient Planet. The recurrence of heptads raises the possibility that Saturn is at issue. Moreover, as posited, the date on stage is 1597.3 ± 0.7, which overlaps the interval when the rings of Saturn are effectively telescopically invisible (table 4.3), and of the many alchemical symbols for the materials gold, silver, and lead from which the caskets are made, some closely resemble the astronomical ones for the Ancient Planets Sun ☉, Moon ☾ and Saturn ♄. The last symbol represents a sickle, which suits Saturn and conforms to the winning casket being leaden. This explains why Bassanio gets to choose the lead casket—because time is of the essence and in his naiveté, he views his prospective relationship with Portia as long term. With these coincidences, it is reasonable to inquire whether a connection exists between Saturn and the ring that Portia has given Bassanio.

NERISSA'S BETROTHAL

No sooner have Portia and Bassanio agreed to wed, than Nerissa announces that the time has come for her to wed too. Graziano names Nerissa as his prospective bride and asks Bassanio for permission to tie the knot. Bassanio gives them his blessing.

Graziano's courtship of Nerissa has no precedence in Shakespeare's sources and appears to be an addition "from his own fertile imagination" (14). Probably, Shakespeare added it for a reason. It has already been argued that, as far as the subtext is concerned, Graziano and Bassanio are linked in a special way, and this second romance will complement the primary one between Bassanio and Portia. Another peculiarity is that Graziano ends up with a ring, but the script does not specify when Nerissa gives it to him. Similar lacunae related to Nerissa's ring occur as the play progresses.

The joyous occasion of the union of Nerissa and Graziano is short-lived, however. Lorenzo, Jessica, and Salerio arrive from Venice with a letter from Antonio to Bassanio, who blanches at its contents. Portia demands an explanation, and Bassanio describes his situation and Antonio's role in it. Then he announces that all six of Antonio's ships have been lost. Apparently, three months have passed and Antonio will have to pay the forfeit, but the further news is that Shylock will entertain no other recompense than the taking of Antonio's flesh.

Jessica opines that unless the law intervenes, "It will go hard with poor Antonio" (3.2.288). Portia heeds her remarks. She inquires of the sum owed and offers to repay the loan at twice, six, or twenty times the owed amount, but Shylock has already dismissed any humane alternative to that which was legally prescribed. His obstinacy does not exhaust Portia's options, however. Speedily she sets in motion a plan whose first step is to rid herself of Bassanio's company. She urges him to attend to the business in Venice, and he complies, vowing to remain true to her.

Whereas these developments promote the literal storyline, Portia will use Bassanio's vow of fidelity to further the subtextual plot.

The Jailer

In Venice, Antonio is in the custody of a jailer who has asked Shylock to take pity on his charge, but the usurer has experienced multiple shocks and is in no mood to be merciful. He says that Antonio "is the fool that lent out money gratis" (3.3.2), thereby undercutting the livelihood of usurers like himself who derive income from interest. Antonio believes that the law cannot be denied, and he hopes that Bassanio will return in time to witness his death.

Bellario

The action shifts back to Belmont, where Lorenzo praises Portia for having encouraged Bassanio to leave for Venice. Lorenzo says that Portia has "a noble and a true conceit / Of godlike amity" (3.4.2–3.4.3), meaning that she has the highest form of love—platonic love (3.4.3n). Portia replies that she has never regretted doing good, and concludes that if Antonio is Bassanio's bosom friend then he must resemble him in spirit. She therefore plans to aid Antonio, and hoodwinks her servants by announcing that she and Nerissa will repair to a monastery two miles off to pray.

She entrusts the management of her house to Lorenzo, and instructs her manservant, Balthazar, to betake himself speedily to Padua and give "this same letter" (3.4.47)—the one describing Antonio's plight—to her cousin Bellario. Through all this, Portia manages to write a note to Bellario, which she hands to Balthazar (3.4.51SD) for him to deliver as well. Balthazar is to "look what notes and garments" Bellario gives him, and return quickly to meet Portia, who will be *en route* to Venice. The garments conform to what Portia tells Nerissa, that the two shall disguise themselves as men. Their mission is urgent and they haste away to cover twenty miles that day.

THE DUKE

Bassanio has returned to Venice, and he and Antonio are present at court. The Duke presides over the proceedings along with a select cadre of magnificoes. The Duke urges Shylock to show mercy toward Antonio, but the moneylender again refuses. He also declines Bassanio's offer of sums of money greater than the amount of the bond. In the face of Shylock's recalcitrance, the Duke prepares to enforce the terms of the bond (4.1.104–4.1.106), that is, unless a certain Bellario arrives to preside over the proceedings. This Bellario is the same person whom Portia has alerted to the pending case and to whom she has appealed for help. His residence in Padua is explained by the fact that civil law was taught at the University there (Brown 3.4.49n), so he is probably the recognized local authority and the Venetian Duke's first choice from whom to seek help.

However, Bellario does not arrive. Instead, and just in time, Nerissa enters clad as a male clerk in clothes that Bellario has supplied. The clerk announces that "he" has just left Bellario's presence, and "he" presents a letter from Bellario excusing himself owing to illness, but commending to the court, "A young and learnèd doctor" (4.1.143). On cue, Portia enters attired as a male Doctor of Laws, also in clothes supplied by Bellario. She says that she knows about the case and asks, "Which is the merchant here, and which the Jew?" (4.1.171). Sometimes in courts of law, accusers and defenders are indistinguishable. She recommends that Shylock "must…be merciful" (4.1.179), but Shylock is defiant. The honorable Doctor then delivers a sermon on the merits of mercy, and Bassanio repeats his offer of cash, but Shylock stands firm. "I crave the law / The penalty and forfeit of my bond," he says (4.1.203–4.1.204). The Doctor concedes that under these circumstances, the terms of the bond are inviolate, and Shylock is gleeful. Still, the Doctor presses Shylock to relent, to no avail.

Shylock's joy evaporates as soon as the Doctor prepares to carry out Shylock's wish. "He" enforces the exact terms of the bond and instructs Shylock (4.1.320–4.1.322) to go ahead and "cut off the flesh," but not to shed any blood in the process. "He" decrees further that if the amount of severed flesh deviates from a pound but "in the estimation of a hair" (4.1.327), then Shylock will die and all his goods confiscated. "He" then indicts Shylock under a law that forbids menacing another's life under penalty of the loss of the culprit's estate and his life.

Even before Shylock appeals this judgment, the Duke spares Shylock's life, and Antonio acts holier than thou in declining to accept his share of Shylock's estate. The Duke too seems to have acted magnanimously, but the fact is that before the Doctor arrived, the Duke was in charge of the court and could have spared Antonio's life by the same elementary arguments as the Doctor used. Moreover, Antonio reveals his prejudice by demanding that Shylock forsake his Jewish faith and "become a Christian" (4.1.383). These threats and slurs transpire in the court of the Duke, who is therefore complicit in them. In Venice, it seems that there is enough depravity to go around, showing that *humanum est errare* (to err is human) is ubiquitously applicable.[2]

Antonio demands further that Shylock bequeath his remaining wealth to his daughter, Jessica, and her boyfriend, Lorenzo. The Court allows this, which may seem well intentioned except that this, in effect, condones and rewards her thievery. Shylock has tried to exact revenge on Antonio, and now Antonio and Graziano in turn have become vengeful and abusive, like Shylock. Shylock has no choice but to agree to these terms, and there is no talk of an appeal to a higher authority.

The Duke dismisses the court and advises Antonio to pay his lawyer, but the Doctor of Laws refuses payment. When Bassanio begs "him" to "Take some remembrance of us as a tribute" (4.1.418), the Doctor asks for Antonio's gloves and Bassanio's ring. Bassanio is in a quandary, but he talks his way out of the jam and keeps his ring.

Broken Vows

After the legal team leaves, Antonio urges Bassanio to comply with the Doctor's wishes, so Bassanio removes his ring and hands it to Graziano with instructions to overtake the Doctor and present the ring to "him."

Graziano is also to invite the legal team to Antonio's house for dinner, following which Bassanio and Antonio plan to leave for Belmont. Evidently, Portia's house will soon see a convergence of principals.

Graziano catches up to the Doctor and "his" clerk. The Doctor declines the invitation to dine, but accepts the ring "most thankfully" (4.2.9). Nerissa adds to the complexity by telling Portia, aside, that in her guise as a law clerk she plans to ask Graziano for his ring. This is the first time it is mentioned that Nerissa has given Graziano a ring, an event that, as remarked earlier, may well have transpired soon after Graziano and Nerissa received permission to wed. Nerissa commits to Portia's plan without knowing fully what it is, but Portia now reveals that with the two rings in hand, they can claim that their husbands have broken their vows to wear them forever. Further, they could insinuate that their husbands were unfaithful.

Portia's strategy is to embarrass Bassanio and Graziano, but to what end? Surely, such a jape is beneath the dignity of this sterling woman (15). Unless, of course, her full nature has not been revealed; for despite her receiving Bassanio enthusiastically and accepting him wholeheartedly as her soul mate, she now loves him platonically (she has "a noble and a true conceit / Of godlike amity," 3.4.2–3.4.3) and not humanly ("is not love," 3.2.4), fearing only the loss of his "company" (3.2.3).

The clerk needs to know the way to Shylock's house in order for the convict to sign court documents. Graziano will show the way, and *en route*, the clerk gulls him into handing over his ring. The clerk then joins the Doctor and they head home to Belmont, discarding their disguises on the way.

CONVERGENCE

Act 5 opens in Belmont. It is nighttime, the Moon shines bright, and Lorenzo and Jessica imagine that on such a night, noteworthy deeds of classical significance occurred. Suddenly, Lorenzo shifts from events of yore to one of immediate concern. On such a night, he says, Jessica filched her father's possessions, and the two trade barbs. Shakespeare is purposeful in contrasting the couple's wrongdoing with the beauty of the night, leading the audience to wonder whether it is his intent to contrast the rancor and sinfulness displayed in Venice with some yet unspecified natural wonder, and in particular to disassociate the couple from its natural beauty.

A servant enters and announces that Portia will soon arrive, and Lorenzo asks Jessica to help him prepare a welcome for her in her house. Portia's house is now the focus of attention. The occasion has special significance, for straightaway the younger Gobbo enters and declares that his new master Bassanio will arrive ere morning.

As mentioned, the significance of Lancelot's brief appearance lies in his hailing Lorenzo with the cry of a falconer (5.1.39n), which seems odd because nothing about Lorenzo warrants such a mode of greeting. To be sure, Gobbo behaves like a zombie and there is no telling what zombies might do, but as posited, the mode of greeting associates him with falconry, specifically with the leonard, or more particularly with Leonard Digges. Plausibly, the interior of Portia's house will be the site of cosmic revelation.

Lorenzo asks Portia's servant to alert the housekeepers of Portia's imminent arrival, and he calls for music. A passage of exquisite beauty follows in which Lorenzo invokes the Pythagorean doctrine of the musical harmony of the heavenly spheres.

> LORENZO How sweet the moonlight sleeps upon this bank!
> Here will we sit and let the sounds of music
> Creep in our ears. Soft stillness and the night
> Become the touches of sweet harmony.
> Sit, Jessica. Look how the floor of heaven
> Is thick inlaid with patens of bright gold.
> There's not the smallest orb which thou behold'st
> But in his motion like an angel sings,
> Still quiring to the young-eyed cherubins. (5.1.54–5.1.62)

Lorenzo mixes Pythagorean doctrine with post-Pythagorean imagery, in which the "floor of heaven" to the deities that reside beyond the eighth sphere of the stars is, to the denizens on Earth, heaven's "ceiling." The last lines of the passage refer to the superior eyesight of youth and especially that of cherubs (5.1.62n), and Lorenzo ends by citing the belief that only departed souls—like those belonging to the zombies in question—can hear the music of the orbs.

The musicians enter to play music audible to mere mortals. Again, Shakespeare contrasts musical richness with spiritual poverty.

> LORENZO The man that hath no music in himself,
> Nor is not moved with concord of sweet sounds,
> Is fit for treasons, stratagems, and spoils.
> The motions of his spirit are dull as night,
> And his affections dark as Erebus. (5.1.83–5.1.87)

Recall the harrowing scenes of prior Acts featuring Venetians bereft of grace, from which Lorenzo concludes, "Let no such man be trusted" (5.1.88).

Portia and Nerissa enter and discover that their husbands have not yet arrived. Portia furthers her plan by demanding that all present vouch that she and Nerissa were present at home in Belmont. "Fear you not," says Lorenzo, "We are no tell-tales" (5.1.123).

Night pales into twilight, and the husbands arrive accompanied by Antonio. Stage directions (5.1.126SD) indicate that Nerissa and Graziano embrace and fall into conversation, but there are no similar directions for Portia and Bassanio. This is to be expected, as it appears that their relationship is not normal. Instead of expressing delight, as one might expect from a lover reunited, Bassanio speaks of the brightening twilight. "We should hold day with the Antipodes, / If you would walk in absence of the Sun" he says, which serves as a "courtly compliment [that] eases him into conversation with his wife" (5.1.127–5.1.128n). Why there should be any tension between the two is puzzling. Portia replies, "Let me give light, but let me not be light," which may well pun on "light" meaning "wanton" (5.1.129n), but could mean that she is merely the facilitator of a pending enlightenment. Nevertheless, she welcomes Bassanio home and gives credit where credit is due. "God sort all!" she exclaims, meaning that matters will "be as God wishes"

(5.1.132n). Portia believes that a higher power supervises her beneficence, and her altered relationship to Bassanio echoes the one between Innogen and Posthumus in *Cymbeline*, which similarly connotes service to lofty ideals.

After Bassanio introduces Antonio to Portia, Nerissa and Graziano's conversation takes center stage. It seems that Nerissa must already have expressed mock surprise that the ring she gave Graziano is missing from his finger because the script suddenly has Graziano explaining that he "gave it to the judge's clerk" (5.1.143) (i.e., to Nerissa in disguise). It has already been pointed out that in scene 3.2, Shakespeare does not specify when Nerissa gave a ring to Graziano, and here in scene 5.1, the script does not specify exactly when Nerissa notices its absence.

Portia butts in. "A quarrel, ho, already! What's the matter?" she asks, and Graziano replies that it concerns "a hoop of gold, a paltry ring" (5.1.146–5.1.147). Portia condemns Graziano for not honoring his vow to wear his ring forever, and as if to promote feelings of guilt, she reminds everyone that she too gave a ring to Bassanio, the presumption being that of course *her* beau Bassanio would never part with *his* ring (5.1.184–5.1.185). But Graziano cannot contain himself and says that Bassanio gave his ring away too, which Bassanio grudgingly admits. Portia says that she will decline to sleep with Bassanio, and Nerissa threatens the same reprisal against Graziano.

Bassanio attempts to defend himself. Of his first six complete lines, the first five end with "ring." The first four lines of Portia's response also end with "ring," and Portia ends her speech with "ring" (5.1.208). In total, ten of the sixteen lines spoken in the two passages by Bassanio and Portia end in "ring," and Bassanio adds to this by using "ring" two more times (5.1.212, 5.1.222). Another peculiarity is that clauses frequently begin with the same word. The unusual poetical construction (5.1.193–5.1.197, 5.1.199–5.1.202n) alerts playgoers to the significance of "ring" and when Portia promptly calls the ring she gave Bassanio a "jewel" (5.1.224) she refers to Saturn, which as mentioned, many today regard as the "jewel of the solar system."

Nerissa and Graziano exchange a few sharp words, prompting Antonio to proclaim (5.1.238), "I am th'unhappy subject of these quarrels." Recall that at the opening of the play, Antonio pronounced in an abbreviated line (1.1.5), "I am to learn," and one expects that, in the end, these quarrels will lead to increased knowledge.

Bassanio and Portia continue to argue and Antonia defends Bassanio. Portia forgives Bassanio and gives him a ring (5.1.254–5.1.256), and Bassanio is astonished to see that it is the very one that Portia had first given him and that he had subsequently given to the Doctor of Laws (i.e., Portia in disguise). Portia says that she slept with the Doctor who then gave it to her, and Nerissa chimes in with a parallel story of dalliance with the law clerk. Here again the script is silent on whether Nerissa and Graziano's ring changes hands, but parallelism implies that Nerissa gave Graziano the same ring that Graziano gave to her when she was disguised as a law clerk. Graziano is amazed, "What, are we cuckolds ere we have deserved it?" (5.1.265).

ANOTHER HEPTAD

Portia produces a letter from Bellario that she says identifies her as the Doctor of Laws and Nerissa as "his" clerk. She cites Lorenzo as a witness to her absence and to her subsequent return to Belmont, which she qualifies by saying, "I have not yet / Entered my house" (5.1.272–5.1.273). She then makes a point of welcoming Antonio to her home. There are six newlyweds, leaving Antonio as the odd man out. Halio remarks that Antonio is "a seventh wheel among the newly-wed couples," and he wonders what Antonio "will do" (31). No evidence suggests that Antonio overcomes his melancholy, yet as gathered from lines 1.1.73–76, he is also "marvellously changed." Thus, he still has saturnine symptoms, and as a "seventh wheel" he is again associated with the seventh and melancholic Ancient Planet, Saturn.

Portia and Bassanio have not restored their idyllic relationship, and because Antonio and Bassanio were hitherto the best of friends, perhaps by inviting Antonio into her house, Portia welcomes Antonio as someone who can fill the lacuna created by her altered relationship to Bassanio. What else might Antonio learn? He is the saturnine merchant of Venice, and two years hence when Saturn's rings open up sufficiently, Antonio will see the benefits of Portia's strategy, as will Lorenzo and other interested parties.

Outdoors

Portia is still outdoors and her remaining remarks concern the subtext. She says that she has news for Antonio that is better than he expects. "Unseal this letter soon" (5.1.275), she says, whereupon she divulges its content, which is that three of Antonio's ships have miraculously returned to port. She does not explain how she knows beforehand the contents of the sealed letter (5.1.278–5.1.279n), nor does she identify its author, and there is no explanation of why previous reports had said that all of Antonio's ships were lost. Instead, she tells Antonio, "You shall not know by what strange accident / I chancèd on this letter" (5.1.278–5.1.279). This sentence has been heavily criticized (5.1.278–5.1.279n), but it is Shakespeare's clear hint that one should read *MV* with a paranormal subtext in mind.

Although Antonio opens and reads the letter, he does not divulge the identity of the sender. I posit that Portia came upon this letter by a means that even she as an accomplished miracle worker terms a "strange accident,"[3] as if through metaphysical influence, which would explain how she knows the contents. Portia introduces this letter right after she produces Bellario's letter addressed to her, and the simplest inference is that the author of the second letter is Bellario as well. This would turn Bellario into a superhuman figure who becomes a player at the behest of Portia and who, like her, resolves to do good. Similarly, Bellario's knowledge of the survival of three of Antonio's ships can be ascribed to paranormal action. The rigors of acting as an agent of the Good also explain Portia's altered relationship to Bassanio.

Superior management has succeeded in rescuing the merchant of Venice, but the plot resolution of the subtext is not yet forthcoming, and in fact Portia announces that she feels sure that the accumulated guests are "not satisfied / Of these events in full."

> PORTIA Let us go in,
> And charge us there upon inter'gatories,
> And we will answer all things faithfully. (5.1.296–5.1.297)

The erstwhile Doctor of Laws resorts to legal jargon as she announces that inside her house, witnesses will be required under oath to answer questions (interrogatories). Unfortunately, the play ends before she enters and thus before a full explanation is delivered, so the audience must shift for itself.

SAFEKEEPING

From evidence described, I argue for descriptions and events that associate rings with the planet Saturn. A chief feature of the two rings in question is that the diamond ring in the alliance of Portia and Bassanio is primary, and the "hoop of gold" and the "paltry ring" (5.1.146–5.1.147)" of the alliance of Nerissa and Graziano is secondary. The script is consistent in this regard. As already remarked, the latter alliance has no precedent in Shakespeare's sources; it specifies when Bassanio receives his ring (4.2.9) but not when Graziano receives his; and when Portia discovers that Bassanio no longer wears his ring (5.1.184–5.1.185), it is not stated precisely when Nerissa notices the absence of Graziano's (5.1.265). These events must all be inferred from the parallelism between Portia's and Nerissa's actions, but what stands out is that the script is lacking for a third time, as it does not specify when Graziano receives his ring back. Again, one must infer that event, but this time Shakespeare places the information prominently at the very end of the play:

> GRAZIANO Well, while I live I'll fear no other thing
> So sore as keeping safe Nerissa's ring. (5.1.306–5.1.307)

So "Nerissa's ring" is indeed again in Graziano's possession, and when Shakespeare makes "Nerissa's ring" the last two words of the play, he establishes it as the crux of the subtext. He couches the crux in risqué language that serves to distract, but those who seek reality behind appearances are not misled.

Discreteness

The histories of the two "betrothal rings" are parallel, just as the two physical rings are concentric. Both follow the course: wives to husbands, husbands to wives (in disguise), and wives (out of disguise) to husbands. Portia's ring attracts more attention in part because it has a diamond, and as such is a cut above Nerissa's ring. Telescopy supports the distinction as well. Saturn's inner ring (now known as the B-ring) is very bright and represents the ring that Portia gave Bassanio. The outer ring (now known as the A-ring), is only moderately bright and as such represents the ring that Nerissa gave Graziano. The combination accords with the view of Saturn seen under moderately good resolution, in which there is a gap, named for Cassini, cleanly separating the two rings, with the diamond—bathed and sparkling in the sunlight—representing the planet proper (figure 1.12).

In a likely reconstruction of events, seen through early telescopes the inner brighter B-ring would dominate the earliest telescopic ring image (figure 1.11c). Resolution of a one-ring image is sufficient to earn Saturn the epithet "jewel," but another ring, the "paltry hoop" of gold, is in the picture too. The discrete rings and their relative brightnesses are detectable when looking through a telescope of modest aperture, provided its optical quality and weather conditions are good. The Gainer reproduction (figure 1.9) has magnification sufficient to detect a ring but too low to separate it into discrete rings, which suggests that the Diggeses used a perspective glass of longer focal length (see chapter 7).

Magnificoes

In consequence of the subtext, the question arises as to whether there is any metaphorical connection between the magnificoes of Venice and the magnificence of Saturn. The answer may lie in the history of optics.

The earliest evidence for the study of lenses and optics in England comes from reports by Roger Bacon's predecessor, Robert Grosseteste (*c.*1170–1253), who taught at Oxford and who wrote that people knew the Milky Way to be an agglomeration of closely spaced stars. In a treatise on rainbows, he stated that lenses show how to "make things a very long distance off appear as if placed very close, and large near things appear very small," with the result that observers can "read the smallest letters at incredible distances."

At about the same time, lens grinding became an important industrial activity at the leading center of glass making, Venice, whence itinerant peddlers sold glass artifacts throughout Europe. As mentioned in chapter 1, Bourne tells of lenses made of fine Venice glass in England in about 1585, and it is fair to surmise that in the sixteenth century as well, lenses of good optical quality existed in England. Perhaps their quality was sufficient to enable details of the Saturn ring system to be seen, but if the Venetian industry is so laudable, why did Shakespeare cast Venice is such a poor light, and why did he set the happy finale in neighboring Belmont?

As mentioned in chapter 1, Bourne did not know all there was to know about English telescopy, and perhaps Venice glass did not play an important role there. The Diggeses were capable craftsmen, and just as Tycho made his own paper by which to publish his books, so I suggest, the Diggeses made their own glass.

The fact that the "doubling of the ring business" (16) will conclude in the house where Portia's "ever virtuous" and "holy" father once resided is testimony to the association between heavenly harmony and the Good. Perhaps, even, her father's spirit devised the hurdle of the three caskets in order to effect Good in the world after he was gone, which ultimately led to introducing an unknowing public to the stunning spectacle of a doubly ringed Saturn.

CHAPTER 6

THE WINTER'S TALE

Singularity is almost invariably a clue.

—Arthur Conan Doyle

The First Folio of 1623 contains the first printing of Shakespeare's *The Winter's Tale* (*WT*). It is one of a few plays with a *dramatis personae*, and of the sixteen named parts in the original printing, over two-thirds are from Plutarch's *Lives of the Noble Grecians and Romans*, which he wrote early in the second century AD. These are Antigonus, Archidamus, Autolycus, Camillo, Cleomenes, Dion, Emilia (Aemylia), Hermione, Leontes, Paulina (Paulinus), and Polixenes (Polyxemus). Autolycus also appears in Plutarch's *Life of Lucullus*, and Leontes may derive from the city of Leontium in Plutarch's *Life of Cleomenes* of AD 75 (Snyder and Curren-Aquino 80). (In this Chapter, unattributed notes, lineation, and pagination refer to Snyder and Curren-Aquino's edition of *The Winter's Tale*, and other editions of Shakespeare's plays are also identified by editors' names.) Like Plutarch, Shakespeare chronicles events in the lives of eminent personages, except that in *WT*, his concern is with individuals who contributed in one way or another to the scientific awakening in the later Renaissance. These relationships emerge through the god of Time in his role as a chorus and through numerous precise specifications of intervals of time. Bizarre features include a bear hunting down and eating a nobleman and the petrifaction and reviviscence of a Queen after a passage of sixteen years.

SEASONS

The Bohemian Lord Archidamus is in Sicilia conversing with his Sicilian counterpart, Camillo. Archidamus is an aide to Polixenes, the King of Bohemia. Camillo is an aide to Leontes, the King of Sicilia. Leontes is hosting a visit from Polixenes, with whom he has been friends since childhood.

Directors sometimes omit the short opening scene 1.1, which deprives playgoers of an early opportunity to appreciate the differences between Bohemia and Sicilia. The opening words belong to Archidamus. "If you shall chance, Camillo, to visit Bohemia...you shall see, as I have said, great differences betwixt our Bohemia and your Sicilia" (1.1.1–1.1.4). The play's first word—"If"—sets a subjunctive mood later amplified by dreams and improbabilities, giving events a surreal quality in which nothing is what it seems. Once again, Shakespeare invites the audience to distinguish between appearance and reality, and if challenges encountered in previous chapters are any guide, one might wonder whether *WT* has a subtext that relates to real phenomena and events.

Camillo's reply, "this coming summer the king of Sicilia means to pay Bohemia the visitation which he justly owes him" (1.1.5–1.1.6), introduces the play's concern with time. It sounds as if it is now winter or spring, which makes sense, as the northerly King probably would go south in the winter and return home in the spring.

POLIXENES AND LEONTES

Shakespeare adapts the name "Polyxemus"—occurring in Plutarch's *Lives*—to resemble the Greek word "polyxenos," meaning "hospitable and much visited" (81), derived from *poly-* (meaning several or many) and *xenos* (meaning guest or foreigner). This raises the question of why Leontes, who is Sicilian, is playing the host rather than the supposedly hospitable Polixenes. Camillo is the trusted aide to Leontes (1.2.232–1.2.236), yet the first sentence of the play (1.1.1–1.1.4) makes clear that Camillo has never visited Bohemia. If the Bohemian Archidamus accompanies his King to Sicilia, his Sicilian peer Camillo would have accompanied his King to Bohemia. The implication is that Leontes has never visited Bohemia, and thus the nominally hospitable Polixenes has never hosted a visit from his best and lifelong friend. A further inference is that Polixenes has entertained others, but who might they be?

The association of King Polixenes with Bohemia suggests that Shakespeare attributes to him a characteristic of Emperor Rudolph II (1552–1612), who attracted all manner of intellectuals (and charlatans) to his court in Prague. Another attribution is evident when the Bohemian monarch enters, and Camillo remarks, "Happy star reign now! / Here comes Bohemia" (1.2.359–1.2.360). The "happy star" is probably *Polos*, the Pole Star, reigning like a celestial tsar over northern sites.

Polos provides a means for navigators on any meridian in the northern hemisphere to set their bearing, and to that extent, like Polixenes, *Polos* is "much visited," or much observed. The meridian line that passes through the Bohemian capital, Prague, runs due south to Sicily, and in so doing cuts across Italia, whose ancient name Saturnia derives from the Roman god of Time. Moving north from Prague, the meridian reaches the Earth's North Pole, where all the meridians converge. *Polos* lies nearly above that singular spot. In the sixteenth century, the likely referent for *Polos* was the star α in the constellation Ursa Minor (Little Bear). Attending this celestial whelp is the constellation Ursa Major (Big Bear).

After Camillo broaches the possibility of a Sicilian visit to Bohemia, Archidamus apologizes for Bohemia's lack of hospitality and hopes that the goodwill of its people will compensate (1.1.7–1.1.8n). "We cannot with such magnificence—in so rare—I know not what to say," he stammers (1.1.11–1.1.12). Polixenes excels at hospitality yet his aide feels obliged to apologize for lack of it. Camillo does not press for an explanation, which keeps us guessing. A further oddity is that Archidamus appears in the first scene, yet later when his king has cause to flee Sicilia, he is out of sight. If Archidamus knows of this rare Sicilian commodity, then Polixenes must know of it too and must surely have condoned his aide's dropping out of sight. One wonders whether Archidamus is dismayed because he knows that Polixenes is up to something and feels it prudent to make himself scarce.

Camillo explains that Leontes and Polixenes "were trained together" as children. "Trained" can mean either "instructed" or forced to grow in a particular direction, like a plant (1.1.19n). Camillo asserts that the Kings' mutual affection is "rooted" in childhood, from which as adults they "cannot choose but [to] branch." In short, Polixenes and Leontes have diverged "from a central beginning" (1.1.21n), as if a discarnate botanical hand guided them fatefully. The horticultural conceits signify that the grounds of knowledge are at issue, and that a matter of "central" significance is at stake.

According to Camillo, after Polixenes and Leontes had matured and had gone their separate ways, they remained close in spirit, embracing one another "as it were" from "the ends of opposed winds" (1.1.26). The synonymity of "wind" with a compass bearing resembles that of the nearly opposite winds occurring in *Hamlet* that influence Tycho Brahe's hybrid cosmology (figure 1.6). Camillo, never dreaming that opposing winds might confound the kingly kinship, hopes that the "heavens" will "continue their loves" (1.1.27). The passage foretells that the once-kindred monarchs will come to oppose one another, and that—like Hamlet's winds—one will represent a beneficent influence and the other will not.

The conversation shifts from love to Leontes. Archidamus believes that Leontes' only child, Mamillius, "is a gentleman of the greatest promise" (1.1.30), and Camillo agrees. He calls Mamillius "a gallant child" who "physics the subject" and "makes old hearts fresh," adding that those who "went on crutches ere he was born desire yet their life to see him a man" (1.1.32–1.1.35). "Gallant" is "a general term of admiration or praise," and refers specifically to one who is "chivalrously brave" (*OED*), and indeed the lad will soon assure his father, "I'll fight"

(1.2.160). The allusion to crutches means that a man would find a reason to live even if he had no heir (1.1.36–1.1.40n). A sense of foreboding persists, for as the first scene ends, Archidamus repeats the crutch trope, "If the King had no son, they would desire to live on crutches till he had one" (1.1.39–1.1.40). Archidamus opines that the followers of Leontes ("they") would limp along in a debilitated state until a new heir emerged. However, Leontes actually does have an heir, Mamillius, who now it seems is not long for this world.

Timing

Polixenes opens scene 1.2 by telling Leontes, "Nine changes of the wat'ry star hath been the shepherd's note since we have left our throne" (1.2.1–1.2.2). A lunation is the shepherd's unit of time, which is the interval between successive identical phases of the Moon and equals on average $29^d12^h44^m$ (29.53 ± 0.27 days). Members of the audience who happen to have an abacus handy will find that nine lunations is 265¾ days, which in round numbers is 38 weeks (or 8 months, 3 weeks, and 1 day, counting an average month as approximately 30½ days). This establishes the zero point of the chronology of table 6.1 (*entry i*).

When scene 1.2 opens, and Polixenes specifies the length of his stay, 266 days have already passed (*entry iv*, table 6.1), and it is now day 267 after Polixenes left Bohemia. It will be revealed that on this day, Polixenes promises to stay one more week (*entry v*). The action in scene 1.2 is continuous, and at scene's end, when Polixenes leaves court, he does so late and on the same day as he opened the scene (*entry vi*). Through the night, he makes his way to his ships, and it is reasonable to suppose that he sails the next day, number 268 (*entry vii*). Later developments lead to the remaining entries in table 6.1.

At court on day 267, Polixenes expresses gratitude for his host's hospitality in sophisticated terms, using place values to base 10 ("I multiply… many thousands more," 1.2.6–1.2.9). His aptitude for mathematics and preference for shepherd's time are traits associated with the god Apollo,

TABLE 6.1. Chronology of Hermione's pregnancy (± ½ day) in *The Winter's Tale.*

Entry	Day	Scene	Event
i	1	—	Polixenes departs Bohemia
ii	16½	—	Polixenes arrives in Sicilia
iii	?	—	Hermione conceives
iv	267	1.2	9 lunations have passed
v	267½	1.2	Polixenes will stay 1 week longer
vi	end 267	1.2	Polixenes departs Leontes' court
vii	267–268	1.2–2.1	Polixenes leaves Sicilia
viii	268	2.1	Ambassadors depart for Delphos
ix	268	2.1	Hermione goes to jail
x	291	2.2	Hermione delivers
xi	291	2.3	Ambassadors return after 23 days

whose name occurs in *WT* about as often as all other plays combined, and his poetic exponentiation forewarns auditors to keep their abacuses at the ready.

HERMIONE

Polixenes announces he must depart on the morrow (day 268), but Leontes tries to persuade him to remain another week. Polixenes acknowledges that "no tongue that moves, none, none i'th'world" (1.2.20), other than Leontes' tongue, could persuade him to change his mind. Yet Polixenes remains adamant, so Leontes invites Hermione to try. Does Leontes think that Hermione is not of this world?

In response to the King's invitation, Hermione explains that she was waiting for her husband to finish trying before she tried. Polixenes refuses her entreaty too, but she insists, saying that even if Polixenes' resolve were strong enough to dislodge stars from their spheres, "Verily, / You shall not go" (1.2.48–1.2.49). The old category "stars" includes Ancient Planets, and according to the Old Astronomy, "spheres" are their magical supports. Apparently, Hermione's power of persuasion is so powerful that it exceeds even the capacity of stars to adhere to their spheres. The hyperbole emphasizes Hermione's superb and perhaps otherworldly qualities, and indeed she does participate in unseating the stars and planets—of the Old Astronomy.

With such powerful credentials, it is not surprising that Hermione wins the contest of wills. Sometime during day 267, therefore, Polixenes grants her the "borrow of a week" (1.2.39). For convenience, table 6.1 lists the day value as 267½ (*entry v*).

Polixenes and Hermione continue to socialize in the presence of Leontes. Polixenes tells her of times when he and Leontes were young and had not yet won their spouses. In those days, says Polixenes, "Your precious self had then not crossed the eyes / Of my young playfellow" (1.2.78–1.2.79) and Hermione responds, "Grace to boot!" (1.2.79), meaning "grace as well." "Grace" is an elegance of proportion and a pleasing quality, a divine influence that operates in humans to regenerate and sanctify. It is a force with qualities of excellence, clemency, and forgiveness, with its seat in the soul.

Leontes has been inattentive to the interchange between Polixenes and Hermione, and now he asks whether she has succeeded in persuading Polixenes to stay. When she responds in the affirmative, Leontes immediately states the obvious, that she had succeeded where he had failed. This should intrigue the audience as Polixenes had said that no human could accomplish this.

The King is pleased with the outcome and tells his wife that this is the second time she has spoken well. Hermione responds with a tinge of sarcasm, "What, have I twice said well? When was't before?" (1.2.89). She feels it necessary to recall such a memorable occasion because "One good deed dying tongueless / Slaughters a thousand waiting upon that" (1.2.91–1.2.92). She calls the second occasion of her speaking well, an "elder sister…O, would her name were Grace!" (1.2.97–1.2.98). "Elder" means having lived or existed longer, and it may occur in quasi-personification, and "sister" can signify something with characteristics similar to something else (*OED*). Leontes explains that the first time she spoke well was when she accepted his offer of marriage, which, Hermione says again with a trace of humor, was "grace indeed" (1.2.104). Thus, she has spoken "to the purpose" twice (1.2.105)—once when she agreed to marry Leontes, and once when she persuaded Polixenes to stay another week. The grace that blesses both occasions is Shakespeare's word for beneficence that is unexpected, unearned, and granted by Providence or the gods (*OED*; Dolan xli, 1.2.81n).

Hermione's quality of grace concerns her having gained "for some while a friend" (1.2.107), but the trouble is that "friend" has the connotation of "lover." Leontes is unsure when Hermione became pregnant, but he knows that Polixenes has been in Sicilia long enough to have planted his seed in her, if not immediately upon arrival then soon thereafter. This could explain Hermione's amazing success in persuading Polixenes to grant the King's wish that Polixenes stay another week. Leontes is concerned about whether he is the father of the unborn child and begins now to harbor doubts about the paternity of Mamillius as well.

Hermione realizes that Polixenes "longs to see his son" (1.2.34), who is in Bohemia, but there is no mention of Polixenes' wife. If she is estranged or deceased, this could fuel Leontes' fears as well.

Leontes might have dismissed these connections as coincidental and unproven were it not that he sees Hermione give her hand to Polixenes in a tender manner. "Too hot, too hot!" (1.2.108) he exclaims. He is beset by "tremor cordis" (1.2.109n), complaining that his "heart dances / But not for joy" (1.2.109–1.2.110). Later, Camillo describes his ailment as a "sickness" whose name he does not know, although *Diagnostic* (393–444) may have remedied the deficiency by now. Perhaps the handholding was the trigger for Leontes' anxiety, and he has reason to worry as Hermione's supernal allies slowly enmesh him. In his anguish, he imagines that Hermione and Polixenes are lovers whose sighs are like the hunter's horn announcing the death of the deer. The slaying of a deer normally would be cause for celebration, but neither Leontes' bosom nor his brow shows signs of joy. "Brow" commonly refers to appearance, countenance, or forehead, and when Leontes refers to a "hard'ning of his brows," he could mean a hardening of the imagined horns of a cuckold, or else a muscular contraction of his eyebrows brought on by the infection of his brain. Later, Hermione tells Leontes that he holds "a brow of much distraction" (1.2.148), which is a reference to horns and an outward manifestation of worry (1.2.145n).

Leontes is not completely out of touch because at least he allows for the possibility that Hermione is friendly toward Polixenes because she is simply a graceful hostess. Leontes' has a smidgen of mental flexibility, which hints that there is hope for him, but for the time being he is inclined to believe that Polixenes and Hermione have something going between them.

Solace and Affection

Leontes seeks solace from Mamillius, who pledges devotion to his father. Leontes remarks that women consider the two of them almost "as like as eggs" (1.2.129), but he adds that women are fickle and will say anything. Leontes is now unsure that he sired Mamillius. "Art thou my boy?" (1.2.119), "Art thou my calf?" (1.2.126), he asks. He tries to reassure himself, "They say [thy nose] is a copy out of mine," (1.2.121) and he addresses Mamillius as "my collop" (1.2.136), but the terms are "contorted" and "perhaps not meant to carry a clear meaning," so that "a single coherent interpretation remains...elusive" (1.2.136–1.2.145n).

Consider the first part of his address.

> LEONTES Come, sir page,
> Look on me with your welkin eye. Sweet villain,
> Most dear'st, my collop! Can thy dam, may't be—
> Affection! Thy intention stabs the centre. (1.2.134–1.2.137)

The first three lines of the passage refer unambiguously to Mamillius. Leontes proclaims his son "most dearest," where the double superlative measures the depth of his love. The lines establish the mutual affinity of father and son, but Leontes is so suspicious of Hermione that he now wonders about her fidelity and her influence on Mamillius.

The third line in the passage contains two questions which Leontes does not fully articulate, as if he were merely musing, and this uncertainty is confounded by unclear punctuation (1.2.137nn). Leontes has just finished calling his son "my collop" (i.e., of his own flesh), whereupon thoughts about the nature of the boy's dam suddenly assail him, and just as suddenly, in the fourth line he hits upon the answer, "Affection!" Critics focus attention on this word whose common meaning is tender regard, emotion, attraction, or in reference to substances and essences, "a mode of being" (*OED*).

Next comes "intention," which can mean intensity and which Leontes specifies as directed toward "the centre" ("intention stabs the centre"). "Centre" is supposed to refer to the center or core of being, and "to stab" means to wound or kill, as in "to stab the heart"—the physical and spiritual center of human beings (*OED*). Leontes' conceives of a type of force killing a mode of being by stabbing its center, but "centre" also can mean "the Earth" (*OED*, 2.1.100–2.1.103n; Dolan 2.1.102n). Primitive cosmologies are geocentric, and Leontes obsesses that a forceful purpose will stab his perception of the World's center.

Leontes rambles on. "Thou dost make possible things not so held," he says (1.2.138), referring to the forceful purpose that dislodges stars and planets from their appointed rounds. Equally unsettling to Leontes is that he has no idea from where his thoughts originate (1.2.138–1.2.135), and he fears that he is dreaming. "How can this be?" (1.2.139) he asks.

> LEONTES With what's unreal thou coactive art
> And fellow'st nothing. Then 'tis very credent
> Thou may'st co-join with something, and thou dost,
> And that beyond commission, and I find it,
> And that to the infection of my brains
> And hardening of my brows. (1.2.140–1.2.145)

"Thou" continues to refer to the forceful purpose "affection," which is now "acting in concert" (1.2.140n) with "what's unreal" (i.e., with something immaterial, invisible, or insubstantial), which "fellow'st nothing," or rather is not physically part of the fellow or body in question (1.2.141n1). Leontes fears that Hermione has influenced Mamillius adversely by inculcating the falseness of the Ptolemaic machinery that holds the Ancient Planets in orbit, and replacing it with the idea of intangible action. Then, concludes Leontes, it is most credible that "Thou may'st co-join with something," meaning that the invisible interaction between fellows or bodies acts to "co-join" them, or bind them to one another. To Leontes, this is absurd because he thinks that celestial bodies cannot circle the Earth without spheres and swivels to guide them.

At the suggested writing time of 1610, the language could refer to the first stirrings of the concept of an invisible interaction between celestial bodies. Kepler's *Mysterium Cosmographicum* ("Mystery of the Universe") of 1596 and Gilbert's *De Magnete* address the concept of magnetic attraction at a distance, which is a concept far advanced over views expressed by Aristotle. Leontes is concerned that such concepts are "beyond commission," or not "authorized" (1.2.143n2), where authorities are often a barrier to original thought. He speaks of results not yet established, but although the magnetic force is supposed to act tangentially, it nonetheless shares the idea of invisible action and reaction with Newtonian gravity. Instead of containing a line fit for excision (1.2.143n1), the passage in question is significant to the understanding of the development of the New Physics because it pertains to the then-contemporary problem of how the Sun holds planets in orbit.

Unsettling thoughts would put a frown on any husband's face, and in this case, the King's delusions cause him to feel that his brow is hardened with the worry of being a cuckold and thus being without a legitimate heir. *WT* appears to be another instance of Shakespeare couching cosmic advances in terms of human sexuality. He often paints conflicts in extreme ways, and so the audience is not dismayed that Leontes becomes increasingly displeased with what he perceives subliminally to be his erstwhile friend Polixenes bedding his wife and threatening his kingdom and his worldview. The irony in Leontes' speech is that in a mere ten lines addressed to his young son and in the presence of Polixenes, he glosses the progress of astronomy from pre-Pythagorean times to the turn of the seventeenth century, when magnetism and its intangible action were coming into vogue.

After Leontes finishes, it is not Mamillius who professes bafflement, but Polixenes. Hermione is present too and adds that Leontes seems unsettled. She asks pointedly, "Are you moved, my lord?" and Leontes replies, "No in good earnest" (1.2.149). She is asking whether the earthling Leontes feels displaced from his perceived center, and whether he and the Earth are in motion. The Kings are polar opposites, the alpha and the omega of cosmological thinking, and if Leontes believes he is cosmically unmoved, one should expect Polixenes to hold the opposite view. As well he might, because he hails from Bohemia whose capital, Prague, was the foremost center of scientific learning in Europe in the first decade of the seventeenth century, and whose emperor was hospitable to new thinking in the sciences and allied arts.

AGE LIMITS

Leontes says that Mamillius reminds him of when he was as old as his son is now.

> LEONTES Looking on the lines
> Of my boy's face, methoughts I did recoil
> Twenty-three years, and saw myself unbreeched. (1.2.152–1.2.154)

By 1.2.154, Leontes is 23 years older than Mamillius. This is *entry* α of table 6.2. When he was as old as his son is now, he was not yet "breeched," meaning not dressed in trousers. Breeching is a rite of passage that occurs usually before age 7 (1.2.155n), so it follows that Mamillius is less than 7 years old, and Leontes is less than 30 years old (*entry* β). The upper limit to Mamillius' age agrees with Othello's age of 7, when his grooming for war begins (*Othello* 1.3.85), and with Mamillius' resolve to fight (1.2.160). Remaining entries in table 6.2 await developments.

TABLE 6.2. Ages and epochs on stage.

entry	SCENE	AGES AND EPOCHS ON STAGE
α	1.2	Leontes is 23 years older than his son
β	1.2	Mamillius is < 7 years old, and Leontes < 30 years old
γ	3.3	The epoch of Acts 1,2 and 3 is 1588
δ	—	In 1588 both Kings are 28 years old and their sons are 5
ε	4.1	The epoch of Acts 4 and 5 is 1604
ζ	—	In 1604 the Kings are 44 years old, Florizel is 21, Perdita 16

INSTRUMENT

Leontes tells Hermione and Polixenes that he will keep company with Mamillius. He calls his son a "squash," which is both an unripe peapod (1.2.159n) and a term of derision (*OED*). The disintegration of the King's relationships continues as he says that he will leave Hermione and Polixenes to their "graver steps" (1.2.171). His choice of words reveals the depth of his resentment. Hermione announces that they will be "i'th'garden," so the weather must not be too chilly. Hermione asks whether they can expect Leontes to join them, but Leontes begs off and adds (seemingly gratuitously) that they can be found "beneath the sky" (1.2.178). Aside, Leontes announces, "I am angling now, / Though you perceive me not how I give line" (1.2.178–1.2.179). He is on a fishing expedition, and he will give scope or latitude like an angler who plays a fish (1.2.178n, 1.2.179n).

Leontes questions Camillo and tries to win him over to his side, but Camillo demurs. They argue, and Leontes orders Camillo to poison Polixenes. Leontes leaves and passes Polixenes offstage. Polixenes enters and expresses concern that Leontes averted his gaze and did not speak to him. Camillo explains that Polixenes' "distemper" (1.2.380) has triggered Leontes' anxiety. This is the "disease" (1.2.381) that the boreal visitor is supposed to have brought to Sicilia.

Camillo volunteers that Leontes appointed him to kill Polixenes, and he explains why.

> CAMILLO ...he swears,
> As he had seen't, or been an instrument
> To vice you to't, that you have touched his queen
> Forbiddenly. (1.2.409–1.2.412)

"Vice" as a verb connotes illicit persuasion. "Instrument" is a "tool" that Camillo says Polixenes used to touch Hermione "forbiddenly" (1.2.411n), and introduced later are "instruments that feel," which supposedly are fingers (2.1.154n). Broadly speaking, from the fourteenth century an instrument is "anything that serves or contributes to the accomplishment of a purpose or end," specifically a device "whose primary function is to respond to a physical quantity or phenomenon...by registering or measuring it...and which may be of a complicated design or construction" (*OED*). It can be argued that Camillo and Leontes refer unwittingly to the spiritual transferal of the New Philosophy to the unsuspecting Hermione and her unborn child.

Toward the end of day 267, Camillo convinces Polixenes to flee (*entry vi*, table 6.1). He offers to switch allegiance and abet the escape on condition that Polixenes brings him along.

> CAMILLO If therefore you dare trust my honesty,
> That lies enclosed in this trunk which you
> Shall bear along impawned, away tonight! (1.2.429–1.2.431)

They will leave court late that day and Camillo will open the posterns that would otherwise impede their flight. It is reasonable to suppose, therefore, that Polixenes departs Sicilia early the next day, number 268 (*entry vii*, table 6.1).

Camillo envisions that his honesty lies "in this trunk" (1.2.430). "Trunk" could mean his body sans head and limbs, where his heart is; or, as professed earlier, it could mean a perspective glass. Figuratively, via a conflation of an optical instrument used to explore the stars and a physiological instrument used to impregnate, Shakespeare identifies the tool with which

Polixenes infected Hermione. Polixenes opines that Leontes is jealous because Hermione is "rare" and "precious" (1.2.447), so much so that one must suspect that her spirit is as receptive to impregnation as her soma.

Camillo, having tried to dissuade Leontes of his presumption of cuckoldry and having disobeyed his order to commit murder, is now squarely in the camp of the righteous.

> CAMILLO For myself, I'll put
> My fortunes to your service, which are here
> By this discovery lost. (1.2.434–1.2.436)

By "discovery," Camillo means his disclosures to Polixenes, who responds to Camillo's commitment by pledging to respect Camillo as long as "Thou bear'st my life off hence" (i.e., as long as Camillo gets him out of Sicilia) (1.2.457n). Both aforementioned passages contain the verb "bear"—Camillo's in 1.2.431 and Polixenes' in 1.2.457—and Hermione is bearing a child. All bear responsibility for moving the plot along, and in the next Act, the verb "bear" transmogrifies into a noun.

Treason

Act 2 opens with Hermione expressing exasperation with her son, who "so troubles me / 'Tis past enduring" (2.1.1–2.1.2). Poor Mamillius! His father suspects his lineage and his mother cannot bear his presence. Perhaps her impatience is a symptom of pregnancy, which is now learned about formally from the ladies in attendance, who announce that she "rounds apace" and has "spread of late / Into a goodly bulk" (2.1.19–2.1.20). In response, Hermione asks, "What wisdom stirs amongst you?" (2.1.21) in other words, to what wise matter do the women refer? (Dolan 2.1.21n). Hermione knows that the life stirring in her womb is synonymous with "wisdom," and the contrast suggests that her son is not so blessed.

Hermione's mood swings and she asks Mamillius to sit beside her and tell her a story. Mamillius asks whether the tale shall be merry or sad, to which his mother responds "As merry as you will" (2.1.24). Mamillius decides, "A sad tale's best for winter. I have one / Of sprites and goblins"

(2.1.25–2.1.26). Yet concerning the season currently prevailing, his meaning is doubly ambiguous. The word "goblin" derives from the Greek for a wicked sprite, and from the fourteenth century, it has come to mean an ugly, mischievous demon (*OED*). From the same century, the word "sprite" occurs in various senses ranging from wicked to holy, so in principle, Mamillius' tale could be happy or sad. *WT* opened with Camillo saying that Leontes plans to visit Polixenes in the coming summer, from which it appears that the season is either winter or spring; however, Mamillius never tells his story, so his meaning remains unclear.

A Lord tells Leontes that he had observed Polixenes and Camillo fleeing, and that he "eyed" them to their waiting ships. The amount of traveling involved to this point—Polixenes leaving late and passing through posterns to his ship and the spy's journey to court from his lookout—suggests that Act 2 opens the next day 268 (*entry viii*, table 6.1), which would be the same day as Polixenes sets sail for home (*entry vii*).

Leontes takes the flight of Polixenes as evidence of guilt, "How blest am I / In my just censure, in my true opinion!" (2.1.36–2.1.37) Shakespeare illustrates how an erroneous assumption can lead to a verisimilar sequela, the one in this case being a reaction that appears to verify the assumption.

The King takes Mamillius from his mother and remarks that he is glad that Hermione did not breastfeed him because Mamillius already possesses too much of her. This follows from the belief that babies inherit characteristics through suckling (2.1.56–2.1.58n). Leontes tells Hermione that Polixenes has made her "swell thus" and must "sport herself / With that she's big with" (2.1.60–2.1.62). He accuses her of treason, which she denies, and he tells her,

> LEONTES If I mistake
> In those foundations which I build upon,
> The centre is not big enough to bear
> A schoolboy's top. (2.1.100–2.1.103)[1]

He is certain of the correctness of his logic because it emanates from the "centre" of his being and figuratively from the center of the Earth.

The same day (268), Leontes sends Hermione to prison (*entry ix*, table 6.1). She concludes that, "some ill planet reigns" (2.1.105), but pledges to "be patient till the heavens look / With an aspect more favourable" (2.1.106–2.1.107). She speaks confidently of a happy outcome, "This action I now go on / Is for my better grace" (2.1.121–2.1.122), recalling that grace is a divine influence that regenerates and sanctifies. She departs with her ladies-in-waiting.

Antigonus and a courtier try to convince Leontes of the Queen's innocence, but the King dismisses their arguments, claiming it is his prerogative to disregard the counsel of his advisors because he acts by virtue of "natural goodness" (2.1.164). Leontes announces that he has dispatched ambassadors, Cleomenes and Dion, "in post" (2.1.182) (i.e., post haste, immediately) to the sacred shrine of "Delphos" (2.1.183). "Delphos" combines Delos in the Aegean Sea (where Apollo was born) with Delphi on the Greek mainland (where Apollo's oracle is actually located) (2.1.183n), but for purposes of the play, the oracle lies on the island Delos. Such is the strength of his conviction that Leontes is sure that the oracle will support his beliefs and condone his actions. Leontes might have ordered their departure shortly after he snubbed Polixenes in scene 1.2, on day 267 (*entries v* and *vi*, table 6.1), but that scene drags on with Leontes present three-quarters of the time, and it is more likely that the ambassadors left the next day, 268 (*entry viii*, table 6.1).

Paulina

The next scene (2.2) opens with Lady Paulina visiting Queen Hermione in jail. It is not yet clear when this occurs, but until such time as it is revealed what day that is, let this be called day "Δ" (I will posit later that Δ = 291; see *entry x*, table 6.1). Paulina's lady-in-waiting, Emilia, announces that Hermione "is something before her time delivered" (2.2.24). The news of a momentous event like the birth of a Princess ought to spread rapidly regardless of its venue, yet Paulina knows nothing of it when she arrives. It seems likely that the birth has occurred that very same day, Δ.

The child is a lusty girl, and apparently, the mother is doing well. Paulina offers to take the babe and show her to the King in hope that he will relent. She is confident in her mission. Let it "not be doubted / I shall do good" (2.2.52–2.2.53), she promises. The tension is between good and evil, as Paulina's resolve to do well by the child contrasts with the King's "natural goodness" to kill it. Meanwhile, Mamillius becomes despondent, loses his appetite, and cannot sleep. The King cannot sleep either, and wonders whether he might get some rest if Hermione were "given to the fire" (2.3.8). Persecution is addictive; once started, it is hard to stop.

Paulina loses no time in bringing the babe to the King—she arrives at court on the same day Δ as Hermione gave birth. She finds Leontes in a foul mood. Leontes blames Paulina's husband, Antigonus, for the intrusion, but he protests that he told her not to visit the King. Leontes sneers, "What, canst not rule her?" (2.3.46) His pinched worldview precludes self-willed wives.

Paulina argues the babe's case eloquently, but Leontes orders that mother and child be burned to death. Paulina is undeterred and the King threatens to have her burnt as well. Her retort is memorable, "It is an heretic that makes the fire, / Not she which burns in't" (2.3.113–2.3.115).[2]

When Paulina removed the baby girl from her mother, she guaranteed that she would do well by her, but now it would seem she has failed. However, before departing Paulina prays, "Jove send her / A better guiding spirit!" (2.3.126), so in fact, Paulina does do well by the infant because the deity will hear her prayer and heed her plea.

Leontes calls Antigonus a traitor for allowing his wife Paulina's visit, and commands him to take the child and burn it. Attendant courtiers intercede and the King relents, but orders Antigonus on pain of death to "bear" the "female bastard" to some remote and desolate place (2.3.174), there to leave it to the mercy of fate. Leontes argues that because "strange fortune" brought the child to him, "chance" must dictate its survival (2.3.182). The "better guiding spirit" that Paulina prayed for may have shielded the child temporarily, but its ultimate fate is still uncertain.

ANTIGONUS

The name Antigonus probably has roots in *Antigone* of Sophocles (*c.*496–*c.*406 BC). Both names mean "against generation, against child-bearing," and "in place of a mother" (80; Graves 2: 380). In mythology, Queen Juno changes Antigone into a stork (Ovid 4: 132), which in fairy tales is the delivery vehicle for babies. Antigonus too will deliver a baby, albeit a born one, and he mothers it to the extent of hoping that a "powerful spirit" will "instruct the kites and ravens" to nurse and save her (2.3.184–2.3.186). Like his wife, he pins hope of the babe's salvation on a powerful spirit, but "kites and ravens" are the wrong kinds of bird. Fear and pity mount as he notes that bears and wolves have occasionally cast aside their savageness and have raised human infants.

Speedy messengers arrive announcing that the King's minions have returned from Delphos and are "Hasting to th'court" (2.3.196). The day is still Δ. Leontes orders preparations for Hermione's arraignment, while the audience waits on tenterhooks to learn of the oracle's pronouncement. Apollo is a champion of righteousness, and one would expect that the news from Delphos is good, but with almost unbearable pathos, it has not arrived in time to forestall the departure of Antigonus and his charge.

TWENTY-THREE DAYS

The audience learns that Cleomenes and Dion have been gone for 23 days (*entry xi*, table 6.1), and they have accomplished their mission in record time (2.3.197n). Because the ambassadors left on day 268 (*entry viii*), they arrive back on day 291 (*entry xi*). From preceding sections, this is the value of Δ, the same day that Hermione delivers (*entry x*), and the same day that Antigonus departs with the babe.

The couriers' 23-day absence probably includes time spent at Delphos interacting with the oracle, and for the sake of argument, suppose this lasts a day. If so, the pair spends a total of 22 days on the round trip, or 11 days each way. The journey was "rare, pleasant, speedy" (3.1.13), and because the messengers do not complain about the weather, it is fair to assume that it was clement. It makes sense that they sailed the entire way through the Aegean archipelago.

These circumstances enable an estimate of the duration of Polixenes' sail from Bohemia (*entry ii*, table 6.1). It is safe to assume that the Sicilian and Bohemian ships are equally fleet, as they both have a royal mission and royalty would demand capable crews and ships that are up to speed. The distance from Sicily to modern Delos is about two-thirds of the distance from Sicily to Trieste, which I posit is approximately the site of the play's Bohemian port. Thus, a reasonable estimate for the time that it took Polixenes to sail from Bohemia to Sicilia is 1½ times 11 days, or about 16½ days (*entry ii*, table 6.1). This is a rough estimate, but it is supported by the fact that it agrees well with contemporary sailing times in the Mediterranean (Sohmer, *Wise* 125n55).

With the exception of *entry iii*, which is the topic of ensuing sections, table 6.1 is complete. Column 3 indicates the flow of information that the script provides through Acts 1 and 2, and *entry ii* follows from an inference from the final *entry xi*.

Two Coincidences

There is now enough information in table 6.1 to shed light on the question of whether Polixenes or Leontes impregnated Hermione (*entry iii*). The data reveal two remarkable coincidences.

Consider *entry v* first, which says that Polixenes plans to stay to day 274½. With an average month of 30½ days, 274½ days is almost exactly 9 months, which is the nominal length for human gestation!

This means that Polixenes' departure for Sicilia has initiated a figurative incubation period for the introduction of the New Philosophy to Sicilia. On day 267½, the figurative mother to be wants the figurative father to be at her side 7 days hence, when she expects to deliver (figuratively). This suggests that the neonate is figuratively the spirit of the

New Philosophy that has emerged in Sicilia, the irony being that Leontes, who opposes it, suggested the extra week's stay to begin with.

A corollary of the first coincidence is that when Hermione makes her request on day 267½ (*entry v*, table 6.1), Polixenes has been in Sicilia for the interval *entry v* minus *entry ii* (i.e., for only 251 days, or 8 months and 1 week). Even if he had impregnated his hostess immediately on his arrival, on day 267½, Hermione would not expect to deliver after only 258 days (8 months and 2 weeks).

Next, consider *entry x*, which specifies when Hermione actually did deliver. The script infers that she did so prematurely, but *entry x* minus *entry ii* equals 274½ days, which is 9 months. In other words, the nominal period that passes from the time Polixenes first sets foot in Sicilia to when Hermione actually delivers again equals the human gestation period!

However, Polixenes is not the father, as the oracle will pronounce that the baby girl is the child of Leontes. Because she was born prematurely, it can be concluded that Leontes impregnated Hermione after day 16½ (*entry iii*, table 6.1), the exact value of which is of no concern other than to ensure that the newborn is strong enough to survive.

Note that this latter coincidence uses *entry ii*, which is an inference derived from information delivered indirectly by the script, whereas the first coincidence relies on data supplied directly by the script. The two coincidences are therefore independent.

In short, Hermione's spiritual delivery occurs 9 months after Polixenes left Bohemia, and her physical delivery occurs 9 months after he set foot in Sicilia. Shakespeare associates the physical and spiritual existence of the baby with the prospective dawn of the New Philosophy in Sicilia.

Trial

In scene 3.1, Cleomenes and Dion are on their way back to court, and they recount their visit to Delphos (Delos). The isle is fertile, the climate delicate, and the air "most sweet" (3.1.1). Delos has roughly the same latitude as Sicily, so the two climates are roughly similar, and "delicate" could well describe the climate in Sicilia at the time of the pilgrimage. According to Plutarch, the dissolute Dionysus rules Delphi for the three winter months and Apollo for the remaining nine (Jones and Pennick 45), which suggests that it is now springtime. In consequence, the tale that Mamillius wanted to tell his mother in scene 2.1 is not a "sad" one fit for winter, but a happy one as befits the start of a new seasonal cycle and the advent of the New Philosophy in Sicilia.

Leontes is utterly convinced of his wife's guilt and starts the trial even before his minions have arrived at court. By the sequence of events described in table 6.1, this is almost immediately after Hermione has undergone the throes of childbirth, and indeed she protests that she was hurried "Here, to this place, i'th'open air, before / I have got strength of limit" (3.2.103–3.2.104). Contemporary medicine held that fresh air was inimical to postpartum recovery, and "limit" refers to the typically one-month period of bed rest (3.2.103n, 3.2.104n).

The King professes, "we so openly / Proceed in justice" (3.2.5–3.2.6), but in Sicilia, the judiciary is not an independent branch of government, inviting bias. The indictment accuses Hermione of the crimes of adultery, aiding and abetting the flight of Polixenes and Camillo, and conspiring with them to kill the King. Hermione recognizes the futility of proclaiming innocence, but as she is ordered to speak, she expresses confidence that her innocence will prevail because "powers divine / Behold our human actions (as they do)" and "shall make / false accusation blush" (3.2.26–3.2.29). The parenthetical "(as they do)" (3.2.27) supports the notion of divine intervention.

Hermione asserts that her past life has been continent, chaste, and true, where chastity in this case refers to her marital relation with the King, and she protests, "With what encounter so uncurrent I / Have strained t'appear thus" (3.2.47–3.2.48). An "encounter" can mean a liaison or amorous affair, "uncurrent" means "exceptional, aberrant, out of the ordinary," and to "strain" can mean to embrace, hold, or hug. Thus, she asks, what limits of behavior has she exceeded that she should now find herself on trial (3.2.47–3.2.48n).

She convinces the assembled witnesses of her innocence, but the King is deaf to her protestations. Leontes' reply is a tangle of negatives that could well appeal to lawyers at the Inns of Court. In effect, Leontes turns Hermione's denial into an argument for her guilt by alleging that if someone has the temerity to commit a crime, that person would be capable of denying it. By this logic, denial is evidence of guilt. Hermione protests that she cannot accept responsibility for faults that are not hers, and she denies being part of a conspiracy and of attending to Polixenes more than Leontes commanded her to do. She says that she does not understand the language that Leontes uses, but realizes she is powerless to effect change and will give up voluntarily to save Leontes the trouble of having to "shoot [her] down" (3.2.79–3.2.80n).

To Hermione, the trial seems like a bad dream. "My life stands at the level of your dreams," she tells Leontes (3.2.79). Yet Leontes blames her for his dreams. Dreams abound in *WT*, and it is hard to know where they end and reality begins.

Outcomes

Cleomenes and Dion arrive at court and deliver Apollo's verdict. As expected, it exonerates Hermione and establishes that the newborn is Leontes' truly begotten child. The oracle predicts that the King shall live without an heir if the babe remains missing, which implies either that Mamillius is not his heir or that the Prince is about to die. Apollo exculpates Polixenes and Camillo, and all present rejoice except the King who brands the testimony false. In a trice, an attendant enters and announces that Mamillius is dead.

Leontes decides that Apollo is angry and concludes that the heavens strike at his injustice. He apologizes and is thankful that Camillo disobeyed his order to poison Polixenes. He seems relatively immune to grief over the death of his son, but Hermione faints. Paulina opines that the news "is mortal" to Hermione (3.2.145), yet Leontes predicts, "she will recover" (3.2.147). Paulina's response is curiously ambivalent, "I say she's dead; I'll swear't…go and see" (3.2.200). As far as anyone can tell, Hermione has undergone a transformation from vitality to a sort of petrified hibernation, there to remain until a wellspring of understanding graces the shores of Sicilia.

Abandonment

The boat bearing Antigonus and the Princess reaches Bohemia, where Antigonus plans to abandon the child. Miraculously, the infant has survived. The ship's captain warns Antigonus not to stray too far inland owing to predatory animals. The sound of thunder foretells the ire of Jupiter, son of Saturn, which bodes well for the survival of the babe, as it is expected that divine intervention will save her.

Before deserting the infant, Antigonus announces that an image of her mother appeared to him in a dream that was so realistic that it seemed as if he were awake. "Dreams are toys," he says, "Yet for this once, yea superstitiously, / I will be squared by this" (3.3.38–3.3.40). Somehow, he realizes that he must not discount this dream. The spirit asked him to call the child "Perdita," meaning "she who is lost" (3.3.32n), and he further prophesizes that he will never see his wife Paulina again.

Antigonus is sure that Hermione "hath suffered death" (3.3.41) and that the babe is the issue of Polixenes. Consequently, he believes that Apollo would sanction laying it "Either for life or death, upon the earth / Of its right father" (3.3.44–3.3.45). Antigonus' position contrasts sharply with that of his fellow Lord Camillo, who chose to flee rather than commit murder. Antigonus ignores the clamor of the weather and lays down the child. The claps of thunder are not sounds of heavenly applause but of Jovian censure, and they are especially ominous as they precede the roar of a bear. The animal lopes onto the stage and the script directs that Antigonus must "Exit pursued by a bear" (3.3.57SD1). This famous stage direction calls for a short digression.

Ursus

Nicholai Reymers Bär (Baer) (*c.*1550–1600), known as Ursus (Latin for "bear"), was a native of Ditmarschen in Holstein in northern Germany, which (with Schleswig) became a state with close ties to Denmark. His family was impoverished, but through ability and industry, he acquired proficiency in Latin, mathematics, and surveying. In September of 1584, under the aegis of another Danish nobleman, he paid a visit to Tycho at his island observatory on Hven, but Tycho treated him shabbily and deliberately excluded him from meetings that discussed his World model. Others on site taunted and harassed him. Writing later from exile in Prague, Tycho called Ursus "that bear-like and Dithmarsian brute," but Tycho may have wished that he had "let the *bear* alone" (Dreyer 185, 288; italics original).

Against that, Ursus behaved generally in an uncouth manner, as befitted his nickname, because in English from at least 1390, a "bear" was a "rough unmannerly or uncouth person" (*OED*). He snooped about in Tycho's library and read papers that he found lying around. One of Tycho's assistants discovered items in his possession that in modern times would qualify as intellectual property. Eventually, Tycho expelled him from Hven, but as Thoren puts it, "his precautions were in vain" (255).

In 1587 Tycho's *Liber Secundus* was nearing completion, and when the book appeared early in 1588, it included a section on his World model. In that same year, Ursus published *Fundamentum astronomicum* ("Fundamental Astronomy") which contains a system of the World in which, unlike Tycho, he assumed that the Earth rotated and that the sphere of stars was stationary. By modern standards, this was an admirable departure as it indicates that Ursus had at least overcome the objection that the Earth was too inert to move, a view to which Tycho's chief disciple, Longomontanus, later subscribed (Thoren 255n47).

Another departure in Ursus' book was the supposition that Mars' orbit wholly encloses the orbit of the Sun, where in 1588—when *Liber Secundus* appeared—Tycho believed that the orbits of Mars and the Sun should intersect (figure 1.6) (Thoren 253–256). Tycho's model dates to 1583 and was in use at Hven in 1584 when Ursus visited, at which time Tycho still believed "by a mistake" that the orbits of Mars and Sun did not cross (Dreyer 185). After seeing Ursus' book, Tycho leapt to the conclusion that the inquisitive, acquisitive visitor had stumbled upon the earlier diagram while on Tycho's island, and copied it, but the evidence for this is circumstantial and "slight." Ursus claimed that he developed the idea independently a year or two after his sojourn on Hven, and the quarrel set the stage for a dispute that lasted until 1601, when Death terminated it.

Shakespeare, writing several years after Tycho's death and fully cognizant of the latest World theories, probably saw this saga as much ado about trifles because from a purely geometrical standpoint Tycho's model was a simple inversion of the Copernican (Thoren 251–253). If anything, Shakespeare would have favored Ursus' model because at least Ursus had the sense to resurrect the Pythagorean notions of a rotating Earth and the possibility of stars lying at different distances and even possibly stretching to infinity (Johnson and Larkey 97n2). Tycho's outright dismissal of even the possibility of a spinning Earth and his failure to profess an attempt to measure stellar parallax (which would show that the Earth revolved; see figures 1.3a, 1.3b) showed that despite his claims to the contrary, he was still fundamentally an Aristotelian. To Shakespeare, the backwardness of Tycho's thinking and the Dane's haughty attitude to social inferiors would be a sufficient reason to invent a vengeful bear.

GRACELESSNESS

The powerful Lord of Uraniborg might have avoided the dispute with Ursus had he not ostracized him to begin with, but Tycho may have been incapable of gracious behavior, as evidenced by numerous other altercations throughout his life. For example, Paul Wittich (*c.*1546–1587) was Tycho's most capable mathematics assistant, but after he had left Hven, he related advances that Tycho had made in instrumentation. Tycho believed that Wittich tried to rob him of fame and mounted a campaign against him. Later in 1588, Tycho asserted that he would have written more kindly about Wittich had he known he was dead (Dreyer 121).

Another instance arose from a visit to Hven by Duncan Liddell (1561–1617). It is not certain whether Tycho mentioned his system to him, but the visitor nonetheless learned enough about it that he went to great lengths through correspondence to confirm his understanding of it. Liddell was the first to lecture on all three planetary system models that were known to him (the Ptolemaic, Copernican, and Tychonic models). His teacher, Wittich, remained true to geocentricism, but Liddell's pedagogy resembled Wittich's, which "juggled circles just for the sport of considering alternative possibilities" (Thoren 456–457).

Tycho complained of scholars who borrow without attribution and worried that Liddell was not crediting him. Even after one of Liddell's former students showed Tycho the notes he had taken of Liddell's lectures that completely exonerated him, Tycho remained unconvinced and instigated a private inquiry. This not only verified the student's account, but also showed that Liddell had expressed disapproval of what he considered to be Ursus' impudence toward Tycho. Notwithstanding the strength of contrary evidence, Tycho still found something to complain about. He discovered that Liddell had not referred to the system as "Tychonic" but merely as one that Tycho "mentions in his book" (Thoren 456), which he felt left open the possibility that Liddell was simply referring to a preexisting model instead of the one that he considered the child of his own invention. To make matters worse, Tycho heard by word of mouth that Liddell took credit for discovering the system, which however Liddell vehemently denied (Dreyer 184n3), but the net result was that Tycho planned to sue Liddell as soon as he had settled his score with Ursus.

Tycho mistreated his underlings as well. Hans Crol (?–1591) worked on Hven for about six years as Tycho's chief instrument maker, during which time he built all of the now-famous instruments of his observatory. Crol had sharp vision and could measure angular sizes accurately. When the Wittenberg graduate Georg Ludwig Frobenius (1566–1645) visited Hven in 1591, Crol complained to him that he had received no compensation other than board, lodging, and clothes, and that he had no means to get off the island. Frobenius also became disillusioned with conditions and hatched a successful escape plan (Christianson 270, 280).

In granting Tycho use of Hven, King Frederick decreed that Tycho not injure any tenant "against the law or by any new impost or other unusual tax" (Dreyer 86–87), but Tycho made ever-increasing demands on them. According to Thoren, "class ethics had undoubtedly conditioned Tycho to expect as his due the right to harvest the fruits of the labors of his social inferiors" (293). The serfs complained to Frederick, who felt obliged to appoint a commission to investigate. Despite a reprimand from Frederick, the abuses continued until 1597, when the new 20-year-old King Christian IV—now old enough to serve as King after the death of his father in 1588—started to take matters into his own hands. Tycho wrote a subservient and self-serving letter to him, but Christian's reply left no doubt about his low opinion of Tycho and the fact that he had abused the vassals of Hven (Dreyer 243–245). Soon thereafter, Tycho composed an elegy to Denmark, whose opening line asks disingenuously, "what is my offense?" (Christianson 216). The noble Dane complained bitterly about those who did "all the harm they could" to him (Dreyer 254). He claimed that it was the peasants who abused him—through their "contrary and disobedient" behavior—despite his being "more kind to them than they deserve[d]" (Dreyer 267). Nevertheless, Tycho could see the writing on the wall and thought it best to leave Denmark. The irony is that when Tycho cast the firstborn Prince's horoscope in 1577, he did not foresee that he and the future King Christian IV would later have a falling out. Tycho left Hven in 1597 and arrived in Prague in 1599, where Ursus had in the meantime acquired a professorship in mathematics.

EXILE

In 1597, having read Tycho's choice derogations of him, Ursus published *De Astronomicis Hypothesibus* ("On Astronomical Hypotheses"), in which he vented his spleen against his old nemesis. On hearing of Tycho's plans to come to Prague, Ursus beat a hasty retreat, and sure enough, Tycho instigated legal proceedings against him. In the summer of 1600, Ursus became seriously ill, but Tycho "persisted in having the poor wretch punished" (Dreyer 304). He arranged for a commission to try Ursus for libel, but late in the summer before the trial opened, Tycho announced that Death had "struck that wild beast...and saved him from a thoroughly deserved punishment" (Thoren 454). Tycho believed that if Ursus had lived, the court would have found him guilty, and he expected that it would have exacted a penalty of quartering or beheading in accordance with Bohemian Law. Tycho's patron, Rudolph II, ordered all copies of Ursus' book burned, whereupon the noble Dane prepared to write a refutation of Ursus' claims. However, Tycho died before completing it,[3] and he was unable to prosecute his vendetta against Liddell.

Evening

Returning to the *WT* script, no sooner have predator and prey left the scene than a Shepherd enters, followed a short time later by his son, the Clown, who reports that he had seen a bear tear out a man's shoulder bone. The victim called for help and cried that he was a nobleman, and the consecution of events implies that Antigonus is the delectable comestible in question. He also cried out that "the bear mocked him" (3.3.91), injuring the noble's ego as well.

Tycho had called Ursus a wild beast and had mocked him, and Shakespeare uses *WT* to even the score. He is not content to leave the Dane to the mercy of the Furies, so he lets Ursus be the predator and Tycho the prey. When the bear tore out the victim's shoulder, the beast could well have been devouring its prey by quarters, just as Tycho had relished the prospect of quartering Ursus. Tycho had once remarked in his presence that Germans "are all half-cracked" (Rosen, as cited by Christianson 89), and the audience learns that by the time the Clown arrives on the scene the animal had "half dined on the gentleman" (3.2.95–3.2.96), implying the bear had consumed the second of two bony quarters. The fact that the bear targets Antigonus and not the tender morsel Perdita shows that Jove is answering Paulina's prayer.

Coincidentally, the title page of Ursus' book, *De Astronomicis Hypothesibus*, quotes the beginning of *Hosea* 13:8, "I will meet them as a bear bereaved of her whelps" (Dreyer 273). Shakespeare's bear behaves as a she-bear separated from her whelp, which is to say, ferociously. When Tycho threatened Ursus' brainchild—that hybrid model of the Universe that he considered the whelp of his inventiveness—Ursus too became ferocious. The rest of the verse reads, "and will rend the caul [covering] of their heart, and there I will devour them like a lion: the wild beast shall tear them," and in line 3.3.51, Antigonus bemoans his lot and cries, "my heart bleeds, and most accursed am I." In *Hamlet,* Shakespeare dispatches Tycho's model by killing Rosencrantz and Guildenstern, and in *WT,* he dispatches Tycho himself. Moreover, he does so with finality, for as he flees the stage, Antigonus proclaims, "I am gone forever!" (3.3.57)

Shakespeare never shows hostility to genuine learning or spiritual life and is wont to praise the higher virtues of forgiveness, compassion, humility, and reverence; but in casting Tycho in an unfavorable light, does he ignore the dictum to speak only well of the dead? Normally, the aphorism applies only to the recently deceased, as Tycho was during the times in which *WT* and *Hamlet* were composed. Yet at the same time, Shakespeare knew that he could comply with the dictum because, in the short run, few if any would understand the extent of his natural-philosophical subtext. He could then discourse with impunity on contemporary facts that, by the time understanding dawned, would be historical.

STAGE AGE

In his dialogue with Leontes in scene 2.1, Antigonus inserts the ages of his daughters seemingly with little bearing on the matter at hand, "I have three daughters: the eldest is eleven, / The second and the third, nine and some five" (2.1.144–2.1.145). There is no indication that Antigonus sired children out of wedlock or that he had an earlier wife, so the girls must be Paulina's too, yet nowhere does Paulina mention them. Her concern is with the well-being of another's daughter. This peculiarity casts doubt upon the nature of the couple's marital relationship, as does Antigonus' indifference to the prospect of a permanent separation from his wife.

The three girls likely serve some dramatic purpose, and may well relate to one of the principals. If Antigonus represents Tycho Brahe, then Tycho is a likely place to begin because—judging by his eight offspring—Tycho loves children every bit as much as Antigonus does.

Tycho's children, three boys and five girls, were born between 1573 and 1583. Of these, two boys and four girls survived past 1577 into adulthood, so the possibility that Antigonus' three girls relate to a particular three of Tycho's four surviving daughters is unsustainable. On the other hand, his sponsor King Frederick's offspring—three boys and four girls, born also between 1573 and 1583—survived into the seventeenth century, but again there is no correspondence to Antigonus' three girls.

The only numerical coincidence with Antigonus' three daughters is Frederick's three sons—Christian, Ulrich, and Johan (Hans)— who were born in 1577, 1579, and 1583. Adding the girls' ages of 11, 9, and 5 to these birth dates produces the same answer—1588 (table 6.3). If the girls are this old in Act 2, and because there is essentially no break in first three Acts of *WT*, one can presume that the corresponding epoch on stage is 1588. This is *entry* γ in table 6.2.

However, the correspondences of table 6.3 are gender disjunctive (3 girls versus 3 boys). Perhaps Shakespeare chose Antigonus' children to be daughters rather than sons lest the choice of sons be too obvious a reference to King Frederick's offspring, but if the mooted correspondence is to be strong, there should be an extra set of three reasons to neuter each of the gender disjunctions. Given the existence of a scientific subtext, and Tycho's identification with Antigonus, one might wonder whether *Antigonus' girls* represent *Tycho's brainchildren*. If so, the gender disjunction is beside the point, as brainchildren are gender free.

Table 6.3. Determination of the epoch on stage of Acts 1 to 3 of *The Winter's Tale*.

Frederick's Sons	Christian	Ulrich	Hans
Princes' birth year	1577	1579	1583
Tycho's horoscopes	1577	1579	1583
Antigonus' "daughters'" ages	11	9	5
epoch on stage	1588	1588	1588

The search settles on horoscopes that King Frederick II called upon Tycho to cast for his three sons, which he presented in the years of their birth (Dreyer 144–153; Thoren 216). In constructing the horoscopes, Tycho used principles that he had outlined in a series of orations in 1574. He devoted much effort to calculating planetary positions according to the Alfonsine (geocentric) and Prutenic (heliocentric) tables and used the best astronomical data available. These included some of his own, so the horoscopes were quasi-scientific and deserving of mention in a play with a scientific subtext.

In Prince Christian's horoscope, Tycho supplied data that covered 27 pages, and the remaining 46 pages he devoted to fortune telling (Thoren 119–122). Tycho presented the horoscope to the King early in July 1577. Tycho predicted that the comet of 1577 "had important implications for Danish national security" (Christianson 64), the irony being that Tycho did not predict Christian's later antipathy toward him. Two years later in 1579, Tycho worked out a horoscope for Ulrich and presented it as a handsome volume of 300 handwritten pages. In July 1583, Hans was born, and Tycho prepared a horoscope for him in the same manner. It had about the same length as the second one and was similarly bound, but this time Tycho seemed much less certain of the infant's fortune than he was of the first Prince's in 1577 (Dreyer 154).

Opera

If intellectual property can masquerade as brainchildren, one should examine other manuscripts that might pertain to the subtext. In late winter of 1577–1578, Tycho wrote a document for the King, but researchers discovered its existence only in the twentieth century (Thoren 129), so it probably is irrelevant to the present discussion. In the spring of 1578, Tycho wrote a preliminary treatise on the comet, which he presented to the King (Christianson 64). Its significance lies in the fact that 1578 marks Tycho's first bona fide scientific report after his *De Stella Nova* of 1573. Subsequently, Tycho collected, analyzed, and integrated his data with those of others and printed the synthesized data in January

1588 in *Liber Secundus*, to which he added at the last minute his hybrid World model.

Two more reasons for the significance of the year 1588 are that in writing *WT*, Shakespeare borrowed extensively from Robert Green's *Pandosto* of that year and, as uncovered in the following sections, he alludes to the defeat of the Armada. Of course, many other important events occurred in 1588, but I can find none that bears systematically on the subplot other than that this was the year that Ursus published *Fundamentum Astronomicum* (see table 6.4). As mentioned, this contains a description of a World model very similar to Tycho's, and the dissemination of the competing World models in the same year parallels the stage age and justifies the pursuit.

Drivel

The Shepherd delivers what on a first reading could pass as drivel.

> SHEPHERD I would there were no age between ten and three-and-twenty, or that youth would sleep out the rest; for there is nothing in the between but getting wenches with child, wronging the ancientry, stealing, fighting — hark you now! Would any but these boiled brains of nineteen and two-and-twenty hunt this weather? They have scared away two of my best sheep, which I fear the wolf would sooner find than the master. If anywhere I find them, 'tis by the seaside browsing ivy. (3.3.58–3.3.65)

A paraphrase is useful.

{A} The Shepherd wishes that "youth" did not exist between the age of 10 and 23, or at least,
{B} that "youth" would sleep through that passage of time because
{C} there are no activities in that interval except (i) impregnating women, (ii) wronging elders, (iii) stealing, and (iv) fighting.
{D} The Shepherd is wonderstruck ("hark you now!") that
{E} two boiled brains (maniacs, hotheads) of 19 and 22 would hunt in inclement weather.

{F} The two scared away two of the Shepherd's best sheep, which he fears the wolf would sooner find than he would.
{G} If the shepherd were to find the sheep, they would be by the seaside browsing ivy.

The bear (as Ursus) has just chased Antigonus (as Tycho) from the stage, so the passage probably refers to them.

INTEGERS

Concerning {A}, "age" can have the primary meaning of a "period of existence," but it can also mean "any particular length of life which naturally or conventionally qualifies for anything," or in short, "a period or stage of life" (*OED*). Thus, "no age between ten and three-and-twenty" need not refer to youths whose ages lie between 10 to 23, but to a portion or portions of life lasting that long. Thus, the audience is relieved of the burden of having prepubescent boys impregnating wenches, as {C} (i) might imply.

With reference to {A} and {E}, and the corresponding integer pairs [10, 23] and [19, 22], and in accordance with the meanings previously stated, let "10" represent the decade from 1578 to the stage age 1588. This covers Tycho's first bona fide scientific document—his preliminary account of the Comet of 1577—up to the dissemination of his *Liber Secundus*, and up to the printing of Ursus' first book, *Fundamentum Astronomicum* (table 6.4). The specification ignores Tycho's *De Nova Stella* of 1573, which ranks in quality and importance alongside Thomas Digges' work on the same topic, both of which Shakespeare dealt with at length in *Hamlet*, and both of which possess scientific content that makes them sufficiently worthy of respect.

Because 1588 is significant, the sum involving the second integer of the pair [10, 23] may be significant too, and indeed, adding 23 to the year 1578 results in 1601—the year of Tycho's death. The second pair [19, 22] yields parallel results for Ursus; the sum of 1578 and 19 producing the year 1597, when Ursus published *De Astronomicis Hypothesibus*, and likewise adding 1578 and 22 yields 1600, the year he died (table 6.4). Therefore,

Table 6.4. Role of integer pairs [10, 23] and [19, 22] for basis year 1578.

1578 +	Year	Tycho	Ursus
10 – –	1588	*Liber Secundus*	*Fundamentum Astronomicum*
– – 19	1597		*De astronomicis hypothesibus*
– – 22	1600		† dies
23 – –	1601	† dies	

one can obtain an aesthetically pleasing correspondence between each principal and each part of each pair, [10, 23] and [19, 22].

The results indicate that the Shepherd specifies Ursus' second brainchild, *De Astronomicis Hypothesibus*, but not Tycho's second brainchild, *Epistolarum Astronomicarum* ("Astronomical Letters") of 1596 because the former is a polemic against Tycho and is especially germane to the run in on stage. Shakespeare will account for the latter in another way, as will be discussed. Note also that at the stage time of 1588, the Shepherd cannot know of the future events listed in table 6.4 unless, as posited, he is an unwitting instrument of a transcendent power that is all knowing and for whom violations of causality are everyday occurrences.

Returning to {A}, the Shepherd wishes that "youth" had not existed between the ages of 10 and 23. From table 6.4, this corresponds to the interval 1588 to 1601. It will soon materialize that from 1588 to 1601, the Shepherd ages from his late 60s to about 80, compared to which Tycho's age goes from 42 to 55. Thus, the Shepherd may justifiably regard Tycho as comparatively young, and because "youth" could be plural, the same goes for Ursus, who is about 4 years older than Tycho.

Concerning {B}, "that youth would sleep out the rest," applies to the remainder of their lives after 1588. For example, the Shepherd wishes that Tycho had not published *Liber Secundus* of 1588 as well as later works that include *Epistolarum Astronomicarum* and the handsome *Astronomiae Instaurata Mechanica* ("Mechanics of the Instauration of Astronomy") of 1598.

{C} (i), "getting wenches with child," refers again to the brainchildren identified in table 6.4.[4] {C} (ii), "wronging the ancientry," could cover a multitude of sins, of which the most inglorious in this context are Tycho's outright dismissal of the Copernican model and his myopic adherence to visual sightings to the neglect of contemporary advances in optics. In addition, the possibility that it refers back to {C} (i) should not be dismissed. The third part, {C} (iii), "stealing," might well refer to Ursus' World model and its resemblance to Tycho's. Although historians do not attribute the skill of stealing to Tycho, he systematically undervalued, and thus purloined, the sweat of others' brows, and Thoren wonders whether others' data on the New Star played a role in Tycho's data.[5] In {C} (iv), "fighting" applies to both principals, who were notoriously disputatious.

From {D} and {E}, the Shepherd marvels that the boiled brains of 19 and 22 would hunt under present conditions. From table 6.4, the integers 19 and 22 pertain to Ursus the bear, whose pursuit of Antigonus (3.3.57) occurs right after the "storm begins" (3.3.48). Apollo's representative, the Shepherd, marvels that pursuer and prey can allow petty rivalry and revenge to dominate their existence without respect for mighty nature and the thunder of godly opinion.

{F} tells that the two hotheads scared away two of the Shepherd's best sheep. This refers to two players who, in the real world of Tycho's circle, upheld at least some principles of the New Philosophy. Wittich is an unlikely candidate because he was primarily a mathematician whose models were geocentric, and other visitors to Hven are unlikely for the same reason. Two others stand out. As already discussed, Duncan Liddell promoted dialogue by teaching simultaneously all three of the World models known to him, and he presented theories and weighed evidence objectively in accordance with the methodological principles of the New Philosophy, but Tycho targeted him, and as a result, he was "lost" to the cause of heliocentricism. The other prospect is Christopher Rothmann (*fl.* 1580), who visited Hven in 1590 at which time Tycho persuaded him never to publish anything in support of Copernican theory (Christianson 349). In these ways, like sheep straying from a flock, Liddell and Rothmann were lost to Apollo's fold. In {G}, the lost sheep browsing ivy may refer to *Pandosto*'s "sea ivy," which sheep are known to eat (3.3.65n), where, as posited, the two "sheep" are Liddell and Rothmann.

Bairn

The Shepherd spies the little Princess and exclaims, "Mercy on's, a barne, a very pretty barne!" (3.3.66) In the Scottish dialect, "barne" means "bairn," and indeed, Perdita is *a wee bairn under the vast lift* (Merriam-Webster). Repetition of "barne" stresses its special relationship to the Shepherd (3.3.67n) and is another play on "bear." Within a few lines, the Shepherd expresses both empathy and possessiveness as he remarks, "they were warmer that [be]got this than the poor thing is here. I'll take it up for pity" (3.3.71–3.3.72).

The irony is that events have fulfilled Leontes' wish that chance should decree the fate of the child, except that the outcome is not quite what Leontes expected. The audience shall soon ponder the soundness of *1 Corinthians* 1:27 that "God hath chosen the weak things of the world to confound the things that are mighty," for the child's astonishing hardiness in the face of adversity signify the blessedness of her nature. Among interpretations of *WT*, there is the summary suggestion that "Perdita…*is* the play" (Bloom 1).

The Shepherd's son, Clown, arrives and reports that he witnessed a ship swallowed by the frothy main, the ship being presumably the one that brought Antigonus and the babe to Bohemia. It is revealed later that the sinking occurs in sight of "the" shepherd (5.2.60), suggesting that either the Clown or his father is the witness. In addition, when it sinks it takes with it "the instruments which aided to expose the child" (5.2.60–5.2.61). Tycho brought "instruments" with him to Bohemia (Dreyer 279, 285), but optical science and technology has outmoded them, and Shakespeare stores them figuratively in Davy Jones' locker. Perhaps Tycho's 1598 book on instruments, *Astronomiae Instaurata Mechanica*, receives the same fate. The two "lost sheep" of {G} would enjoy the spectacle of a ship of the old scientific state foundering in a welter of godly censure.

HALLOO

Returning to {D} of the paraphrase, in the midst of his speech, and for no apparent reason, the Shepherd interjects "hark you now!" (3.3.61). Perhaps he heard Perdita crying, but it is more likely that he heard his son calling him from afar because at the end of his speech he says that his son "hallooed" him (3.3.73). The Shepherd's response is noteworthy. Perhaps instead of "hark you now" he meant to say, "hawk you now" because he responds with a hawker's cry—"Whoa-ho-hoa!"—and the Clown responds in kind, "Hilloa, loa!" (3.3.73–3.3.74). The interchange brings to mind hawkish references encountered in *Love's Labour's Lost*, *Hamlet*, *Cymbeline*, and *The Merchant of Venice*, and focuses attention on the telescope designed by that falcon of philosophy, Leonard Digges. In *WT*, the spirit of Leonard resides in Bohemia, and it forebodes that the topic of optical resolution will soon emerge fully fledged (chapter 7).

Hawk-eyed vision enables the Shepherd to spy Perdita, and right before he and his son hail one another, he speculates on conditions that might have brought the newborn to her current state. "This has been some stair-work, some trunk-work, some behind-door work," he alleges. Taken at face value, this implies illicit impregnation (3.3.70–3.3.71n), but one might argue equally here as twice before, that "trunk" refers to Leonard Digges' perspective glass. "Behind-door work" may then refer to the secretive research and development that went into the design and construction of the optical trunk. "Stair-work" may refer to steps necessary for a celestial observer to ascend Plato's ladder of understanding, or if the trunk is a long-focus reflector as Gainer has already foreseen, it refers to the steps necessary to access the prime focus.

Support for the proposition that Digges used telescopes of long focal length comes from the ladders ("scalae") in the title *Alae seu Scalae Mathematicae* ("Mathematical Wings or Ladders"). "Wings or ladders" may refer to a high-resolution telescope designed by lanneresque Leonard, which would have a prime focus accessible by ladders, or (so to speak) by that particular bird's wings. Johnston notes Thomas' desire to fly on wings or climb on ladders to attain a height not achieved by others.

Shakespeare fashions the Shepherd in the guise of Leonard Digges, in accordance with the description of Ovid, "Apollo, tricked out as a rustic, / now dressed in feathers, now a lion's skin / or as a shepherd…" (4: 175–177). Leonards have "feathers," "lion" is "leo" as in Leonard, and the Shepherd is a rustic endowed with Apollonian qualities. The textual evidence suggests that the topic of optics is the chief subject of the subtext and that Leonard Digges is the prime focus. Act 3 ends as the Shepherd tells his son, "'Tis a lucky day, boy, and we'll do good deeds on't" (3.3.120).

Chorus

Act 4 of *WT* opens with the chorus Time apologizing for advancing time by 16 years. One wonders why the leap in years has this value and not some other, just as Johannes Kepler searched zealously for reasons why things in nature are "the way that they were and not otherwise" (Berry 181).

The chorus speaks as if it were both the playwright and an abstraction as it begs forgiveness for not examining events in the interim, saying merely that they were either "good or bad" (4.1.2). It speaks of Bohemia as "fair" (4.1.21), which is a more complimentary description than the ones encountered previously, wherein Bohemia is pictured as relatively deficient (1.1.11, 2.3.175, 3.3.2). "Fair" may even signify a close approximation to perfection, and it is not surprising that such an upgrade has occurred after the arrival of Perdita.

The chorus states that Polixenes' son is named Florizel, and that Perdita has "grown in grace" to "a degree demanding admiration" (4.1.25n). It explains that Perdita now serves as the Shepherd's daughter who, however, remains ignorant of her true identity, and it concludes by stating that the rest of *WT* concerns her fate.

> CHORUS A shepherd's daughter
> And what to her adheres, which follows after,
> Is th'argument of Time. (4.1.27–4.1.29)

To understand Time's argument, it helps to know the ages of the characters in the present stage epoch, which is now 1604 (*entry* ε, table 6.2).

According to 1.1.18–21, it is reasonable to assume that Leontes and Polixenes are effectively the same age ("trained together in their childhoods"), and lines 5.1.117–5.1.118 show that there is "not a full month" between the births of their sons Mamillius and Florizel. In 1604 Leontes tells Florizel,

> LEONTES Were I but twenty-one,
> Your father's image is so hit in you,
> His very air, that I should call you brother,
> As I did him… (5.1.125–5.1.128)

This means that Florizel "now reminds Leontes of himself and Polixenes when *they* were twenty-one" (italics added), which suggests that Florizel is now 21 (*entry* ζ). Therefore, "the passage of sixteen years…indicates that in the early scenes [Florizel and] Mamillius would have been five" (5.1.125n) (*entry* δ). In 1588 (Acts 1–3), Leontes is 23 years older than Mamillius (*entries* α and γ), and because Leontes and Polixenes are the same age, both must be 28 years old (*entry* δ). Sixteen years later, in 1604, Leontes and Polixenes are 44 years old, and as Perdita was newborn in 1588, she is now 16 (*entry* ζ).

The Chorus speaks of unspecified "bad" events (4.1.2) that occurred during the 16-year gap, of which no doubt there were many. One would be Thomas' death in 1595. However, on stage in 1604, neither the Shepherd's alter ego Leonard Digges, whose death Thomas Digges reported 33 years earlier, nor Thomas himself, whom the Clown represents, was alive. How then can the Shepherd and his son seem alive on stage in 1604?

It is apparent that events occur in Act 3 in a supernatural milieu, and it does not stretch credulity to believe that in Acts 4 and 5, the Shepherd and the Clown have a ghostlike quality. In Act 5, it appears that Hermione, too, will seem both dead and alive.

EIGHTY-THREE

A test of plausibility is to see whether Shakespeare delivers an alternative to the cosmologies that he disparages, and a clue to that effect occurs at line 4.4.433 when the Shepherd tells Florizel, "You have undone a man of fourscore-three." In Acts 4 and 5, therefore, the Shepherd is 83 years old (table 6.5), so that within the accuracy of truncated ages, the Shepherd was born in 1521. This is a remarkable coincidence as most biographical sources state that Leonard Digges was born in about that year. This supports the prior identification of the Shepherd with Leonard, and explains why in 1588 (in Act 3), when the Shepherd is 67 years old, he complained about "youth," which refers to the relatively young ages (42 and 46) of Tycho and Ursus. Thus, Shakespeare's positive alternative to the negative connotations of table 6.3 is to associate Leonard Digges with the Apollonian qualities of a shepherd. In so doing, he brings to the fore what is, or should be, the source of progress in astronomy and cosmology in the sixteenth and the first decade of the seventeenth century, particularly optics and the development of the telescope. Further temporal coincidences indicate the manner of this emphasis.

Lineage

The two years 1588 and 1604 of table 6.5 coincide with two significant dates in the life of Johannes Kepler. He graduated Bachelor of Arts in 1588, and 16 years later in 1604, he published a book that historians regard as the foundation of modern optics (Caspar 142–146).[6] In other words, the 16-year leap announced by the Chorus spans the time from

Table 6.5. Principal dates in Act 4 of *The Winter's Tale.*

Year	L. Digges	Shepherd	Kepler	Perdita
1521	born	born	–	–
1571	*Pantometria*; dies †	–	born	–
1588		rescues Perdita	graduates BA	born
1604		is 83 years old	*Optica*	initiates a new age

Note. † as reported in *Pantometria.*

when Kepler first achieved professional status to when his seminal work on optics dawned upon the world (table 6.5).[7,8] This book's title—*Ad Vitellionem Paralipomina quibus Astronomiae Pars Optica Traditur*—loosely translated means, "Advances on Witelo's Work on Astronomical Optics," and is often abbreviated *Optica*. It builds on work from 1270 by Erazmus Witelo (*c.*1230–*c.*1275), whereas Leonard Digges traces his research lineage to Roger Bacon from virtually the same time, 1267. Table 6.5 shows further that Leonard's work on optics appeared in *Pantometria* in the same year as Thomas Digges reported the death of his father, in 1571, which is also when Kepler was born. Table 6.5 presents these data in such a way as to unite the Bacon-Digges and Witelo-Kepler research lineages; and just as Kepler initiated the dawn of optical research on the Continent, so Perdita will introduce the new age of science to Sicilia.

FLORIZEL

Scene 4.2 opens at the court of King Polixenes. Recall that sixteen years earlier, the Sicilian nobleman, Camillo, had rescued Polixenes from the clutches of Leontes and fled to Bohemia, but now he wishes to return and spend his last days in Sicilia. He says that it has been "fifteen years" since he has seen his fatherland, but critics agree that "fifteen" is an error (4.2.3n) because 5.3.31 confirms the value 16.

Polixenes begs Camillo to stay and help him deal with Florizel, who is spending time away from court visiting a certain shepherd's daughter. Polixenes decides that they will pay a visit to the shepherd, who turns out to be the selfsame Shepherd that rescued Perdita. At age 16, Perdita is "the angle that plucks [Florizel] thither" (4.2.37–4.2.38), and in the surreal setting of Act 4, "angle" could as well be "angel."

The Bohemian King says that the shepherd built himself up "from very nothing" and has "grown into an unspeakable estate" (4.2.32–4.2.33). Apparently, this refers to the Shepherd's remarkable professional success (4.2.33n), although from 1568, a rare usage of "unspeakable" occurs in *2 Corinthians* 12:4, "unspeakable words...not lawful to utter" (*OED*).

The identification of the Shepherd with Leonard Digges suggests that Shakespeare regards Digges' estate as having been rebuilt "from very nothing" following Digges' conviction and pardon for participating in Wyatt's rebellion, and even as late as 1604, his estate probably still held information not meant for public consumption.

Autolycus

Scene 4.3 opens with a ne'er-do-well, Autolycus, singing a song of seasons stretching from winter's end to summertime. Antigonus had warned of wolves and bears, and Shakespeare—having introduced a bear in the previous scene—now introduces a character whose name literally means "the wolf himself" (81). Autolycus allows as how he was "littered under Mercury" (4.3.24–4.3.25), or under the aegis of a Roman god whose Greek counterpart Hermes is the messenger of the Olympian pantheon. Mercury was cunning, fleet of foot, and his propensity to lie, steal, and cheat aided his survival. These properties also qualified Mercury to serve as the god of science (as it was regarded in days of yore), and indeed the mercurial properties of Autolycus enable him to abet its progress. Autolycus is an accomplished pickpocket, and the Clown is *en route* to purchase provisions for an upcoming sheep-shearing feast. The audience fears that Autolycus is a wolf in sheep's clothing, and their worries are realized when Clown enters and the swindler exclaims, "A prize, a prize!" (4.3.30) Autolycus feigns misfortune and separates the Clown from his cash.

Sheep Shearing

Next to scene 5.2 of *Love's Labour's Lost*, scene 4.4 of *WT* is the longest in the Canon. In both cases, metaphysical connections exist to Saturn and to Time.

Preparations for the sheep-shearing party are underway. Florizel is present and Perdita plays hostess. The name "Florizel" has a nominal relation to Flora—the Roman goddess of flowers—and springtime

(4.4.2n2), who is "Peering in April's front" (4.4.3), which means that it is either April or the guise of April (4.4.3n; Dolan 4.4.3n).

Even though spring may officially have begun, up north in Bohemia, the weather can turn chilly. Astronomy helps pin down the time. From the opening scene (1.1.5–1.1.6), it appears that summer is approaching, and in scene 1.2, Hermione and Polixenes offer to stroll outdoors, so it is then comfortably warm in Sicilia. Mamillius' offer to tell a tale in scene 2.1 suggests that the season of spring is a possibility, and in scene 3.1, Cleomenes and Dion describe weather that it spring-like. It is also known that in scene 2.3, Paulina appealed to Jupiter, and Leontes' name probably refers inter alia to the zodiacal constellation Leo—the Lion (80). The brightest star in Leo is Regulus, so named by Copernicus as a diminutive form of Rex the King (Allen 225). The year on stage is 1588, and on about May 31 of that year, Jupiter crosses the meridian at about 16:30 hours (4:30 p.m.), followed by Regulus about 45 minutes later. Jupiter is about to enter Leo, just as Leontes will soon feel the influence of the Jovian deity. Thus, a time in late spring (the guise of April) appears consistent with the storyline and with Paulina's appeals to Leontes and Jupiter.

The association is ominous because Regulus anchors the handle of the Sickle asterism, and at about 17:00 hours (5 p.m.), this configuration stands erect above the southern horizon as if poised to strike a southern target. Recall that Leontes and Polixenes were trained to grow in a particular direction like plants and that their mutual affection is "rooted" in childhood from which as adults they "branched" out. Context dictates that Leontes is the target of the reaping, but who shall wield the blade? It is worth noting is that in classical times—as in England in 1552 (Allen 256)—Regulus was called the Lion's Heart (the heart of the constellation Leo), or poetically, "Leon[h]ard." Shall Apollo and his retinue cut Leontes down to size?

Florizel tells Perdita that he blesses the time when his "good falcon made her flight" across her father's ground, and Perdita responds, "Now Jove afford you cause!" (4.4.15–4.4.16) In *Cymbeline*, the Bird of Jupiter soars, and a bird of prey soars in *WT* as well. In both cases, the allusion is to the new order emerging from the efforts of Leonard Digges in the guise of a leonard falcon.

Florizel disguises himself as a "poor humble swain" and likens himself to "Golden Apollo" (4.4.30). He reassures Perdita, saying that he is "most constant" to his purpose (4.4.45). Like a turtledove that never leaves its mate, he pledges himself to her (4.4.154–4.4.155n) and plans to take her to Sicilia. Perdita worries that Polixenes might put a stop to the relationship, but for reasons known only to the gods, Florizel is unfazed.

The Shepherd and his entourage enter, as do Polixenes and Camillo, both in disguise. Perdita welcomes the strangers with gifts of flowers chosen in accordance with traditional symbolism. She begins with rosemary and rue, which she says keep all winter long. To which Polixenes comments, "well you fit our ages / With flowers of winter" (4.4.78–4.4.79), implying that she gives them old flowers because they are getting on in years. Perdita declares diplomatically that it is not the two Kings who are growing old, but the year, which however is not yet "on summer's death nor on the birth / Of trembling winter" (4.4.80–4.4.81).

She has plenty of flowers "Of middle summer" which she says "are given / To men of middle age" (4.4.107–4.4.108). It is calculated from *entry ζ* of table 6.2 that Polixenes is only 44 years old, which even by seventeenth-century standards is not quite "old." The date on stage is probably late July (4.3.35–4.3.46n), and although sheep shearing occurs generally in the spring, in colder climes as might obtain in Bohemia, the wool harvest should not occur too early for fear of cold snaps. According to the *The Book of Days*, "sheepshearing commences" in England in July, followed by "sheep-shearing feasts," which are "of ancient origin" (Chambers 2). The hypothesis that it is summertime receives support from Autolycus, who refers to the passage of time from winter to summer in his opening song. In addition, Florizel is a guest of the Shepherd, and owing to Florizel's youth (a mere 21 years old; *entry ζ*), he would be a deserving recipient of flowers that bloom in the early part of the Roman year (springtime), but Perdita apologizes for not having any such flowers.

Perdita apologizes too for lacking "carnations and streaked gillyvors," which she calls "nature's bastards" (4.4.82–4.4.83) and prefers not to cultivate. Polixenes takes her to task for her bias against cultivated hybrids because they are as much part of nature as beauty that occurs naturally. "The art itself is nature," he says (4.4.97). Florizel is romancing a lass who he believes is a shepherdess, which indicates that the prospect of mixing royal and commoner blood is no less wonderful than if couples were to wed within their social class.

The dialogue introduces the contemporary nature-versus-art debate, which will soon resurface as players and the audience struggle to understand Hermione's hibernation. The imagery brings to mind a Greek myth in which Pygmalion falls in love with Aphrodite, who spurns him, so he sculpts a statue and wishes that it would come alive. Aphrodite, the Greek goddess of love and generation (whom the Romans called Venus), grants his wish and brings the sculpture to life.

Perdita has offered the visitors flowers of winter and summer, but not of spring or autumn, whereupon Shakespeare invokes the Roman deity Prosperina and her husband, Pluto. In Greek mythology, Persephone (mispronounced "Prosperina" by the Romans) is the daughter of Demeter ("Ceres" to the Romans). She is the goddess of agriculture who spends winter with her abductor, Hades (Dis or Pluto), the god of the underworld, during which times the fields of Earth lie fallow, whereas in summer she rises to ensure a bountiful harvest. The imagery is one of passing seasons, "underscoring time as cyclic" (4.4.79n). Celestial cycles, like solar days, lunations, and the tropical year, define common units of time on Earth. These phenomena are oscillatory like the movement of a clock, and in the present case, the floral conceits refer directly to the tropical year with its cycle of birth and death. The concept of cyclic regeneration is ingrained in *WT* and relates directly to the Roman god of the harvest, who doubles as the god of Time and whose cyclic nature is inherent in the planet Saturn itself (table 4.3).

Floral and other clues pin down the epoch of the festival as middle to late summer of 1604. This comports with the fact that in January of 1604, Kepler presented Rudolph II with the completed manuscript of

Optica, which appeared in print in Frankfurt in time for the great book fair held there in the autumn (Voelkel 62). Specifically, Kepler dated his dedication for August 1, 1604, which is practically midsummer. Astronomy and mythology confirm the dating. Leontes is associated with Regulus (the "little king" in Leo), which in turn is associated with the Sun because in antiquity, the Sun lay in that direction at the peak of summer's heat. In the middle of August of 1604, Regulus and the Sun lie virtually in the same direction, implying that the Sun (which is the true symbol of royalty) will bestow bona fides upon the formerly backward Sicilian King.

Perdita (as the Shepherd's daughter) impresses Polixenes, but when in the midst of the festivities she and Florizel prepare to wed, Polixenes (still in disguise) asks the groom purposefully, "Have you a father?" (4.4.372). Florizel does not recognize his father and asserts that he will make his own decisions. Polixenes presses him to come clean, but Florizel refuses for reasons that he believes are "not fit for you to know" (4.4.392). Polixenes then whips off his disguise and threatens to disown Florizel and to torture Perdita to death. The irony is that Polixenes initiated the process of bringing change to Sicilia, and now that his own son is undergoing change, he acts unwittingly against it. A battle of wills ensues as Florizel vows never to break off his relationship with Perdita.

Florizel chooses to flee rather than return to his father's court, and once again, Camillo comes to the rescue. He advises Florizel to marry Perdita and take her to Sicilia in order to present her to Leontes on the pretext that Polixenes sent them to greet and comfort him. Camillo will send letters to Leontes that will precede their arrival, and feels sure that when Leontes replies, his letters will placate Polixenes and restore the eloping couple to the grace of the court. To protect his identity *en route*, Florizel swaps clothes with Autolycus, and Perdita disguises herself too. As the couple departs, Camillo reveals (out of their earshot) that he plans to tell Polixenes of their escape and of their destination in the hope that this will persuade him to follow. Autolycus abets the progress of Science by vowing to withhold from Polixenes news of the pair's escape, and the Shepherd and Clown resolve to tell all to Polixenes.

Autolycus acts as if he is a courtier and (for a price) offers to facilitate the shepherds' access to court.

SOLUTION

Act 5 opens in Sicilia with Leontes again seeking solace, this time from his counselors. He agrees never to remarry without the consent of Paulina, who anticipates that Hermione will be "again in breath" (5.1.83). A servant enters and announces the arrival of Perdita and Florizel. Leontes welcomes them, and Florizel delivers greetings from his father as planned. A courtier arrives and announces that Polixenes has arrived in town, having previously encountered the Shepherd and the Clown, and having named Florizel as a refugee from court. Florizel entreats Leontes to show understanding of the plight of young men in love, and Leontes—now mellowed and contrite—obliges him.

The next scene (5.2) is odd in the sense that the expected climax of Perdita reuniting with her father does not play out on stage. Instead, the audience learns of this much-anticipated event secondhand from three gentlemen whose roles seem contrived for the purpose. The construction suggests that the subtext carries the day. First Gentleman thinks he overheard the Shepherd describe the manner of baby Perdita's discovery, and Second Gentlemen announces the discovery of the King's daughter (5.2.20). Third Gentleman confirms the finding and adds that the news has affected even the most stoic onlooker. In fact, witnesses who were like marble "changed colour" (5.2.77). The allusion to the animation of a stony-faced individual anticipates events in the next and final scene. This helps explain why the father/daughter reunion is not front and center and is not the major resolution of the play.

Another peculiarity is that First Gentleman and Third Gentleman are nameless, whereas Second Gentleman has the name Rogero. The reason that Second Gentleman is so favored lies in the significance of his announcement that "the King's daughter is found." Recall that Perdita's role is to survive and bring the New Philosophy to Sicilia, specifically advancing the role of Kepler's *Optica* (table 6.5). As noted, Kepler's topical descent from Witelo is expressed in the full title of *Optica*, and although that work honors Witelo's work in the thirteenth century, in temporal sequence, Kepler follows Leonard Digges. Leonard, however, sees himself the successor to Roger Bacon, so Shakespeare selects "Rogero" in order to trace that lineage.

The dissemination of the theory and practice of optics would not be complete without the figurative instigation of the new art in Sicilia. For the past 16 years, Paulina has kept Hermione—who is in a magical condition akin to suspended animation—concealed at a remote location. She takes the form of a marble statue carved by a renowned sculptor named Giulio Romano. It is so lifelike as to seem capable of movement, but when Perdita tries to grasp its hand, Paulina warns that its color is not yet dry, which implies that the sculptor has only just completed his work. The sculptor probably takes his name from the sixteenth-century artist Giulio Romano (*c.*1492–1546), best known for his paintings (5.2.83n), and indeed, it seems that the finishing touches are his application of color.

Magically, Hermione has undergone a series of metamorphoses, and she is now in her penultimate state. Leontes expresses astonishment at the lifelike appearance of the statue, but Paulina cautions him not to gaze upon it, lest he think it moves. Paulina worries that onlookers will think that she commands "wicked powers" (5.3.91), so she invites skeptics to leave. No doubt, early telescopists felt the same way about detractors who refused to believe that the images they saw were representations of reality. She advises onlookers to awaken their faith, and with the help of music, she wills Hermione to life. In Shakespeare's ontology, it is difficult to know whether the reviviscence of Hermione is a case of life imitating art, or the reverse.

The Queen's first words are to thank the gods for saving her daughter.

> HERMIONE   You gods look down,
> And from your sacred vials pour your graces
> Upon my daughter's head!...
> Knowing by Paulina that the oracle
> Gave hope thou wast in being, have preserved
> Myself to see the issue. (5.3.121–5.3.128)

Higher powers have kept their promise. Their unexpected beneficence has restored her to grace, rescued Paulina, and advanced Apollo's agenda.

DATING WINTER'S TALE

After Tycho arrived in Prague in January of 1600, he sought Kepler's aid in working out details of his World model, but Kepler was more interested in Tycho's data on Mars than in promoting what he considered a false model. Mars was the ideal object to study because its orbital diameter is larger than Earth's but only by about 60%, making it favorably and frequently observable in the middle of the night around the time of Opposition (figure 1.5). Data on Mars are therefore comparatively plentiful, and its orbit provides the best challenge to orbital theory because it deviates from a circle more than any superior planet known at the time. Shakespeare knew this in 1591–1592 when he wrote in *Henry VI part 1*, "Mars his true moving, even as in the heavens, / So in the earth, to this day is not known" (Bevington 1.2.1–1.2.2).

Having appropriated Tycho's data on Mars, Kepler proceeded with analysis and published his findings in *Astronomia Nova* in July or August of 1609. Thomas Harriot corresponded with Kepler in the interval 1606–1609, and in the winter of 1609–1610 (a few months after its publication), he and a coterie of natural philosophers were studying Kepler's discoveries. Shakespeare would have known of the work too, and if he wrote *WT* after that date, he would have been in a position to include these two results. However as far as I can tell, *WT* contains no such reference. *WT* especially contains no reference to Kepler's *Dioptrice* (a word that Kepler coined), which appeared in 1611. This describes a design for a telescope using two convex lenses, which was a notable improvement on Galilei's design of 1609–1610 (Berry 183) and would surely have warranted comment if Shakespeare had had a mind to do so.

Customary dating puts the writing of *WT* at late 1610 or early 1611, before an account from May 15, 1611 by the medical astrologer Simon Forman (1552–1611). Recent opinion has tended toward an earlier date, perhaps 1609, and before the writing of *Cymbeline* (62–63). Absence of reference to *Astronomia Nova* and *Dioptrice* supports the trend but does not necessarily imply that Shakespeare wrote *WT* prior to those times, merely that he chose to terminate coverage in 1604. *WT* could still qualify as one of Shakespeare's late plays (*c.* 1609–1611), and the reason for the five-to-seven-year gap between 1604 and the late-play stage is that Shakespeare's chief concern is to document the important role that optics and telescopy played in establishing the new Worldview.

CHAPTER 7

THE RESOLUTION REVOLUTION

> The truly influential and pervasive aspects of modern science are not its facts at all, but rather its method of inquiry and its criterion of truth.
>
> —Stillman Drake

Chapters 2–6 establish that contemporary cosmology played a role in Shakespeare's worldview and raise the question of the genesis of the bard's recognition of telescopic development and achievements. His persistent allusions to birds of prey—notably the leonard—in the context of cosmic discovery suggest that a suitable place to begin a preliminary inquiry is with the life and times of Leonard Digges, the inventor of a two-element telescopic device that enabled humankind to see celestial objects with unparalleled resolution.

EARLY YEARS

Leonard Digges was born in about 1521, the second son of James Digges (d.1535) of Digges Court in Barham (near Canterbury) in Kent, England.[1]

The Diggeses came from a prominent Kentish family with roots dating to the reign of Richard I (1157–1199). Adomarus de Digge was a judge during the reign of Edward II (1284–1327), and other forbears held important positions during the reigns of Henry III (1207–1272), Edward III (1312–1377), Edward IV (1442–1483), Henry VII, and Henry VIII (1491–1547) (Everitt 138; Kippis 5: 238; Philipott 60–1; *DNB*).

Leonard may not have attended a university, and he certainly did not graduate from one. Instead, in 1537 (at the age of about 16), he matriculated to Lincoln's Inn in central London, the oldest of the four Inns of Court (Kippis 5: 238–239; *DNB*). He married Sarah Bridget Wilford (d.*c*.1550)[2] by whom he had his son, Thomas, in about 1546. The next year, he sold the property that he had inherited and purchased the manor of Wotton, about six miles from Canterbury on the road to Dover.

Like Isaac Newton in the seventeenth century, Leonard was both an able mathematician and a keen experimentalist. In Leonard's case, his personal wealth and status enabled him to engage in research privately. After 1550, most scientists distanced themselves from the chief academic centers of learning (Johnson 138) and developed their own culture free of interference, but in Leonard's case his pursuit of learning did not go smoothly. In 1553, when he was in his early thirties, Mary I ascended the throne and announced plans to marry the King of Spain, Philip II. As mentioned, Mary and Philip were religious zealots, and the prospect of such a union did not sit well in certain counties—notably Kent—that would likely bear the brunt of an invasion from the Continent.

In November of 1553, a cadre of nobles and gentry hatched a plot to march on London—ostensibly, as they later claimed, to oust Mary's councilors. Thomas Wyatt Jr. (d.1554), the son and namesake of the Kentish poet Thomas Wyatt Sr. (*c*.1503–1542), led the rebels and Leonard was part of the Kentish contingent. Forces from some counties failed to materialize, and in the ensuing confusion, the uprising faltered and failed. Leonard and nearly 500 others were captured along with the ringleader, who was convicted of treason and hanged. Leonard was convicted and sentenced to die on February 24, 1554, but an act of the Privy

Council reprieved him. Cool heads must have prevailed on that council because all told, only seventy-five rebels met their death.

On April 1, 1554, Leonard received a pardon along with many of his comrades in arms, and although history records the identity of many of the interceders, there is no such entry in Leonard's case. The Governor of the Tower of London, Edward Fiennes Lord Clinton (1512–1585), played a key role in suppressing Wyatt's rebellion and had a reputation for diplomacy and common sense. Some believe that he was the one who arranged the pardon (*DNB*); in fact, in his 1555 edition of *A Prognostication of Right Good Effect*, Leonard thanks Fiennes for preserving him during his "troubles."[3] On May 31, 1554, Leonard entered into an obligation of £49 17s 8d for the redemption of his movables, and on February 20, 1555, he entered a second recognizance of £266 13s 4d for the redemption of his lands, which he finished paying on May 7, 1558. This was the same year in which Elizabeth ascended the throne, and in "the fifth year of Elizabeth," Leonard reclaimed his original status, which enabled his descendants to inherit his estate (*DNB*).

In 1556 Leonard added new material to his almanac and renamed it *A Prognostication Euerlasting*. Thomas continued to publish it, having reported his father's death in 1571, and as noted, he added *A Perfit Description* in 1576 (at age 30). The penultimate edition appeared in 1596, the year after Thomas died, and the last edition was released in 1605, ten years after his death.

Also in 1556, Leonard published his *Tectonicon*, which dealt with "exact measuring, and speedie reckoning [of] all manner of Land, Squares, Timber, Stone…" (*DNB*). *Pantometria* appeared in 1571, and critics believe Leonard was the chief author (table 6.5). In it, Thomas describes his father's endeavors in optics and, as mentioned, announces his father's death. In describing the power of resolution of the perspective glass, he limits himself to terrestrial targets such as coins, letters, and the rooms of houses. There is no mention of celestial targets, including the stars that Leonard's icon, Roger Bacon, had remarked on.

In 1579 Thomas coauthored *Stratioticos*, also with his father. This book is noteworthy for a list of works that he planned to publish, including one mentioned previously and entitled:

> Commentaries vpon the Reuolutions of Copernicus, by euidente Demonstrations grounded vpon late Observations, to ratifye and confirm hys Theorikes and Hypothesis, wherein also Demonstratiuelie shall be discussed, whether it bee possible vpon the vulgare Thesis of the Earthes Stability, to delyer any true Theorike voyde of such irregular Motions, and other absurdities, as repugne the whole Princples of Philosophie naturall, and apparent groundes of common Reason. (Johnson 174)

In essence, the title proclaims that recent observations confirm the heliocentric hypothesis, but there is neither any evidence of the existence of this work nor any clue as to the nature of the observations. The work would have elaborated on *A Perfit Description*, which had appeared three years earlier and seems retrospectively to be a stopgap measure in the overall course of inquiry (Johnson and Larkey 95).

William Bourne's *A treatise on the properties and qualities of glasses* was commissioned by Burghley. Like Bourne's first book, this addresses various aspects of optical imagery and magnification, although the author divulges little that is new. The novel device would surely have deserved a full exposition, but Bourne refers back to Thomas Digges and his *parente altero*, John Dee, who he says know more than he does. This completes the circle of obfuscation.

Spyglasses had obvious military uses, and the lack of printed detail on optical devices no doubt stemmed from a need for a nominally Protestant England to guard its scientific and technical expertise against the Catholic theocracies that menaced her. By 1598, for example, Raffael Gualterotti (1548–1639) had constructed a spyglass for military use and not, as he disclaimed over a decade later, for the study of the stars. In England, the bloody-minded innovators from Kent would not have hesitated to aim their optical invention above the horizon at night, particularly if they had had Burghley's backing and did so out of sight of schoolmen. They epitomized the dictum *Nullius in Verba* ("at the dictate of no-one"), which is the bedrock of scientific objectivity and which would become the motto of the newly founded Royal Society in 1662.

Knowledge of stars would have had little or no military value, but it would affect national security indirectly as discoveries that flew in the face of scholastic certainty could only inflame anti-Elizabethan sentiment at home and abroad and strengthen the resolve of theocracies to stamp out heresy on the island nation. Queen Elizabeth lived under constant threat to her life, beginning when she became a rallying point for anti-Marian Protestants, and continuing in 1570 with the decree *Regnans in Excelsis*, in which Pope Pius V (1504–1572) excommunicated her. The bull was so ambiguous that some interpreted it as endorsing her assassination, and by 1584, Burghley had collaborated with Sir Francis Walsingham (1532–1590) to create a clandestine service to protect her and to act in the event of an emergency.

Despite the secrecy, someone with a poetical bent and a reverence for nature might be tempted to write of the new information in ways incomprehensible to unprepared minds. It has been argued in the preceding chapters that Shakespeare was such a poet. It is less surprising that, in the sixteenth and early seventeenth centuries, Shakespeare's inventory of interests included contemporary astronomy, than it is that before the nominal date of 1610—the year during which the world's first telescopic observations were recorded—he already had at his command most of the very data that such telescopy provided. Moreover, it seems likely that the telescope in question was the perspective glass invented by Leonard Digges.

Later Years

When the inventor of the perspective glass died, he did not go gently into the night. Table 7.1 shows a distribution of years of Leonard's purported death. Of the 20 values used, sources quote 15 as uncertain. Values are not all independent, but suffice to show the wide span of dates (17 years) that biographers have considered and the bi-modality of the sample. The early years (1558–1559) coincide with the time when Leonard had

TABLE 7.1. A sample of published years of death of Leonard Digges the Elder.

1558	1559	1560–1569	1570	1571	1572	1573	1574
1	7	0	1	5	0	4	2

completed payment for the redemption of his lands, and the second peak is compatible with the publication year of *Pantometria* (in 1571) when Thomas announced his father's death.

The later dates 1573–1574 in table 7.1 owe their genesis at least in part to a report in 1860 (Cockle) of an annotation in a copy of *Pantometria* in which the annotator gives 1574 as the year of Leonard's death. In November of 1884, in the journal *Notes and Queries*, Lynn ("Leonard Digges" 368) pointed out the incompatibility of the annotation with the accepted fact that *Pantometria* had appeared three years earlier, and he requests more information. This he himself supplies the next month (Lynn, "Leonard Digges" 515). He refers to Wood, who lists Leonard's three publications—*Prognostication*, *Pantometria*, and *Stratioticus*—and adds:

> What else he wrote I find not, nor certainly when he died…or [where he died]. There is some memory of him and his family… on a monumant in Chilham church in Kent, not to shew that he was buried there but to shew the geneology of his family…. (Lynn, "Leonard Digges" 515)

Lynn again appeals for more information, but no reader of *Notes and Queries* sheds further light,[4] so in 1894 he concluded that the death date of Leonard Digges "is not exactly known" (Lynn, "Thomas Digges" 186).

Leonard was well connected. Lord Clinton, the dedicatee of Leonard's almanac, was a diplomat *par excellence* who served four monarchs (Henry VIII, Edward VI, Mary, and Elizabeth) and, up to the time he died in 1585, never wavered in his friendship with Burghley. Burghley himself presided over the state and had a hand in designing or controlling all significant policies over many decades. He and his family exerted their power through a network of patronage and obligation.

Nicholas Bacon (1510–1579), the dedicatee of Leonard's *Pantometria*, became part of the reticulation when he became Burghley's brother-in-law by his second marriage in 1553. In addition, Robert Dudley (*c.*1532–1588), Earl of Leicester (the dedicatee of Leonard's nominally last work *Stratioticus*), was for a while Elizabeth's constant companion and suitor.[5] Thomas named his eldest son Dudley after him. Yet despite Leonard's contributions to military studies, his reputation as an architect and mathematician, and his invention of a device of value to England's defense, his survivors saw fit to inter him without so much as a grave marker.

Conjectures

One possibility, having the merit of simplicity, is that Leonard did not die as his son reported. The conjecture is worth entertaining, although of course one must not allow an untrammeled careening of conjecture to degenerate into a gloss. A theory must be self-consistent, and it must light the way to testable predictions, or at least to ones that have the prospect of such.

To begin with, one can well imagine why Leonard might want to fade into obscurity. In 1554 (at age about 33), he bore the burden of treason and faced death at the hands of a fanatical Queen, added to which was the prospect of leaving his family destitute. Even after his pardon, he faced the difficulty of reconstituting his life and estate beginning during the reign of the monarch that he had conspired to unseat. As indicated, he could not have accomplished this without the aid of individuals in power who had the health of the realm at heart. These included Lord Clinton, Burghley (after Elizabeth's coronation), and by inference, the Tudor Queen herself.

The early peak around 1558–1559 in the distribution of table 7.1 may signify the first occasion when society lost track of Leonard. He may or may not have abetted the process. His child, Thomas, was then barely a teenager, but in 1571, Thomas was 25 and mature enough to coauthor his father's work and to announce his father's death.

Leonard published *Prognostication*, *Pantometria*, and *Stratioticus* "lest it should be thought that he studied only for himself, and not for the benefit of others" (Wood 415), but if he remained productive after 1570–1574, what else did he produce? It took eight years after *Pantometria* for son (and father) to deliver *Stratioticus*, and if—as announced in *Pantometria*—Leonard had an interest in perspective glasses, one might wonder whether he had pursued that topic further in the interim. A further consideration is that Leonard may have contributed to the research that led to the treatise *Alae seu* ("Mathematical Wings or Ladders") in 1573, a piece ostensibly written by Thomas alone.

Concerning the final entry in table 7.1, it is possible that 1574 was the last time that the public caught wind of Leonard, and Lynn's sleuthing in the late nineteenth century would indicate that in the public's mind, Leonard successfully completed his vanishing act at that time. This suggests that Leonard used his son's name as a front for some or all of his scientific work.

Of interest in this regard is the work of Francis Galton (1822–1911), cousin of Charles Darwin (1809–1882), who discovered the phenomenon of regression to the mean, whereby gifted persons tend to have less gifted offspring, and although Thomas' career was certainly illustrious, he may well have played a second fiddle to his father. Johnston discusses the identity of Thomas Digges and concludes that he seemed "to be in two different worlds...very deliberately creating separate identities for himself," one "as a gentleman in parliament" and the other "as a mathematician" (63). Leonard may well have been the brains behind most mathematical, architectural, and military advances attributed to both father and son, and perhaps to son alone. *A Perfit Description* was perfectly deceptive, as it not only kept the new Worldview beneath the contempt of pedants who would scorn the medium of an almanac, but it also kept the brains behind the New Astronomy from public view.

If Thomas provided a cover that enabled his father to pursue research past 1571, Leonard would have been sufficiently below the radar so that he could continue his work after 1595 (when Thomas died) without fear of censure. In particular, his activities could account for the subtextual data contained in the plays considered in the chapters previously mentioned. *Love's Labour's Lost* appeared around 1594, a year before Thomas died, but the remaining three are associated with later dates (*Hamlet c.*1601, *The Winter's Tale* in 1604, and *Cymbeline* in 1610) of which *Hamlet* seems timed for release after the death of all four major World-model protagonists (Copernicus, Thomas Digges, Tycho Brahe, and of course, Ptolemy).

Johnson and Larkey write, "it becomes of the utmost importance, in evaluating [Thomas] Digges' work, to determine whether his idea that the Universe was infinite resulted solely from metaphysical considerations, or whether scientific reasoning and observation played the decisive part in the formulation of his theory" (105). Concerning the six works listed in *Stratioticus* but not delivered, the one mentioned earlier is *Commentaries upon the Revolutions of Copernicus*, and four other titles are: *A Treatise of the Arte of Navigation, A brief Treatise of Architecture Nauticall, A Treatise of Great Artillerie and Pyrotechnie*, and *A Treatise of Fortification of Townes, Fortes, and Campes*. The remaining title, *A Booke of Dialling*, concerns the measurement of time, which is a difficult topic and a matter of importance to navigation, as well as to Shakespeare. These six titles impute material that seems weightier than those that Thomas alone produced, and although Thomas took to sea to demonstrate the correctness of advanced methods in navigation, one wonders whether Leonard could have influenced the conception and hypothetical content of the headier tomes. Any chance of the pledged works appearing perished with Thomas' death, but this does not mean that the perspective glass did not exist or that it was not used to study the heavens (Gainer and Usher). If Leonard did contribute past 1571 as posited, the attributions of astronomical knowledge made in previous chapters should allow for it.

Evidence from *Cymbeline* (table 4.3) suggests that Leonard continued to observe Saturn beyond the time of Thomas' death and into the seventeenth century. This raises concern about his age, for if he was born in 1521, at the dates 1594, 1597, 1601, 1604, and 1610 of the plays examined in this work, he would have been about 73, 76, 80, 83, and 89.

The Winter's Tale and *Cymbeline* place the most stringent constraint of an age, and although in the early modern age octogenarians were rare, they were not unheard of (Robine).[6] It is not known when or whether Leonard's mental vigor began to decline, or when he died, but if he did survive into his middle or late eighties as table 6.5 would indicate, he must have been sufficiently *compos mentis* to relay information accurately.[7]

That Leonard figures prominently in Shakespeare's plays finds further support in the fact that the Canon contains five *leo*-related names—Leonatus from *Cymbeline*, Leontes from *The Winter's Tale*, Leonardo from *The Merchant of Venice* (*MV*), Leonato from *Much Ado about Nothing*, and Leonine from *Pericles*—but *no* Leonard or its variant, Lennard. In addition, Shakespeare frequently mentions birds of prey, including the buzzard, cormorant, cuckoo, eagle, eyas-musket, falcon, haggard, hawk, kestrel, kite, osprey, owl, puttock, and vulture—but *no* leonard and *no* related words leonarde, lanner, lannard, or leneret (*OED*). The arguments are more plausible given Shakespeare's proclivity to relay information by omission.

Of course, major questions remain. If Leonard died in 1571 or earlier, why did no one other than his son know about it? Why record this event only in the preface of a technical book and not in church records or on a tombstone? If the present thesis has merit, why did the man who created a revolution in the capacity of humans to resolve heavenly bodies leave no evidence of this breakthrough? What induced at least one commentator to raise vigorously the question of Leonard's seeming evanescence? What possessed two exceptionally gifted persons, Shakespeare and Leonard Digges, to leave no artifactual records of their work? Is it coincidence that Leonard's grandson and namesake penned encomia in the first two Folio editions of the Canon? Why do the lives and burial sites of both Leonard Digges and the actor Shakspere (see p. xxiii) prompt so many questions? There is no information whatsoever regarding Leonard's grave, but at least Shakespeare's tomb and monument are there for all to see. In light of the juxtaposition of these two celebrities, perhaps Shakespeare's memorials will prove enlightening.

Monument

William Shakspere's funerary monument and grave are located inside the Holy Trinity Church, Stratford, where they challenge understanding. The earliest known depiction of the memorial is a sketch made in 1634 by William Dugdale (1605–1686) which shows a seated figure framed by pillars and a cornice. Placed upon the cornice are Shakspere's coat of arms and two wingless cherubs, one holding a spade and representing Labor, the other holding an hourglass and representing Rest. The bust sports a drooping mustache and a sour expression, and the arms of the figure are akimbo with hands resting on a sack of wool, which was one of Shakspere's businesses.

An engraving of 1656, based on Dugdale's original sketch, shows the seated figure again framed by a cornice and pillars. A leopard head adorns the top of each pillar, and the wingless cherubs, or *putti*, perch precariously at each end of the cornice. The *putto* to the effigy's right grasps a spade, and the other holds an hourglass. Resting at the center of the cornice is a bas-relief of Shakspere's coat of arms with a shield depicting a quill, above which stands a falcon with wings like a spread eagle (figure 7.1). Topping off the monument is a skull, "chopless" or jawless as in *Hamlet* (Edwards 5.1.74n).

In 1632 (sixteen years after William Shakspere's death), residents of Stratford still recalled him, but by 1660, he was all but forgotten (Whalen, "Stratford" 19). By 1723–1725, the effigy had morphed into the figure of a writer. Gone was the sour expression and bag of wool, replaced by a contemplative figure with hands resting upon a flat cushion, one hand holding a quill while the other rests upon a blank piece of paper. From the mid-eighteenth to the early nineteenth centuries, the bust and monument underwent further changes, and one imagines that Shakespeare—ever an admirer of Ovid (43 BC–AD 18)—would have marveled at the metamorphoses depicted in Whalen's article.

In 1693 a Mr. Dowdall visited Holy Trinity Church and recorded the existence of rhymes carved into the stone slab over Shakspere's grave.

> Good Frend for Jesus SAKE forbeare
> To diGG T–E Dust Enclo–Ased HE.Re.

FIGURE 7.1. Reproduction of a sketch of Shakspere's coat of arms based on the original drawn in 1596 by the Garter King of Arms (see e.g., Michell 72–73). The punning crest features a falcon shaking a spear, and above are the words *Non, Sanz Droict* ("No, without right"), a variation of the intended *Non Sanz Droict* ("Not without right").

> Bleste be T–E Man TY spares T.Es Stones
> And curst be He TY moves my Bones. (Baxter 522)

A year later, a visitor remarked that the lines carried "something in them which stands in need of a comment" (Michell 58). If the rhyme is in code, no one has deciphered it. The power of the malediction, the strangeness of the doggerel, and the manner of the lettering keep bardolaters guessing, but the solution is in plain sight.

In the modern version of the monument, the leftward cherub holds a spade as before, but the rightmost holds a torch upside down as if to light the way. This is appropriate because, from 1605, cherubs were considered "angels of light" (*Ezekiel* 10, as cited in *OED*) whose powerful eyesight (Halio 5.1.62n) befits their roles, as no doubt they spoke to Shakespeare's insights into the world. Ogburn suggests that the figures on the monument "direct us to dig into and search within the monument" (789–790), but nobody has dared to disturb the tomb. Generally, though, a torch would be useful in any tunnel, vault, or crypt.

The first four lines of the message on the monument encourage passersby to pause and reflect.

> Stay Passenger, why goest thov by so fast?
> read if thov canst, whom enviovs Death hath plast
> with in this monvment Shakspeare: with whome
> quick nature dide: whose name doth deck his Tomb...

"Read if thou canst" challenges observers to understand the site—if they can! Investigators should attend to the advice quoted, meaning that the memorial is for one "whose name doth deck his tomb." The memorial sports a *putto* holding a spade with which to dig, and the second line of the injunction on the tombstone has the lettering "diGG." If these refer to Leonard Digges, the solution lies hidden in the noise of all the distractions, and the entire arrangement serves to mislead passersby who are accustomed to accepting appearances as reality.

The conjecture is grist to the mill of heretics who have collectively offered over fifty alternatives to Shakspere being Shakespeare (Bryson 193). Thus, the present proposition may not rank high on the Richter scale of literary and scientific outrage because, with so many aspirants, surely one more will scarcely make a difference.[8,9] Nevertheless, what is at best a 2% proposition is worth at least a moment's reflection, lest pedantry deny even the possibility of another viewpoint. A full discourse lies beyond the scope of this work, but I must at least proceed heuristically and discuss a few pertinent items.

Coat of Arms

In 1596 the Garter King of Arms granted a coat of arms (figure 7.1) to William's father, who had started an application twenty years before but had let the matter drop. A document containing a sketch of it has a phrase added, effectively, in two versions: Non, Sanz Droict and NON SANZ

DROIT. Removal of the comma transforms "No, Without Right" into "NOT WITHOUT RIGHT." In *Every Man out of his Humour*, Ben Jonson (1572–1637) mocks the motto using a rustic's motto, "Not without Mustard" (Michell 72–74). The comma in "No, Without Right" causes an ambiguity intended to reflect a conflict between fact and expediency, for William Shakspere would have committed himself to lifelong service to an important person but at the same time had no birthright to his elevated status.

In figure 7.1, the pen depicted on the shield signifies the paramountcy of writing, and the crest resembles a falcon, perhaps a leonard. The bird stands on one leg with the other raised in front of it. The talons of the raised leg grasp a spear, which presumably it could shake and thus connote "shake-speare." This may explain the idiosyncratic hyphenation "Shake-speare" which occurs ubiquitously in 15 of 32 plays issued before F1, two of the four dedicatory poems in F1, the *Sonnets*, and a poem that appeared with *The Rape of Lucrece*.

John Webster (1580?–1625?) used the hyphenated name in an appraisal of contemporary playwrights. It occurs also in a tribute of 1610 entitled "Our English Terence Mr. Will. Shake-speare," wherein John Davies compares "Shake-speare" to the Roman writer Terence (*c.*185 or *c.*195 BC–*c.*159 BC) who published literary efforts of noblemen loath to admit their authorship. In 1614, in a tribute that begins "Shake-speare, that nimble Mercury thy brain," Thomas Freeman mentions Terence and his mercurial habit of appropriating the wares of others and adapting them to suit his dramatic ends (Michell 56).

The hyphenation recurs in the epigram *To Mr. William Shake-spear* that appears in the anonymous *Wits Recreation* of 1640,

> Shake-speare, we must be silent in thy praise,
> 'Cause our encomiums will but blast thy bays
> Which envy could not, that thou didst so well;
> Let thine own histories prove thy Chronicle. (Ogburn 9)

The hyphen challenges readers to seek a cause more plausible than a mindless repetition of a typesetting error. As discussed earlier, the hyphenations "sur-addition" and "leo-natus" occurring in *Cymbeline* point *inter alia* to Leonard Digges, and perhaps the hyphen in Shake-spear(e) does as well. Recurrent too is the injunction of the last line, which urges the audience to seek information on Shakespeare's life by pondering his plays and poems.

The actor William Shakspere's credentials are less than stellar, yet modern biographers have been reluctant to admit that the poet Shakespeare had more than a grammar school education. Evidence points to knowledge of classical authors that the "small Latin and less Greek" (Jonson, as cited in Michell 76) of a Stratford education could not have supplied. For instance, *Hamlet* and *The Winter's Tale* have origins in classical Greek literature, the former in *Oresteia* of Aeschylus (525–456 BC) and the latter in *Alcestis* of Euripides (Showerman, "Orestes"; Showerman, "Look"). The Canon is evidence of Shakespeare's polymathy, whereas his protégé, Jonson, satirized Shakespeare's erudition. In *Every Man out of his Humour*, Jonson ridicules the hyphenated Shake-spear by hyphenating "Cri-spinus" (Michell 72), where *crispus* means "curly-headed" in Latin. Possibly, this refers to Leonard's hair, which I suggested in chapter 3 is curly.

In *Poetaster*, Jonson again references Leonard indirectly.

> Ramp up my genius, be not retrograde
> But boldly nominate a spade a spade. (Jonson 5.3, L4(b))

Leonard was the "genius" who was instrumental in providing the means to confirm that heliocentricism accounted for the "retrograde" motion of the planets, and Jonson advises to "nominate" (i.e., name) him by alluding to an implement for digging. The parody supports the view that "Shakespeare" is a nom-de-plume for Leonard Digges.

In *To the memory of my beloved, The Author, Mr. William Shakespeare: and what he hath left us*, Jonson writes that Shakespeare is "a Monument, without a tombe" (Michell 75 l. 22). The poem is ambiguous in several places, as if Jonson is trying to say something without actually saying it. Near the poem's end, he writes that Shakespeare is the "Sweet

Swan of Avon," but he cautions against believing that Shakespeare has a tomb:

> ...stay, I see thee in the Hemisphere
> Advanced, and made a constellation there!
> Shine forth thou Starre of Poets... (as cited in Michell 75)

In keeping with Neoplatonic doctrine, he instead places Shakespeare among the stars, from whence the Star of Poets shines steadfastly upon the world.

Jonson was the first to use the term "poetaster" to describe an inferior poet (*OED*), but "Starre of Poets," when adapted by combining two words of Greek origin (Michell 75–76), could mean literally "poet-star." Thus, "Starre of Poets" is ambiguous, referring both to Shakespeare and to one so renowned as to occupy an entire celestial constellation after death. Perhaps that constellation is Leo.

Jonson writes further that Shakespeare "was not of an age, but for all time" who "like Apollo...came forth to warme / Our cares" and "to charme" like Mercury (as cited in Michell 75). Hotson also invokes Apollo and Mercury to explain the 1588 portrait by Nicholas Hilliard (1537–1619).[10] Jonson likens Shakespeare to Euripides, Sophocles, Seneca, and other classical writers, but adds that he accomplished all this despite having little knowledge of Latin and even less of Greek. Then, he puns crudely with "shake a Stage," and as another sop to appearances, he goes on to mention "neat Terence," whose literary style has already been encountered.

GREEN'S GROAT

Robert Greene was a penurious pamphleteer who penned a polemic that appeared posthumously in 1592. In *Groats-worth of witte, bought with a million of Repentance...*, Greene relates how a wealthy stranger—an actor who was a dab hand at poesy and who supposedly represented Shakespeare—approached him and proposed that he write plays for his acting company. Greene's account gives the impression of a pretentious, well-off individual asking one in need to provide a service.

This could mean Leonard approaching Shakspere, the service being for the Stratford man to lend his name—and his silence—to Leonard's literary enterprise and desire for anonymity.

Greene values his tract at the comparatively low value of his own worth, a groat (or four pence), and warns fellow dramatists of those who corrupt the theater and its writers. He singles out one who he says is,

> …an upstart crow, beautified with our feathers, that with his *tiger's heart wrapped in a player's hide* supposes he is as well able to bombast out a blank verse as the best of you; and, being an absolute *Johannes Factotum*, is in his own conceit the only Shake-scene in a country. (as cited in Michell 68)

In 1592 the dramatist Shakespeare's play production had only just begun, so Greene would feel justified in calling him an "upstart," but he buttresses his abuse by likening him to a crow. In Renaissance literature, the crow symbolized thievery because of its "power of mimicry but not of invention" (Schoenbaum 152).[11] Many liken Shakespeare to a magpie because they think that he scavenges for literary morsels, but they have entirely the wrong kind of bird.

Greene characterizes the upstart crow as having a "tiger's heart wrapped in a player's hide" (as cited in Michell 68), which is an adaptation of "O, tiger's heart wrapt in a woman's hide" from *Henry VI part 3* of 1590–1592 (Bevington 1.4.137). The tiger is a symbol of duplicity, deception, and "doublenesse" (Hotson, *Hilliard* 145), for which one may substitute another big cat—"leon" the lion—and in accordance with the etymology previously described, substitute "hard" for "heart." Then, as noted, the combination leads to Leon[h]ard from which follows the name Leonard. In short, Greene is describing Shakespeare as Leonard hiding inside the hide of the player-actor Shakspere.

Greene alleges further that Leonard is a Johannes Factotum, a "Jack of all trades" and a "would-be universal genius" (*OED*). Greene is right, for Leonard excels in all manner of disciplines, and his remark that Shakespeare is a Shake-scene refers to the upstart Shake-speare who struts and frets upon the literary scene.

With Greene, it is hard to separate fact from fiction, and his life history is sufficiently suspect that some experts believe he "conveniently died" after serving as a "cover for someone else" (Michell 69). Henry Chettle (*c.*1560–*c.*1607), the publisher of *Groats-worth*, later apologized for Greene's screed, although to whom he apologized is not entirely clear, and then not without deferring to "divers of worship." This refers to diverse, worshipful gentlemen—the social class to which Leonard belonged—and not to noblemen for whom "divers of honour" would apply (Schoenbaum 155). The *Groats-worth*, it would appear, is another red herring.

OTHER CONNECTIONS

Other aspects of Leonard's life deserve attention. In the list of fines imposed upon Wyatt's rebels, Leonard's status appears as "Esq." (Esquire) and not "Gent." (Gentleman) as it is for others (Loades 253). The designation "Esquire" indicates a man belonging to the higher order of English gentry, including barristers-at-law (*OED*), which is the likely designation because Leonard completed schooling as a lawyer at Lincoln's Inn (Bergin and Speake 130). It is well known that the Canon demonstrates Shakespeare's knowledge of the law, helping to understand yet another piece of self-deprecatory humor from *Henry VI part 2*, "The first thing we do, let's kill all the lawyers" (4.2.74). At the Inns of Court, this risible occasion would have put in mind a passage from *Measure for Measure*.

> ANGELO We must not make a scarecrow of the law,
> Setting it up to fear the birds of prey,
> And let it keep one shape till custom make it
> Their perch, and not their terror. (2.1.1–2.1.4)

"Fear" can mean "to frighten" (Lever 2.1.2n), suggesting that the law must not frighten birds of prey, but equally, the law must not "fear the birds of prey," as occurs when society discriminates against scientists.

By the earliest range of dates for a Shakespearean play (1589–1593), Leonard must have already established an understanding with

Shakspere, which may help explain the Stratfordian's "lost years" 1585–1592. Yet in 1597, the thirty-three-year-old Shakspere purchased the second largest house in Stratford for at least £60 (1,200 shillings) despite having defaulted on a 5-shilling tax only a year earlier. Many wonder how a penny-pinching, grain-hoarding, tax-defaulting deerstalker could suddenly have acquired such a sum.

The *Sonnets* are a likely source of personal information, and here they are touched upon briefly. Sonnet 72 makes it clear that the speaker lives in the shadow of shame and the fear of death. He tells a young man—whom he addresses parenthetically as "dear love" (72.3) and who one suspects is Apollo—to forget him after he dies, and he hopes that "My name be buried where my body is" (72.11)—another reference to digging. Further, he writes, it would be inappropriate for the addressee to "lie" (72.5) about the speaker's life, although presumably it would be appropriate for one named Digges to "lie" in a dug grave. In Sonnet 76, the speaker hints at how his writing contains clues to his identity. "That every word doth almost tell my name, / Showing their birth, and where they did proceed" (76.7–76.8) implies that the Canon is littered with hints, and at least in the plays examined here, certain passages do indeed "almost" spell out the names "Leonard Digges."

Sonnets 21, 38, 78, 79, 82, 85, 100, 101, and 103 are Muse poems, matching the number of Apollo's retinue of nine muses. Eight of Apollo's muses cover topics in what would today be called the Arts and Humanities, with the ninth—Urania—representing Astronomy, which in antique times belonged in the same league.[12] Muse poem Sonnet 38 invokes a "tenth Muse" who the speaker says is "ten times more in worth / Than those old nine which rhymers invocate" (38.9–38.10). By stepping beyond the purview of the classical nine, the speaker sets himself apart from common rhymesters in a very specific way. He explains, "he that calls on thee, let him bring forth / Eternal numbers to outlive long date" (38.11–38.12). In other words, the poet amends the old mythology by weighting the worthiness of mathematical prediction ten times more than the oracular prophesy (over which the classical Apollo holds sway). Implicitly, the sonnet writer welcomes a "New Mythology" that incorporates mathematical prediction. This helps explain the glut of numbers in *The Winter's Tale* and again points to Leonard Digges and his ability to predict planetary positions (table 3.5). Recall the observation of John Keats (1795–1821), which Hotson (as cited in *Hilliard* 196) termed the "rarest of insights," that Shakespeare led a life of allegory and that his works are comments on it.

In the eighth Muse poem, Sonnet 101, the speaker informs Urania that she "must behave as a good functionary" and not delay in properly representing the Platonic triad of truth, beauty, and "the good" (Vendler 430–432). In the ninth Muse poem, Sonnet 103, the writer begs forgiveness, "O blame me not if I no more can write!" (103.5). He advises Apollo that nothing he writes can adequately express that god's magnificence, and that the speaker's writing is a poor rendition of the reflection that the god sees when gazing into a mirror. "And more, much more than in my verse can sit, / Your own glass shows you, when you look in it" (103.13–103.14). The interlocking tropes of "glass" or mirror (103.6, 103.14) and Leonard's "blunt invention" (103.7)—which I believe refers to his perspective glass—indicate the writer's regret at his inability to do justice to celestial splendor with mere words. This comes with dignified humility from the poet of poets, a master wordsmith, discoverer of unparalleled beauty and harmony in the Heavens, and advocate of Apollonian ideals.

POST MORTEM

Only in 1623, at the time of F1, is it definitely proclaimed that Shakespeare has died. A poet signing himself "I.M." wrote an 8-line commendation entitled *To the memorie of M. W. Shake-speare.*

> Wee wondred (*Shake-speare*) that thou went'st so soone
> From the Worlds-Stage, to the Graves-Tyring-room.
> Wee thought thee dead, but this thy printed worth,
> Tels thy Spectators, that thou went'st but forthe
> To enter with applause. An Actors Art,
> Can dye, and live, to act a second part.

> That's but an Exit of Mortalitie;
> This, a Re-entrance to a Plaudite. (as cited in Eliot 163)

The name in the first line is parenthetical, hyphenated, and in a different font, forewarning of content pertinent to the disguised dedicatee.

"I.M." may stand for James Mabbe (1572–1642), a scholar of Spanish. He and others wondered why Shake-speare passed away "so soone." According to table 7.1, this was sometime between 1558 and 1574 when Leonard's age lay in the range 37 to 53. Just as an actor can "die" and reappear on stage in another role, so too did Shake-speare vanish and reappear on life's stage, and by 1623, I.M. had learned that F1 was evidence of his second career. "Return from supposed death is a frequent device of the late plays" (Arbery 166), which further points to Leonard's autobiographical dramaturgy, his reclusion, and his return to the "Worlds-Stage," which is a setting of planetary if not cosmic proportions.

Mabbe was a friend of the younger Leonard Digges, who (with Mabbe) has an encomium in F1. Digges writes that Shakespeare's name and works will outlive his tomb, and that:

> ...when that stone is rent
> And Time dissolve the Stratford Moniment,
> Here we alive shall view thee still. (as cited in Michell 88)

One must "view" Shakespeare through his collected works, advice that is not unique. In F1, in a short poem *To the Reader*, Jonson repeats it as he addresses an image of Shakespeare called the "Droeshout engraving." An artist's rendering of this portrait is on the cover to this book. The poem is best appreciated in its entirety.

> This Figure, that thou here seest put,
> It was for gentle Shakespeare cut;
> Wherein the Graver had a strife
> With Nature, to out-doo the life:
> O, could he but have drawne his wit
> As well in brasse, as he hath hit
> His face; the Print would then surpasse

> All, that was ever writ in brasse.
> But, since he cannot, Reader, looke
> Not on his Picture, but his Booke. (as cited in Michell 79, 86)

Penetrating the poem's ambiguities clarifies meaning. The artist, Martin Droeshout (b.1601?), hid the face of the poet, and Jonson advised readers who seek Shakespeare's image to attend to his Canon instead.

Criticism of the Droeshout engraving has persisted and for good reason (Michell 83–88; Ogburn 222–224; Schoenbaum 315–317). The encephalitic face resembles a mask with two right eyes attached to a stiff ruff held rather too high above the shoulders. The mask seems to cover the back of the head, which implies that Shakespeare had two faces, of which only the one depicted was for public viewing. The 1640 edition of *Sonnets* has an engraving based on Droeshout's, except that it is a mirror image, and an appended poem parodies Jonson's by calling the engraving "renowned" and a "Shadow." On top of that, the author is "John Benson," itself a mirror of "Ben Jonson."

A panegyric of outstanding quality by a "friendly admirer" who signed himself "I.M.S." appeared in the Second Folio of 1632. It begins by comparing Shakespeare's mind to a "cleere / And equall surface" that, "reflecting ages past," makes "things appeare," which "[r]owle back the heavens" and are "[d]istant a thousand yeares" (as cited in Campbell 147–148). "Equal" means "precise" or "exact," as in *MV* when Shylock speaks of "an equal pound…of flesh" (1.3.148–1.3.149). Thus, a "clear and equal surface" suggests a lens, and "equal" describes the quality of the "reflecting…surface" by which Leonard rolled back the conceptual barrier to infinite space.[13] The playwright and his instrument represent these celestial things "in their lively colours just extent" (Campbell 147). "Just" means exact or precise, and from 1624, "extent" means the dimensions or size of anything (*OED*). The things are in color, suggesting (in accordance with Gainer's telescopic reproduction) that mirrors, which cause no chromatic aberration, serve as light collectors and help image objects in their true colors. "Extent" can refer also to scope and breadth of comprehension (*OED*), which enriches the imagery and implies that the images are resolved. "This, and much more which cannot be exprest" was "Shakespeares ſreehold," which was improved by "favour of the ninefold traine" of Muses, of which the last mentioned is Urania.

Shakespeare's persistent quest for anonymity and his apparent determination to play a role in accordance with Apollonian tenets suggests that he would credit the principles espoused by that deity and would see no value in the preservation of his mortal remains. Leonard Digges was a master architect and expert in ſortifications, and the arguments presented lead to the conjecture that whosoever would seek the papers and artifacts of the resolution revolution should digg in Kent.

Murmur, a little sadly, how Love fled
And paced upon the mountains overhead
And hid his face amid a crowd of stars.
—William Butler Yeats

Notes

Chapter 1

1. An angle is the difference between two directions. By convention, direction changes by 360 degrees (360°) for a complete rotation. 1° equals 60 minutes of arc (60′) and 1′ equals 60 seconds of arc (60″). Owing to the enormous distances of stars, stellar parallax angles resulting from the revolution of the Earth around the Sun are less than 1″ (see figure 1.3b).
2. To see this, consider the case of a planet like Mars, whose orbit is larger than the Earth's (figure 1.5). Looking down upon the planets from the north, both move in a counter-clockwise direction, but Mars moves more slowly along a larger track. As the Earth on its inside track overtakes Mars, Mars appears to fall behind even though it moves around the Sun in the same sense as the Earth. Thus, seen from Earth, Mars appears to move retrograde relative to the stars. Only when Earth is either well behind or well ahead of Mars, does Mars appear to run in the direct sense.
3. Digges sees himself surrounded by stars, just as a forester may see trees in every direction, but may be unable to see the forest for the trees because nearby ones block the sight of more distant ones. The sight of the stargazer is limited too, but by the sensitivity of the detecting instrument, whether eye or telescope. The forester may posit an edge to the forest or may wonder whether trees cover the entire surface of the world. In like manner, the stellar observer may posit an edge to the distribution or may wonder whether stars fill the World. This was Thomas Digges' vision. It is unclear, however, whether Digges regarded the Sun and solar system as physically unique; although, in keeping with contemporary thinking, he may have regarded humankind as theologically unique. Even if Digges believed that the solar system is unique, this does not imply that he regarded it as a physical center because it is impossible to assign a center to space whose borders are unknown. The stars that Digges supposed filled all of cosmic space were the first "building blocks" of the Universe. It was discovered in 1921—at the time of the debate between Heber Curtis (1872–1942) and Harlow Shapley (1885–1972)—that entire collections of stars called galaxies are larger building blocks of the Universe (Shapley and Curtis). The change of scale from stars to galaxies is enormous—about 100,000,000,000 give or take a factor of 10 or 100—but this pales in comparison to "infinite" space.

The leap from a quasi-spiritual sixteenth-century model to one in the twentieth century merely requires a redefinition of "monad." Today the term includes much larger and more luminous ones, such as clusters of galaxies. The popular misperception that the Universe has a center persists, owing in part to the misleading name ("Big Bang") for a model of the expanding Universe of galaxies. On the grand scale, these monads recede from one another, not from some supposed center as if they were the shards from an explosion. The "Big Bang" is not like a bomb blast in space, but rather an explosion of space itself in which material objects are imbedded.

4. A valid criticism is that, in *A Perfit Description*, Digges states (figure 1.7) that the stars have properties "far excellinge our sonne both in quantitye and qualitye," which sets the Sun apart from the stars. Suppose that in support of his model, Digges attempted to measure the stellar parallax of bright stars by measuring their angular displacement relative to faint stars in the field of view; on average, bright stars are closer than faint ones and are more likely to have detectable parallax angles (figure 1.3a). Like Copernicus and Tycho Brahe, Digges agreed with early estimates that the Sun's radius was about five Earth radii (E.r.): $s_0 = 5$ E.r. (figure 1.8). This requires the distance between Earth and the Sun to be about $a = 1142$ E.r. (van Helden, Measuring; tables 1, 5). Let d be the distance to a star and s be its apparent radius. The parallax angle of a star is a/d, but Digges detected none to the limit of the resolution s/d of his telescope. Then, $a/d < s/d$, or $s > a$, where $a = 1142$ E.r. $= 228.4\ s_0$. Thus $s > 228.4\ s_0$, and he would conclude that stars are over 200 times the size of the Sun. Because the standard notion at the time was that a star's size was the sole determinant of its intrinsic brightness, stellar luminosities would exceed the Sun's as well. However, if this were Digges' train of thought, it would have raised the difficulty that the luminous Sun was unlike luminous stars, making the Sun especially small. One might argue that this violates both Copernican thinking and what today is known as the Cosmological Principle, which says in effect that all parts of the Universe are the same in their gross properties. Digges would know of this idea, which originated in the 7th or 6th centuries BC among thinkers of ancient Milesia (Pannekoek 98), but it appears that he does not address it.
5. Christopher Marlowe was born and educated in Kent, and on occasion, he appeared before the Kentish Justice Manwood, who pronounced him guilty and meted out sentences that Marlowe considered justifiably lenient. Upon Manwood's death in 1592, Marlowe composed a dozen-line elegy (in Latin) entitled *Epitaph to Sir Roger Manwood*. The elegy is "uncharacteristic,"

raising doubts about Marlowe's sincerity (*DNB*). The first two lines attribute four properties to the famous jurist. The first includes (in translation) "terror of the night prowler," and the second compares him to Hercules. The latter suits Manwood, owing perhaps to his patriotic labors, but it is odd that Marlowe would rank Manwood's capacity to terrorize night prowlers ahead of a laudatory comparison to an offspring of Zeus. The third property is "scourge of the profligate," which suits the concerns of the guardian of the nation's wealth, but peculiarities persist with the fourth attribute, which is usually taken to mean "vulture to the stubborn brigand." Surely, Manwood ranks higher than a consumer of carrion. The primacy of the attribute "terror of the night prowler" invites speculation as to the occupation of the prowler. Manwood was only a few years younger than Leonard Digges and, like him, a follower of Roger Bacon's research. He claimed to have invented the perspective glass (Birch 4: 157), and perhaps Marlowe knew that he would not have taken kindly to others who prowl the sky at night.

6. John Donne began writing in about 1590, when Copernicanism had begun to make inroads in England. He addressed issues raised by the New Astronomy and earned the title of the "Copernicus in poetry." In *Devotions upon Emergent Occasions*, he writes, "I am up, and I seem to stand, and I go round; and I am a new Argument of the new Philosophie, that the earth moves round..."; in 1610 he foresaw the upheaval that heliocentricism could potentially inflict, and the following year in *Ignatius his Conclave*, he prophesied the condemnation of Copernicus and Galileo in the afterlife (Simpson 133–136). In *A Valediction: Forbidding Mourning*, he acknowledges that the circle is a symbol of perfection, but notes that the circular motion of Earth brings "harms and fears," and in *Verse-Letter to the Countesse of Bedford* (III), it says that, "new philosophy arrests the Sunne, / And bids the passive earth about it runne" (as cited in Bennett 33, 166). Also in 1611, in an oft-quoted passage from *The First Anniversary, An Anatomy of the World*, he bemoans the confusion that Copernicus caused.

    And new philosophy calls all in doubt,
    The element of fire is quite put out.
    The sun is lost, and th'earth, and no man's wit
    Can well direct him where to look for it... (as cited in Bennett 206)

Donne was familiar with the work of Galileo (Roberts 393, 750, 973), whose announcement that Venus was a dark star in orbit around the Sun tipped the scales of Donne's opinion toward heliocentricism. I have called this realization "Donne's Criterion" (*HU* 211–212), which represents a noncontroversial benchmark by which to calibrate awareness of World-views among writers of that era. In the passage quoted, Donne refers also to the demise of the ancient element "fire." According to geocentric doctrine, all Ancient Planets except the Moon shine by virtue of their motion, which ignites the element Air that is supposed to exist in them. However, Copernicus assigned Mercury, Venus, Earth, Mars, Jupiter, and Saturn to a new class of modern planets, and when Galileo found that Venus was a "dark star" (i.e., non-self-luminous like Earth), he cast doubt on the fiery nature of all the Ancient Planets. The only object for which the ancient doctrine might reasonably survive was the Sun, but in the Copernican model, it was at rest, and therefore (by ancient dictum) it could not shine by Fire. Moreover, because Copernican theory put the Earth in motion and reduced the stars to rest, the stars had no way to ignite Fire either. Donne concluded that the element of Fire "is quite put out." It took over three centuries to discover that the Sun and stars shine primarily by thermonuclear fusion.

7. In Usher ("Cymbeline") and Sohmer (*Wiser*), I regret misstating the manner and time of Galileo's announcement on Venus. Conclusions are unaffected.
8. This official acceptance of heliocentricism in 1822 occurred about one hundred and fifty years after Isaac Newton had formulated a theory of universal gravitational attraction, and about a hundred years after Samuel Molyneux (1689–1728) and James Bradley (1693–1762) had observed a phenomenon called the "aberration of starlight," which is a direct empirical proof of heliocentricism. In 1838–1839, Friedrich Wilhelm Bessel (1784–1846) and Thomas Henderson (1798–1844) were the first to observe the heliocentric parallax of stars. They showed that stars are distant at least 50,000,000,000,000 kilometers (30 trillion miles) and lie at different distances, directly verifying the immensity of Copernicus' *immensum* and the idea of stars scattered through space. In 1851 Jean Foucault (1819–1868) used the conceptually simple apparatus of a freely swinging pendulum to show that the Earth rotates relative to the stars. Thus, experimental verification of the Copernicus-Digges model came more than two-and-a-half centuries after its synthesis in 1576, and over two millennia after a mob had murdered Pythagoras, the father of astronomy and the first proponent of a moving Earth.

CHAPTER 2

1. In the First Quarto of *LLL*, Mote appears as "Moth" (Mowat and Werstine, *Love's* 1.2.0SD), and in Elizabethan times both were pronounced the

same way. Motes are specks, and moths are tiny, yet the diminutive Mote is astute and, in repartee, more than the equal of his employer. Critics theorize that "Moth" is "Thom" spelled backward, which they claim refers to Thomas Nashe (1567–1601), but Thomas was a common name in the sixteenth century and other Thomases might qualify. Costard likens Mote to someone who is "loaded with honors" ("*honorificabilitudinitatibus*"), which neither Mote nor Nashe are, so either Costard is confused or the epithet refers to someone more deserving of accolades. Perhaps Shakespeare named Moth for Thom(as) Digges, for whom *honorificabilitudinitatibus* would be an apt if retrograde compliment for one who (after 1573) vowed to eschew Latin and "glorify his native tongue" (Johnson and Larkey 114).

2. For ease of reference, let ℓ1–ℓ6 identify lines in Holofernes' doggerel.

ℓ1 The preyful princess pierced and pricked a pretty pleasing pricket,
ℓ2 Some say a sore, but not a sore till now made sore with shooting.
ℓ3 The dogs did yell. Put 'l' to 'sore,' then sorel jumps from thicket,
ℓ4 Or pricket sore, or else sorel. The people fall a-hooting.
ℓ5 If sore be sore, then 'L' to 'sore' makes fifty sores o' sorel.
ℓ6 Of one sore I an hundred make by adding one more 'L.' (4.2.67–4.2.78)

If this is more than satire, the puns in ℓ2, ℓ5 on a sore (4-year-old buck) that is injured ("sore") complicate the verse even further. Grasping at straws, suppose that the cascade of buck types (first head, sore, sorel, and pricket: ages 5, 4, 3, 2 years respectively) suggests a tally of them. Mentions up to and including ℓ3 are as follows: a first head in 4.2.10; a pricket once each in ℓ1, 4.2.12, 4.2.21, 4.2.58, and 4.2.62; a sore three times in ℓ2, and a "sore" and a sorel in ℓ3. In ℓ4, "or…or…" means "either…or…" as in *Julius Caesar* 5.5.3, which may mean that one must select either "pricket sore" or "sorel," but if the tally accumulates by mere mentions, then the words could signify an addition of up to three buck types (pricket, sorel, sore), which makes the contribution from ℓ4 either 1, 2, or 3. ℓ5 mentions a sore twice, plus a "sore" and a sorel once each, plus "fifty sores," and ℓ6 mentions one sore plus fifty more sores. ℓ5 has the only plural mention of a buck type ("fifty sores"), repeated in ℓ6 and implying a total of "an hundred" mentions. The cumulative tally is 117 to 119, or 2 less if the "sore" meaning injury is omitted. The amount "an hundred" in ℓ6 is the same as "fivescore," which occurs also in Holofernes' riddle ("Adam… came to fivescore," 4.2.50) where, as mentioned, its units are undefined. Ostensibly, the units here are "deer," but what purpose does a herd of deer serve, and why is the herd so large? In allegories, one thing can represent something else, and here, the accumulation of items for no apparent reason along with the lack of units hints that one is free to choose a unit appropriate to the context. The scene (4.2) in which this gibberish occurs features Costard, Nathaniel, Holofernes, and Dull, but the pedant Armado is conspicuously absent. Armado takes his name from the Spanish Armada, and because Shakespeare's omissions sometimes hint at his meaning, a heuristic approach is to let the sum of the entries represent the number of vessels in the Spanish fleet. This new assignment of units is plausible given the banter in scene 2.1 between Katherine and Boyet, the former speaking of "Two hot sheeps" with the latter replying "And wherefore not ships?" (1.2.229). Elizabethans pronounced "ship" and "sheep" nearly alike (Bevington, *Love's* 2.1.219n), and Katherine could as well mean "two hot ships." The pun refers to the tactic used against the Armada (as it lay anchored in the waters off Calais, France), which was to fill hulks with incendiary material, set them ablaze and let them drift with the wind and the current into the massed ships. The pun recurs in two other early plays (*Two Gentlemen of Verona* of 1589–1593, *The Comedy of Errors* of 1592–1594), and Hibbard argues that thereafter Shakespeare "mercifully" abandoned it (*Love's* 2.1.217n), Yet here it seems that Shakespeare belabors the pun, and for good reason because the failure of the Armada was a watershed event in English history. Shakespeare encourages belief in this interpretation by writing that the Princess "pierced and pricked" a pricket, and further that she inflicted sores. By implication, hunters inflict injuries to game just as the Armada's ships suffered piercings and prickings. The fleeter English warships engaged the Armada from afar, enabling them to pick off a few Spanish ships as if they were sheep in a flock. The oddity of having four different buck types thrust into the dialogue speaks to the variety of vessels that comprised the Armada. Shakespeare's identifications are none too complimentary, and a contrast exists between the divine spirit of the deer and the defeat of the herd of ships. A pricket can mean a little prick, in which case a first head is a pointed reference to the passing of puberty; and a sore can mean *pudendum muliebris*, all of which refer in a vulgar and derogatory way to human genitalia and their afflictions (Ellis 78–79, 94). On top of that, Holofernes' verse is as chaotic as the Armada in retreat. About 127–130 vessels set sail from Spain (Rasor 79–80), but a lesser number arrived in the English Channel, where the first major encounters with English warships took place. Writing in 1988 on the 400th anniversary of the maritime failure, the naval historian Jose Luis Cassado Soto put the size of the Armada off the English coast at 117 (Rasor 79–80, 227). This lies within the range predicted from the previous tally of 115–119, whose uncertainty is close to that in the estimate 127–130 of the original size of the fleet. Thereafter, the number of seaworthy ships becomes increasingly uncertain. At least one-third were lost and those that returned did so haphazardly along 500 miles of the Iberian coast, so it is unlikely that Shakespeare would make much of later numbers. These numerical convolutions may seem contrived were it not for the fact that arguments like these were commonplace in the sixteenth century and beyond. The irony is that on stage in 1572, Holofernes is a poetaster foretelling of events 16 years before they occurred.

3. When Rosaline (as the Princess who represents the Moon) speaks of a clouded face, she probably refers both to her face covered by a mask and to features on the face of the Moon (which are readily visible to the naked eye) that the ancients attributed to foreground clouds. However, in the confusion of swapped identities, does Rosaline speak for herself as Venus when she says her face is clouded? Could Shakespeare have known that planet Venus is cloud-covered? In 1788 Johann Schroeter (1745–1816) is said to have observed that the cusps of the crescent phase extend beyond where the point of the crescent would nominally be, signifying that Venus has clouds that scatter sunlight (Wagner 180), whereas the Moon shows no such effect. The telescope needed to observe this phenomenon would have had to have good optics and high resolution.
4. "Novum" can refer to the Latin *novem* (meaning "nine"), but "novum" is also part of *novus* (meaning "new"). This may refer to the New Star that erupted in November of 1572, a few months after the Paris massacre, which Tycho Brahe and others observed. "Vein" is a pun on Tycho's observatory on Hven, and is also a pun on "vain," a term used by Hotson (*Appoint* 124) to describe Tycho's personality. Berowne says that "the whole world…cannot pick out five such" (5.2.598–5.2.599). If Shakespeare is referring to the New Star of 1572, this may mean that he knows of only four other instances of the phenomenon. It is worth mentioning Tycho Brahe's reference to New Stars, said by Cyprianus Leoviticus (1524–1574) to have appeared in 945 and 1264 (Dreyer 188). In addition, Shakespeare's knowledge of astronomy and of classical Greek writers probably means that he knew of the outburst in the second century BC that occasioned much consternation and prompted Hipparchus to commence studying star positions. The tabulation by Green and Stephenson of historical stellar outbursts occurring in the Common Era suggests that Shakespeare may also have known of those New Stars that appeared in

1006 and 1054, for which Arabic (and many Chinese) records exist. Two European reports exist on the former, which would account for five outbursts, and concerning the other, only one western report exists, given by a doctor in Constantinople.

CHAPTER 3

1. The fact that astronomers call the New Star of 1572 "Tycho's supernova" biases modern readers into believing that Tycho's *De Stella Nova* was the definitive work on it. Recent work (Pumfrey and Riley n.p.) suggests that in 1573, Thomas Digges wrote an anonymous pamphlet, *Letter sent by a gentleman of England*, which was published by John Marsh (Thomas' regular printer who also published Field's *Ephemeris Anni 1557*). The pamphlet opines on the New Star, but Pumfrey and Riley believe Thomas Digges did the research. Thomas published *Alae seu* on the New Star of 1572 in late February of 1573, whereas Tycho's *De Stella Nova*—which was in press before April 16, 1573—had not yet appeared on May 3, 1573. Thoren notes that Tycho's measurements of the position of the New Star are the only ones of his entire career that "were not copied when one of Tycho's assistants later compiled a notebook that constitutes the sole record of his observations from his first efforts in 1563 up to those in December 1577" (55). Thoren suggests that by May 1573, Tycho "had seen some other writings on the star" (69). Perhaps these were the data tabled prominently at the beginning of *Alae seu*, which Johnson terms "surprisingly" accurate (157). By the turn of the twentieth century, the bias against Thomas Digges' work became evident, as Berry regarded *Alae seu* as a treatise "of no great importance" (127) and the entry in the first edition of the *DNB* states that he "made no great addition to science."
2. Differences between Q2 and F1 may arise through human error, but others are intentional, and over two-thirds are important or essential to understanding the cosmic allegory. Passages that are in Q2 but not in F1 and that relate to the subplot are as follows:

   1.1.108–125. Shakespeare describes evil astral influences upon Hamlet and Ophelia ("dews of blood," "disasters in the sun," and "the moist star…sick almost to doomsday with eclipse") that come to pass and have allegorical significance. In addition, an omission occurs in the Q2 passage itself (Edwards 1.1.117–1.1.118n) precisely at the point where its inclusion might have clarified the relationship of the omens to events in the allegory.

1.4.17–38. These lines on the vicious mole of nature and the dram of eale, along with those prior, have been roundly criticized (Edwards 14), yet help elucidate the subtext. The passage immediately precedes the entrance of the Ghost, to which the mole and the dram refer.

3.4.203–211. Edwards (15–16) argues that the passage is an authorial cut, yet it is important in explaining the subtext.

4.4.9–66. This cut is extensive but holds clues vital to understanding the allegory. Perhaps the cut resulted from events that transpired in the interim between Q2 and F1 (i.e., from about 1604 to 1623). In the first two decades of the seventeenth century, and thus before the printing of F1, Johannes Kepler discovered three empirical "laws" for planets, and it was Tycho Brahe's data that made his discoveries possible. Of these, the Harmonic Law of 1619 especially captures the essence of Pythagorean philosophy, and because much of Q2 is devoted to belittling Tychonic contributions and advancing Pythagorean, perhaps the F1 editors saw fit to reduce the cosmic emphasis in general in order to soften the impact. A similar argument pertains to cuts 5.2.100–125 and 127–130. In addition, the repercussions of Galileo's celestial observations of 1610 were widespread and led to increasing interference by ecclesiastical authority. Perhaps the F1 editors, fearing that the purpose of the Norwegian plot against Poland might be as transparent as the plot in Poland is small, may have decided that it was in the national interest to omit lines which describe the allegory with such comparative clarity. The omission allows the editors to have their cake and eat it too because they could plausibly deny heresy, but later readers would surely wonder why Q2 and F1 are as they are and not otherwise.

4.7.67–80. Laertes, freshly returned from Paris, enlists on the side of the geocentricists and thus becomes a legitimate target for Hamlet.

5.2.100–125, 127–130. This "swinging cut" (Edwards 5.2.100–5.2.125n) is thought not to contribute to the literal plot but to shorten the long buildup to the final action. Rather, all items pertain to Thomas Harriot and events in his life prior to the writing of Q2 at about turn of the century. F1 not only omits these lines, but also inserts the explanation 5.2.68–80 (see below, this endnote) that helps counter the parody, and thus Hamlet's other apology 5.2.198–216 remains in F1.

Passages in F1 and not Q2 are as follows:

2.2.229–256. This is probably a cut from Q2 rather than an insert to F1 (Edwards 2.2.229–2.2.256n; Jenkins 2.2.239–2.2.269LN). It relates political restrictions in Denmark to bounded space in the Aristotelian model, and makes the case for infinite space (i.e., for Thomas Digges' break with all past cosmological models).

2.2.313–333. These lines deal with child actors and the "war-of-the-theatres" of *c.*1601 and are unique in the Canon (Edwards 4–5, 2.2.313–2.2.333n). Heretofore, the lines had seemed digressive or the result of error, but their insertion draws attention to the telescopic imagery reported. The imagery is sufficiently obscure for safe inclusion in F1, particularly because by that time, Galileo had publicly revealed the merits of astronomical telescopy.

3.1.32. The two-word extra-metrical phrase draws attention to spying by authorities (Edwards 3.1.32n), who noted the capabilities of the perspective glass as well as of Galileo's "spyglass"

(5.2.57). The line explains that Rosencrantz and Guildenstern were willing agents (Jenkins 5.2.57n) in the promotion of geocentricism.

5.2.68–80. Hamlet expresses regret about his treatment of Laertes, which is tantamount to recognition of Thomas Harriot's post-Q2 accomplishments (see endnote 17 of this chapter) as within about 20 lines, F1 omits lines 5.2.100–125 and 127–130 that are less than complimentary to him.

3. "Ophelia" may derive from the Greek word meaning "to succor" or "to help" (Hoff 173), but some critics deem this inapt, and it is unlikely that it derives from the male character, Ofelia, in Jacopo Sannazaro's (1458–1530) *Arcadia* (Jenkins 163n7). An equally unlikely source is the "aphelion"—the point in a planet's orbit that is farthest from the Sun—as this term came into use only in 1656 (*OED*). Egert lists other suggestions.
4. In *HU*, I suggested that the day immediately following Munosius' class is when the New Star was first observed, but the present analysis is preferable because it uses the two temporal constraints without need of further assumptions.
5. Clocks that keep mean solar time advance steadily in lockstep with a fictitious sun, the "mean sun," that moves at a uniform rate with respect to the stars as if executing a perfectly circular orbit above the Earth's equator. Owing to the inclination of the Earth's equator to the plane of its orbit and the ellipticity of its orbit around the Sun, the real Sun is sometimes behind and sometimes ahead of the mean sun. The difference is the Equation of Time (mean solar time minus apparent solar time = E), which varies through the year. E often appears on modern world globes as a sort of figure 8 (the analemma) placed between the Tropics of Cancer and Capricorn. Observers of celestial phenomena record occurrences on clocks that keep time at a steady pace, and reduce them to a common astronomical time

called Universal Time. For present purposes, Universal Time is the same as Greenwich Mean Time (GMT), or the time shown by a clock on the Greenwich meridian that keeps time at a steady pace regardless of the actual position of the Sun in the sky. Modern ephemerides generated by computers list positions as a function of Universal Time, from which local mean times at any site follow from the fact that the Earth rotates at the rate of one complete turn from west to east every 24 hours. Local mean time at any site is later than local mean time at Greenwich by one hour for every 15° of eastward longitude (4 minutes per degree), and correspondingly earlier for sites west. Mechanical clocks emerged during the second half of the thirteenth century, and by 1500, most towns had clocks mounted on the walls of prominent buildings. These struck the hours, as they do in *Hamlet* ("I think it lacks of twelve…No, it is struck," 1.4.3–1.4.4), and by 1559, clocks on the Continent were sufficiently mechanically sound that they could keep time to about 15 minutes per day (Sobel 35–36). Accuracy continued to improve steadily (Cipolla 41–61) until in 1577, five years after the date of the present chronology, an error of as little as four minutes in the royal clock was for Tycho Brahe a sufficient hedge against astrological error (Thoren 121). This suggests that at and around that time, the best clocks had an expected error not much larger. Some clock keepers set time by observing the stars (Crombie 1: 217–219), whose uniform motion across the sky is a reflection of the steady rotation of the Earth, so that with constant correction using a Nocturnal Clock or its equivalent, these clocks could keep time at a more-or-less steady pace (i.e., keep local mean solar time). This art is manifest in clocks dating to the late Middle Ages placed on major buildings in Europe's capitals, such as the Astronomical Clock of 1572 on Prague's Town Hall that shows solar as well as sidereal and lunar time (Michel 164). In that same year (1572) at Elsinore, one would expect the royal clock to be equally capable. Furthermore, it is reasonable to assume that the celestial visitor appearing from the direction of the stars would take its cue from the very source by which timekeepers keep steady time (the stars), and that in 1572 at Elsinore, when the bell beats "one" it is telling local mean solar time. The division of the hour into minutes was common as early as 1345 (Crombie 1: 219), so when Marcellus explains that he invited Horatio to watch the minutes of the night, he creates an image of sentries watching and waiting minute by minute. On the night in question, therefore, it is reasonable to believe that the chiming clock is accurate to better than 15 minutes.

6. A Möbius strip is a surface constructed by twisting one end of a flat strip through 180° and joining it to the other. The strip now has only one side,

even though it appears still to have two (try coloring only one side). In the current era of green conservation, it is a metaphor for recycling waste material.

7. The first two lines of the doggerel illustrate the difficulty of interpretations taken out of context. The meaning favored by Levy (*Starry* 68–69) is the one favored here. (See also Trimble and Aschwanden 330.)
8. Copernicus made the sphere of stars that bounds the heliocentric model indefinitely large in order to accommodate the observed lack of annual parallax. It is unlikely that "nutshell" refers also to this *immensum* because, even though nothing finite, whether "large" or "small," can compare to the Diggesian *infinitum*, which is indeterminably large, the Copernican model contains too much "empty" space to bear comparison to a seed.
9. Atlas, the first astronomer of myth and father of the Pleiades, held the bounded World on his shoulders. Hercules had once rescued Atlas′ daughters, and Atlas was so grateful that he let Hercules bear this burden for a while and taught him astronomy as well (Graves, sections 39.d, 133.3). In the classical myth, Atlas bears a motionless sphere that represents the sky, although a modern variant that started when Mercator published a collection of world maps (an "atlas") has Atlas supporting the Earth. Nowadays, Atlas carries both "Heaven and Earth on his shoulders" (Sobel 1–2).
10. In *A Perfit Description*, Digges theorized that space was "infinite" and filled with stars, but knowledge of the incidence of naked-eye stars is a necessary precursor to this conclusion. Shakespeare describes this count as a massy wheel to whose huge spokes ten thousand lesser things are mortised and adjoined, where (as posited in the text) the 10,000 lesser things are the number of stars visible to the naked eye. To estimate this number, Digges would count stars in small, randomly selected patches of sky whose projected areas (i.e., solid angles) he could measure. In principle, from these data he could infer the total number across the entire sky by assuming that the sky unseen from England is no different in gross properties from that visible from England. Digges did not mention specifics in *A Perfit Description*, but by the time *Hamlet* came to the fore, he and Shakespeare might have felt free to report the consequences. Having estimated the incidence of stars up to the natural limit set by the physiology of the naked eye, Digges may have taken the next logical step and used a telescope to count stars up to successively fainter thresholds. He would observe that the numbers of stars increases rapidly with ever-increasing faintness, and perhaps this observation alone was sufficient to assert the "infinitude" of stars in space. Johnson writes, somewhat loosely, that the most casual telescopic

observation of the sky "would have revealed…the fact that the stars are *infinitely* more numerous than they seem to the naked eye" (175; italics added). With ever-increasing distance, a star would appear ever fainter, but it is unknown how far below the naked eye threshold Digges quantified his observations. It is not at all obvious that he could translate the numbers of stars he could see telescopically in the 2-dimensional plane of the sky into a relative spatial distribution in 3 dimensions without at least performing experiments to correlate and calibrate physiological sensation with the physical stimulus of light. This seems unlikely, and even if he had done so, he probably could not have mounted arguments with sufficient clarity to convince skeptical poets of the significance of his observations. Instead, it seems likely that Digges' positive assertion of a physically infinite stellar distribution is a quasi-empirical assessment reached in conjunction with theological considerations. A close reading of the *Hamlet* script supports the conclusion that Thomas Digges (alias Hamlet) counted stars because when Hamlet says (italics added), "*I* could be bounded in a nutshell and *count myself* a king of infinite space" (2.2.243–2.2.244), he means *inter alia* that *he himself counts* stars (i.e., in the present interpretive framework, Thomas Digges himself counts stars). This establishes each protagonist as a "king of infinite space." In the real world, Thomas Digges is at the pinnacle of the hierarchy of philosophers and theologians who debate the question of the infinitude of space because, unlike any predecessor, he achieves that vision with the benefit of empirical evidence. Shakespeare's description is fully in accord with his philosophy of celestial cartography, as expressed in *Love's Labour's Lost* (1.1.88–1.1.93), where Berowne advocates selecting stars with a scientific goal in mind. Shakespeare knew that counting stars is pointless unless accompanied by data analysis and theory building. Naming stars is equally pointless unless to import scientific information like object class, position, or (in at least one instance) criteria of selection. In Shakespeare's allusions in *LLL*, he specifically fails to identify the observer, leaving it to *us*, the human inquirers, to supply details.

11. If "Hyperion's curls" refers to curls or ringlets, perhaps the description refers to the man in the Hilliard portrait showing a redhead with curly locks and a penetrating gaze. The subject wears a hat with a hatband sporting some 100 eyelets called "Oes" (Hotson, *Hilliard* 63–78). That the artist intended these bead-like attachments to be eyes is plain when one observes their similarity to the penetrating eyes of the subject. The black centers of the "eyes" add up to a collecting area whose aperture is about 10 cm (4 inches). Perhaps Hilliard painted Thomas Digges in 1588 at age 42, or

perhaps he portrayed Leonard as he appeared in 1558 at about 37 years old, when he had redeemed his lands and reconstituted his life after the debacle of Wyatt's Rebellion. Hotson's interpretation is that the ghostlike hand reaching down from the heavens is that of the Olympian god Apollo, the subject is Mercury, and the portrait's inscription, *Attici amoris ergo*, means "Athenians because of love," where the Athenians in question are Apollo and Mercury. This metaphor is consistent with the proposition that the picture depicts Apollo transmitting the New Philosophy *per manus tradere* (from hand to hand) to a real-life Hermes, who serves as the messenger of the gods.

12. Jupiter's Great Red Spot is a storm of immense size, large enough to hold several Earths, and unusual in that it is an anticyclone, the opposite of the common, low-pressure typhoons of Earth. Jupiter's relatively large size and rapid rotation abet its ferocity. High-pressure systems are associated with fair weather, allowing outsiders to see deep into the atmosphere. Molecules comprised of hydrogen, carbon, nitrogen, oxygen, sulfur, and phosphorus lend color to the spot, which varies from a barely perceptible discoloration through orange-red to red. It appears that this storm has been blowing incessantly from the nominal time of its discovery in 1664–1665, and the present interpretation increases its age to over four centuries.
13. Shakespeare is wont to capitalize on coincidence, and on the morning on February 9, 1591 at 05:10 hours Greenwich Mean Time, Jupiter and the Red Planet Mars lay nearly in the same direction in the sky. Tycho's sharp-eyed assistant, Hans Crol, observed that the image of Mars did not obscure that of Jupiter (Christianson 270), and a modern estimate using Voyager 4 software indicates that Mars, which at that time was about 6″ in diameter, came within 30″ of the direction of Jupiter, which was itself about 30″ in diameter. This angle is at or below the visual resolution of the human eye, and the two objects might well have looked coincident to average onlookers who lacked Crol's acuity. Such a close conjunction of these planets is rare. The two instances of close approach that occurred nearest in time to this one occurred in 1577 and 1597, but their minimum separations were substantially greater (nearly a quarter of a degree, which is half the angular diameter of the Moon) and easily seen by eye.
14. Many wonder whether Hamlet could still be a student at the ripe old age of 30, the argument being that his classmates at Wittenberg would be more than ten years younger, and surely, no one would remain a student that long. However, Copernicus was about 33 when he returned to Frauenburg after studying under da Novara, and Rheticus was about 27 when he returned to

Wittenberg after studying under Copernicus. Thomas Digges was 30 when the almanac containing *A Perfit Description* was published, and because statistical inference begins confidently with sample sizes of three, it may be concluded that in the sixteenth century, 30 was an average age for students at the cutting edge of the New Astronomy. In classical times, maturity began at age 30, which is a good age for fledgling cosmologists to lose their innocence.

15. The common belief is that Field's only notable achievement was that he was a "proto-Copernican," the first in England to recognize and assert the Copernican philosophy (*DNB*), but I assert that Shakespeare knew that Field's *Ephemeris Anni 1557* was an adaptation of the Reinhold tables, and that Leonard's ephemerides were England's first (see table 3.5).
16. Concerning "Georg" and "Johann," note the working relationship between Georg Peurbach (1423–1461) and his student, Johann Müller (1436–1476), known as Regiomontanus. Their work associates astronomy and humanism, which may have inclined Shakespeare to incorporate their names into the allegory even though both remained convinced geocentricists.
17. *HU* (163–171) lists these properties, which are repeated here for the sake of completeness. By line numbers in scene 5.2 from Edwards' *Hamlet*, they are:

100–101. Osric announces, "...here is newly come to court Laertes." Harriot returned from the Virginia colony in 1586 and, after publication of his report, he emerged as a leading intellectual. By 1590, most people recognized him as a foremost English mathematician, along with John Dee and Dee's student, Thomas Digges, at which time his work began to be cited. In c.1601, therefore, Shakespeare could legitimately write that Harriot was "newly come to court" in the sense that he was recognized by Elizabethan society.

101. Osric continues, "...believe me an absolute gentleman." Notes inscribed on Thomas Harriot's matriculation to Oxford list his age as 17 and the social status of his father as "plebian." Thus, he belonged to the social order of the common man, but his social class changed on graduating BA in 1580 at age 20. It changed further when, in 1595, he received a gift from the ninth Earl of Northumberland, Henry Percy, of a life interest in the income of Percy's holdings in Durham, enabling Harriot to become a lifetime member of the landed gentry. As a result, Harriot was entitled to attach the title "Gentleman" to his name.

101–102. "...full of most excellent differences." "Differences" are characteristics or distinctions that are out of the ordinary (Jenkins 5.2.108n) so that a person "excels in a variety of different accomplishments" (Edwards 5.2.102n). By 1601, select circles recognized Harriot's accomplishments in several fields, including navigation, cartography, ethnography, linguistics, meteorology, and his work on atomism, which began with his studies in military science.

102. "...of very soft society and great showing." The phrase "soft society" means "easy sociability," and "great showing" means "excellent appearance" (Edwards 5.2.102n). Harriot's personality was warm and attractive, and Percy admired his affability and learning.

103–104. "Indeed to speak freely of him, he is the card or calendar of gentry." He is the map or guide of gentility. A "calendar" is a registry or directory essential to keeping track of events and time. A "card" can mean a map or a stiff piece of paper containing the points of the compass (*OED*). Harriot is both a model gentleman and an expert cartographer and navigator. He kept the maps of Walter Raleigh (c.1552–1618) up to date, especially those of the New World, and drew maps for the Guiana expedition and for Raleigh's Irish holdings. These lines connect the preceding ones on Harriot's gentility to the next topic, his voyage to the Virginia colony, where Harriot was the official mapmaker and surveyor of the coastline.

104–105. "...for you shall find in him the continent of what part a gentleman would see." On April 9, 1585, a fleet of seven ships, led by Sir Richard Grenville (1542–1591) and his flagship *Tiger*, set sail from Plymouth bound for Roanoke Island. Evidence suggests that Harriot, now a gentleman by virtue of his Oxford education, accompanied Grenville aboard the *Tiger*, and reached the Carolina Outer Banks in late June.

106–107. Hamlet responds, "Sir, his definement suffers no perdition in you, though I know to divide him inventorially would dozy th'arithmetic of memory." Hamlet agrees with Osric, adding that to list all his qualities would make one dizzy. Such an inventory would have to come from memory for want of a significant number of published works. The words "divide" and "arithmetic" refer to Harriot's mathematical prowess.

108. "...and yet but yaw neither in respect of his quick sail." To yaw is to swing off course. The fleet led by Grenville on the *Tiger*, having set sail on April 9, 1585, soon encountered a storm that sank the *Tiger*'s pinnace and scattered the fleet. The *Tiger* sailed on alone, reaching the Canaries five days later. It proceeded westward, reaching Dominica in the Lesser Antilles on May 7 and Puerto Rico on May 10. This was "a rapid crossing" (Shirley, *Harriot* 126) to which "quick sail" refers, and is also a pun on Grenville's "fleet," for at least two meanings of "fleet" were in use at the

end of the sixteenth century (*OED*): "a sea force," as in "flete of schyppys," and "swift," as in "fleeter than arrowes" used in *Love's Labour's Lost*.

108–109. "But in the verity of extolment, I take him to be a soul of great article." In truth, "there would be many articles to list in his inventory" (Edwards 5.2.109n), if only he had published them. In fact, the only work Harriot published in his lifetime was *A briefe and true report*. He had raised expectations that he would publish a full account of his research on Virginia (Sokol 2), but this never appeared, and his textbook on navigation, Arcticon, was never published either. His reputation as a mathematician results from unpublished papers and a draft of a mathematics text in which he "virtually gave to algebra its modern form" (*DNB*).

109–110. "...and his infusion of such dearth and rareness." That which is "poured into him" by nature is dear and rare (Edwards 5.2.109–5.2.111nn). *OED* uses this line to illustrate the meanings of "infusion" and "dearth," but another meaning for "infusion" is the action of infusing some principle or idea into the mind, used as early as c.1450. This fits well with Harriot's dearth of publications, as fewer publications imply less impact on people's thinking.

110. "...as, to make true diction of him." Harriot was generations ahead of his time in creating a way to reduce speech to symbols. Unfortunately, he did not leave a treatise on phonetics nor a key to his symbols.

110–111. "...his semblable is his mirror." "The (only) person like him is his own image in the glass" (Jenkins 5.2.118n). Harriot brought several scientific instruments with him to Virginia, including burning glasses and "a perspective glasse whereby was shewed manie strange sightes" (*Harriot* 27). Harriot therefore possessed and demonstrated the forerunner of the telescope, but it was not until after Hamlet was written that he studied celestial objects telescopically.

111. "...and who else would trace him." Here "trace him" means "follow him closely" (Edwards 5.2.111n). After the storm scattered the fleet, the *Tiger* (presumably with Harriot on board) arrived at the appointed rendezvous eight days ahead of the next fastest vessel. No ship of the fleet followed the flagship closely. The present interpretation supports the conjecture of Shirley that Harriot accompanied Grenville aboard the *Tiger* (*Harriot* 125–126).

111. "...his umbrage." At sea, Harriot observed a partial eclipse of the Sun and was partly in the Moon's "umbrage," or shadow.

112. "...nothing more." Shirley writes, "whether he attempted to use these observations [of the eclipse] to calculate longitude accurately is

doubtful" (*Harriot* 125). Harriot passed up this opportunity to contribute to the solution of a difficult problem.

113. Osric is pleased with Hamlet's flattery. "Your lordship speaks most infallibly of him," he says. Here, as in 5.2.81 and elsewhere, Osric and Horatio address Hamlet as "lord," but Horatio and Osric are addressed as "sir" (see lines 117–118).

114–115. Hamlet seeks relevancy and asks, "The concernancy, sir? Why do we wrap the gentleman in our more rawer breath?" In other words, "How does this concern us? Why do we clothe him in words of ours which can only fall short of his refinement?" (Jenkins 5.2.112n, 5.2.122–5.2.123n).

116. Osric is baffled. "Sir?" he asks.

117. Horatio tries to be helpful and asks Osric, "Is't not possible to understand in another tongue?" The 1584 Amadas-Barlow expedition to Virginia brought back to England two American Indians, from whom Harriot learned Algonquian and then developed a special alphabet by which to record their language. The line may also refer to the facts that (with the help of Richard Hakluyt) Harriot's *A briefe and true report* was published in three languages besides English, and that Harriot's fondness for Latin coincided with a time when English was emerging as the language of scientific literature.

117–118. Horatio mimics Hamlet's balderdash, "You will to't sir, really." Horatio always addresses Hamlet as "lord" (Jenkins 5.2.125–5.2.126n) and here addresses Osric as "sir." As it stands, the remark makes little sense, and because Horatio does not explain to what "it" refers, the audience is free to improvise. At a stretch, perhaps he means, "You [are] Walter Sir, Raleigh," where "will to't" and "really" (suitably slurred) sound like "Walter Raleigh." "You" then suggests that Horatio identifies Osric as Raleigh (see table 3.1). The possibility is credible given that Raleigh championed Harriot just as Osric champions Laertes. Moreover, at the outset of their meeting, Hamlet identifies Osric to Horatio as one who "hath much land and fertile," which fits because Raleigh owned much fertile land (lines 103–104).

119. Hamlet tries again. "What imports the nomination of this gentleman?" he asks. He wants to know "the purpose of naming this gentleman" (Edwards 5.2.119n), referring to the naming of Raleigh. The question answers itself, thanks to a pun on "imports." According to the *OED*, "import" means a commodity brought in from abroad and was in use only by 1690; but the associated verb was in use much earlier, in 1508 and

1548. Perhaps the remark refers to Raleigh's interest in importing natural resources from England's colonies.

120. Osric is still confused. He asks, "Of Laertes?" He does not realize that Hamlet has shifted the conversation briefly to Raleigh (i.e., to Osric himself). By asking this question, Osric shifts attention back to Laertes (i.e., in the present reading, back to Harriot).

121. Horatio responds accordingly, "His purse is empty already, all's golden words are spent." "His purse" refers to H[is]enry P[urse]ercy's coffers, from which gifts of money were made to Harriot. In the early 1590s when Raleigh was in disfavor with Queen Elizabeth, Harriot sought other patronage; and it was natural for him to turn to Raleigh's friend, Percy. Percy's largesse commenced in 1593 with a sizable annual grant of £80. Thus, the names of the two persons who significantly influenced Harriot's life, the close friends Raleigh and Percy, appear in the order in which they patronized him. Although "[H]is [P]urse" refers to Henry Percy, the second half of the sentence suggests that his purse might store "golden words" rather than golden coins, so that in this case "H[is]" also refers to H[arriot], who has a purse full of words at his disposal. As coins are held in a purse before they are spent, so too could "his purse" also connote H(is)arriot's P(urse)ercy. Despite Percy's generosity, by 1601, Harriot had produced only one work and showed no sign of producing more. The line contrasts the cessation of Harriot's output after 1588 with his receipt of a lavish pension from Henry that resulted in no published works. It seemed as if Harriot had depleted his store of words and had none left to commit to paper. Shakespeare may have intended the Osric dialogue to be highly opaque because in 1601, Harriot, Raleigh, and Percy were all still alive.

122. Hamlet says, "Of him sir." The words "of him" have occurred three times before (103–104, 110, 113), and the referent in all cases is Laertes (i.e., H[im]arriot). Horatio's remarks refer to Percy, and Hamlet says that they refer also to Harriot.

123. Says Osric, "I know you are not ignorant..." Osric (also known as Raleigh) is referring to the prologue to Harriot's *A briefe and true report*, where thrice in the space of six paragraphs Harriot calls his compatriots "ignorant" (Harriot 5–6). Perhaps Shakespeare intends to emphasize the word "you," in which case the remark could mean that Osric (Raleigh) knows that Hamlet (Thomas Digges) is not ignorant; the presumption being that Laertes (Harriot) is not ignorant either. Shakespeare's intent is to show that the quoted remark is tantamount to Raleigh (Osric) telling Thomas Digges (Hamlet) that he, Digges, is not ignorant!

124–125. Hamlet interrupts, "I would you did sir, yet in faith if you did, it would not much approve me. Well sir?" Hamlet has no need for fatuous compliments from Osric any more than Thomas Digges needs them from Walter Raleigh. What does Osric have to say about that?

126. Osric tries to recover by explaining, "You are not ignorant of what excellence Laertes is." Osric supposes that Hamlet (also known as Thomas Digges) knows all about the excellence of Laertes (also known as Harriot), but Osric has dug his hole a little deeper.

127–128. Hamlet cannot admit to such a comparison, "I dare not confess that, lest I should compare with him in excellence, but to know a man well were to know himself." For Hamlet to admit knowing of Laertes' excellence would be like claiming such excellence for himself, as to know such excellence one would need to be able to perform such excellence; beyond that, the sentence "is not meant to have much meaning" (Edwards 5.2.127–5.2.128n). On the contrary, the comparison between Laertes and Hamlet is one between Harriot and Thomas Digges, and as far as Hamlet is concerned, it is an unequal comparison.

Shakespeare's redemption of Harriot in 1623 in F1 followed Harriot's scientific activity that occurred and became known only after about 1601–1604, about the same time or shortly after Q2. In 1606 Harriot and his circle vigorously defended the concept of an infinite Universe against arguments presented by Kepler, who mistakenly attributed them to Bruno and Gilbert (Johnson and Larkey 116). In about 1609, well after Q2 had appeared and at about the same time as Galileo developed his spyglass, Harriot acquired a spyglass of his own and (like Galileo) trained it on the Sun, Moon, and planets. He discovered sunspots, deduced the Sun's period of rotation, and made the world's first telescopic drawing of the Moon (a sketch that however was barely better than could be made by eye). He collected data on the Comet of 1607 (Halley's), which Bessel used to write his first scientific paper, and he observed the Comet of 1618. In 1621, at the time of his death and two years before publication of F1, Harriot's scientific reputation was high. The Q2 excision indirectly restores Harriot's reputation in the minds of playgoers, and the cosmic subtext explains the changes. From *Hamlet* 4.3.4, it becomes clear that Hamlet was once "loved of the distracted multitude," but two scenes later, they suddenly switch allegiance to Laertes, whom they see as eligible for the throne (4.5.102–4.5.108). This sequence parallels that between Thomas Harriot and Thomas Digges in the interval of 1573–1588. In 1588 Harriot published his brief report on the Virginia colony, and because he had already established a reputation as a mathematician, his esteem rose to the point where the English public believed he was their preeminent mathematician. In Q2, by making Laertes a latter-day contender for the throne of Denmark, Shakespeare identifies Harriot as a latter-day pretender to the throne of English mathematics (see endnote 2 of this chapter).

18. The King thinks he lives near the Hub of Universe, which makes Elsinore the origin of coordinates from which to measure terrestrial distances, just as in a geocentric Universe, the Earth is the origin from which to measure cosmic distances. Shakespeare equates geocentricism with egocentrism and defeats them both in the impending conflict. Egocentrism is a common affliction that modern cartoonists spoof by depicting how big-city dwellers view the world (Gould and White 22; Steinberg). There is a grain of truth also in the parody of regicentricism, as the Danish King Frederick II regarded Kronborg Castle and the Oresund Sound as the heart of his far-flung realm (Christianson 96–98). Tycho's observatory on Hven lay only a short distance from Kronborg Castle, so the parody extends to that site as well, particularly because Tycho regarded his observatory as central to the glories of Denmark and the world.

Chapter 4

1. Choreography should comply with Galileo's report in *Sidereus Nuncius* that the moons with smaller orbits complete their revolutions in shorter intervals of time (van Helden, *Sidereus* 84). This result anticipates Kepler's quantification for the planets of the solar system—known as Kepler's Third or Harmonic Law—published in 1619. It states that $rv^2 = C$, where (in application to the dance of the ghosts) r is the radius of a circular path, v is the speed of the dancer, and C is a constant whose value suits the size of the stage and the sprightliness of the actors. Presumably, the filial phantoms would execute the smaller orbits at the highest speeds, whereas the parents would move lethargically in larger orbits.
2. Shakespeare picks Orleans as the site of this initial altercation perhaps because the Maid of Orleans, Joan of Arc (*c.*1412–1431), is divine like Innogen. The Maid Joan rallied the French to defeat the English at the siege of Orleans, whereupon Charles VII (1403–1461) awarded her and her family the sur-addition *du Lis* ("of the lily"). Innogen is also associated with a lily (*Cymbeline* 2.2.15), the symbol of purity. After becoming a warrior in the French cause, Joan of Arc adopted male attire, as does Innogen in her role as Fidele. Allegations of heresy against the Maid of Orleans resulted in

a trial by university theologians, during which the ingrate King remained indifferent to her fate. In 1431 she was convicted of heresy and burned at the stake. In the early history play, *Henry VI part 1*, Shakespeare portrays Joan la Pucelle ("Joan of Arc") in less than flattering terms, but perhaps in *Cymbeline*, by intimating an association with the paragon Innogen, Shakespeare rehabilitates Joan's reputation, just as in a new trial in 1456, the Church exculpated her of her alleged sins—posthumously.

3. Galileo states that he has postponed a full "assault" on the problem of counting stars owing to the press of time and their "enormous" and "unfathomable multitude" (van Helden, *Sidereus* 59–62). Galileo's descriptions are qualitative and imprecise, and he did not "assault" the problem any more than Giacomo assaults Innogen. He states that telescopic stars "exceed tenfold the number of old and known ones," but he does not estimate the number of "old and known" stars, therefore it is not known what number he assigns to the incidence of telescopic stars (van Helden, *Sidereus* 35, 59). Ptolemy's *Almagest* of the second century AD was still current, and Galileo probably used its value "1,022" for the number of visible stars (Gilbert 319; van Helden, *Sidereus* 35n18), making his estimate for the number of *telescopic* stars ten times that value, or 10,220. This is virtually the same as Rosencrantz's "lesser things" that he says are "mortised and adjoined" to the sphere of stars, and which can be interpreted as the number of *naked eye* stars. This coincidence is an opportunity that Shakespeare would surely not pass up. Although Galileo made a perfunctory attempt to estimate star numbers, his claim that he saw six magnitudes below the visual threshold is grossly in error.
4. It is known today that Saturn has many concentric rings comprised of a multitude of highly reflective particles approximately 10 centimeters (4 inches) in size. The brightest ring accounts for more than half of their total brightness and (for want of information to the contrary) is probably the one at issue here. For dynamical reasons, the rings lie above Saturn's equator, meaning the planet and the ring particles turn about a common axis. The ring system has an inclination of 27° to the orbital plane of Saturn, which in turn has a small tilt of 2½° to the orbital plane of Earth. As Saturn and its rings revolve about the Sun, their orientations with respect to the stars remain essentially the same. Because the orbit of Saturn is nearly ten times the size of the Earth's orbit, the maximum tilt of the rings (as seen from Earth) is only about 29½°. Saturn's orbital period is 29½ years, so the rings are edge on to the Sun twice every period and are then invisible from Earth. The rings are also invisible when they are edge on to the Earth, and because

the Earth is comparatively close to the Sun and its orbital plane is about the same as Saturn's, these two times of invisibility occur within a matter of months of one another. A third reason for ring invisibility results from the constitution of the rings. Particle separations are sufficiently great to render them virtually transparent to light, which means they are invisible when backlit, particularly when the plane of the rings passes between the Sun and Earth. The season of ring invisibility also depends on the optical quality of the telescope, for when the rings are highly foreshortened, imperfect telescope optics hinder visibility. Before and after a period of invisibility, viewers on Earth see one side of the rings and then the other for slightly unequal periods. If + and – represent these two viewing aspects, then a cycle of 29½ years comprises any two successive subcycles from the series ...+ – + – +... (see table 4.3).

5. Galileo sometimes receives credit for the discovery of Saturn's rings, but in reality, he explained what he saw as an audial planet or one with stationary moons (figure 1.11a, 1.11b).
6. In 1675 Cassini noticed a dark marking in Saturn's ring, which later observations revealed was a gap dividing the ring concentrically (see figures 1.12, 1.13). The gap ("Cassini's Division") is 3,500 km (2,200 miles) wide (Wagner 314) and is sharply defined. It results from a 2:1 gravitational resonance with Saturn's innermost major moon, Mimas.

Chapter 5

1. Graziano's promise at the time of the epoch on stage of 1597.3 ± 0.7 amounts to a scientific prediction. Coincidentally, and unbeknownst to the playwright at the time, the phenomenon referred to (that Saturn's rings will become visible 2 years hence after the start of subcycle V of table 4.3) is the reverse of one that actually occurred at the end of that subcycle. On July 30, 1610, Galileo announced that Saturn presented an arrangement of three collinear images (figure 1.11b), and not quite 2 years later, near the end of subcycle V, he found that the satellite images had disappeared.
2. The scene illustrates how human frailty can propagate cycles of oppression and revenge. Oppressors may be unable to change their habits, or (fearing reprisals) may be loath to do so, and thus end up trapped by their own iniquity. As a result, the oppressed bear a double and doubly difficult burden. In order to free themselves from the chains of their oppression (and not copy them when the yoke of oppression is lifted), they must act not just in their own interests but also in the interests of the oppressors (Friere 28). Most

often, however, the vicious cycle perpetuates itself. Asimov describes another "vicious cycle that is constantly used to plague minorities... Jews were forced to into becoming usurers and then the fact that they were usurers was used to prove how villainous and hateful they were" (1: 515).

3. The "strange accident" of the letter brings to mind a passage from Thomas Digges' *Stratioticos* of 1579 that relates how Leonard Digges' interest in optics "grew by the aide he had by one old written booke...that by straunge adventure, or rather Destinie, came to his hands" (van Helden, *Transactions* 30). Meanings common to Portia's strange "accident" and Digges' strange "adventure" and "destiny" are "chance" and "fortune" (*OED*).

CHAPTER 6

1. Among all the choices available to exemplify a tiny object, Shakespeare chooses a top because it can gyrate when spinning. The association of ideas relates to the phenomenon of the Precession of the Equinoxes. The shape of the Earth deviates from a sphere by an equatorial bulge of about 43 kilometers (27 miles). Because the Moon and Sun generally do not lie in the plane of the Earth's equator, their gravitational attraction exerts a torque on the bulge of the spinning Earth, which causes the Earth to gyrate much as a spinning top supported against gravity does when its axis of rotation is inclined to the direction of gravity. Shakespeare would not need to know anything about gravitational attraction or the shape of the Earth merely to ascribe the phenomenon of Precession of the Equinoxes to a wobbling Earth. This supports the earlier contention (*HU*) that Shakespeare had a correct phenomenological understanding of all three basic motions of the Earth—rotation, revolution, and precession.
2. See endnote 2, chapter 5.
3. See Andersen for a review of evidence that Tycho died of mercury poisoning.
4. The interval "10" could pertain to the 10 years (1573–1583) over which Tycho sired his eight children. However, the script speaks of pregnancies of a plural number of wenches, followed by "wronging the ancientry." I could find no evidence that Ursus married or had children, but coincidentally, Frederick's children were born over the same decade as Tycho's. For further discussion, see Andersen.
5. See endnote 1 of chapter 3.
6. Conspicuously absent from *The Winter's Tale* is reference to Kepler's earlier work of 1596, *Mysterium Cosmographicum*. Although founded on

heliocentricism, its numerology might have struck Shakespeare as fanciful, and as Shakespeare was apparently privy to accurate ephemerides, he knew that data did not support the text. Nevertheless, this work established Kepler's reputation, which increased further with *Optica* of 1604.

7. The time on stage of Acts 4 and 5 (1604) may also be relevant because a New Star emerged in the constellation Ophiuchus in the evening twilight of October 9, 1604. Several observers spotted it, and after about a week, Kepler saw it too. In the same year, Kepler produced two pamphlets describing it, which (with *Optica*) Shakespeare may have regarded as notable achievements. Kepler's supernova (as it is popularly called) was 1½ magnitudes (four times) brighter than the brightest star in the sky and was a spectacular sight as it lay "in dramatic array" with Saturn, Jupiter, and Mars, which were then close together in the sky (Christianson 368). In 1606 Kepler produced his own *De Stella Nova* on the 1604 supernova. In *All's Well that Ends Well*, Helena may refer to this object when she says, "I should love a bright particular star" (i.e., Bertram) who is "in his sphere" and "so above me" (1.1.84–1.1.87). Hunter suggests a date of 1603–1604 for the writing of *AWW* (Hunter xxv), which is a range that the first sightings of Kepler's New Star could help to establish more accurately.
8. Also in about 1604, Kepler must have attracted the attention of a mysterious individual who went by the name "Edmondio Brutius" or some variation thereof. Editors know little about him, other than from a contemporary account by Galileo's friend Paolo Gualdo (1553–1621) in *Vita Joannis Vincentii Pinelli* (Life of Pinelli), published in Augsburg in 1607, that describes a certain "Edmund Bruce" as "a noble Englishman knowing much of the mathematical disciplines, of military matters and of herbaria" (Singer, *Bruno* 190n23). Koyré writes that "Edward Bruce" was an Englishman known also as "Brutus" (*Infinite* 73n24). Drake (*Work* 46) describes Bruce as Scottish, and Koestler describes him as "a sentimental traveler in Italy, amateur philosopher and science snob, who loved to rub shoulders with scholars and to spread gossip about them" (180). Bruce probably encountered Kepler sometime after 1594, when Kepler was teaching mathematics at Graz. In 1597 Bruce moved to Padua where Galileo held a professorship (Drake, *Work* 442). Bruce was not on friendly terms with Galileo, but in 1599, he acted as a go-between for Kepler, who had asked Galileo to measure the angle of magnetic dip at Padua (Drake, *Work* 46, 48). In 1603 Bruce wrote to Kepler supporting the idea of an infinite Universe of stars that were like the Sun and surrounded by planets (Koyré, *Infinite* 73n24), and with regard to the putative relationship between planetary speed and

distance from the Sun, Bruce warned Kepler that, "Galileo teaches your discoveries as his own" (Drake, *Work* 63–65). Bruce left Padua in about 1605, and no one knows what became of him (Drake, *Work* 442).

CHAPTER 7

1. The Oxford *DNB* entry gives a birth date *c.*1515 based on a letter from 1531 written by Leonard's father, James, that refers to the maintenance of "his son at Oxford" and his grandchildren. The grandchildren are those of James' first son, but the entry draws the groundless conclusion that the "son at Oxford" was the younger son Leonard, who therefore (the argument goes) must have been born about 1515. Furthermore, the note states that the information in that letter is the best indicator of Leonard's birth year. The entry also disregards Thomas Digges' announcement in 1571 of Leonard's death. A simpler explanation that fits the facts as stated is that in 1531, Leonard was about 10 years old and not a drain on family finances, and at the time, his father would be more concerned with the cost of maintaining his older son and his issue.
2. Bridget Wilford's first cousin, Phillippa Wilford, married Sir Thomas Smith (1513–1577) (Hughes 107n2, 108n8), whom Gabriel Harvey proclaimed in a memorial volume in 1578 was "greater than Ptolemy and a hundred Alfonsos" (Johnson 89). Smith's library contained Ptolemy's *Almagest*, Copernicus' *De Revolutionibus*, and the works of Euclid (*fl.c.*300 BC), Peter Apian (1495–1552), and the heliocentric Prutenic Tables (Johnson 89–90). Smith tutored Edward de Vere from ages 4 to 12, and de Vere may have learned about the Diggesian Worldview through Smith's connection to the Wilford women. This connection, along with that between de Vere and the younger Leonard Digges (Whalen, "Cross-examining"), bolsters the case for de Vere as the author of the Canon. The astronomical dating of four of the five plays analyzed here is consistent with de Vere's year of death in 1604, the exception being *Cymbeline*, which dates to 1610. However, a 1611 summary of *Cymbeline* by the necromancer Forman fails to mention both the circling specters and the memorable epiphany, which raises "doubts [about] Shakespeare's authorship of Posthumus' vision" and fuels speculation that it was a later addition "by another hand" (Warren 54–55, 73).
3. Lord Clinton had an uncanny ability to serve monarchs regardless of their religious affiliation, which he did in many capacities, not the least of which being that as a young man, he married a mistress of Henry VIII by whom the King had had a son. In 1572, shortly after he had become Earl of

Lincoln, he traveled to France where he attended the marriage of the King of Navarre six days before the Saint Bartholomew's Day massacre (*DNB*).

4. The sleuthing scholar Hotson published *I, William Shakespeare…* in 1938, and a search of *Notes and Queries* to that year yields no further mention of the Diggeses.
5. Perhaps Leonard's connections reached the monarch herself. For example, John Dee researched the New Star and served as Thomas Digges' tutor. Sometime in the second interval (1570–1574) of table 7.1, Dee made "a very faithful and inviolable promise to the Queene," and coincidentally, these years are also "the least documented of Dee's life" (Deacon 79).
6. Octogenarians who were born in the sixteenth century include the astronomer John Dee (81); Richard Norwood (1590–1675) (85), who was the first in England to measure the length of a terrestrial degree; and William Cavendish Duke of Newcastle (1592–1676) (84), who made contributions to philosophy and translated a Galilean tract on mechanics (*DNB*). Others of note include composer William Byrd (1543–1623) (80), judge Sir Edward Coke (1552–1634) (82), and Thomas Hobbes (1588–1679) (91), who helped transmit Galileo's revival of the distinction between "primary qualities" (innate to things themselves) and "secondary qualities" (their measurable effects) into general philosophy (Singer, *History* 210). On the continent, Joost Bürgi (1552–1632) (80) was an assistant to Tycho Brahe, Gerard Mercator (82) projected a spherical surface onto a flat plane, Cesare Cremonini (81) was a Galilean disputant, Francesco Barberini (82) was Galileo's pupil and advocate during his trial, and (before Galileo's stint at Pisa) Johan Groot (1554–1640) (88) (and Simon Stevin, 1548–1620) demonstrated the incorrectness of Aristotle's theory of falling bodies (Dijksterhuis 4: 69, 81).
7. From the chronology of Shakespeare's plays (Crystal and Crystal 592), the year 1613 associated with the writing of the last two plays in the Canon (*Henry VIII* and *The Two Noble Kinsmen*) attracts attention if only because the date is 10 years before F1. Shakespeare wrote these two plays in collaboration with John Fletcher (1579–1625), with whom, sometime in the interval 1612–1613, he "clearly" worked "for at least 12 months" (*DNB*). Perhaps F1 celebrates a tenth anniversary.
8. It turns out that a study of William Shakespeare's words and sentences, started in 1996, conforms to the precept 291262 (where W_iS_i, $i = 1$, 2…9…12).
9. Sonnet 71 warns skeptics not to raise the question of authorship, "Lest the wise world should look into your moan, / And mock you with me after I am

gone" (71.13–71.14). No one mocks Shakespeare, but noted lyricist William Schwenck Gilbert (1836–1911) is among many who mock attempts to identify Shakespeare's other professions: "The bard play-writing in his room, / The bard a humble clerk, / The bard, a lawyer, parson, groom, / The bard, deer-stalking after dark," etc. (Matus 264). A facetious addendum might be, "That's not the end of it by far, / Another skill is on the list / By which we learn of orb and star,/ — Perspective glass telescopist."

10. See endnote 11, chapter 3.
11. The association of a crow and plagiarism occurs in works by Macrobius (*fl.* AD 430) and others (Schoenbaum 152–153), and was common in the early modern age. In 1589 Tycho Brahe became concerned that he was not receiving proper credit for his model of the Universe and complained of scholars who "deck themselves in borrowed plumes like Aesop's crow" (Thoren 456).
12. Shakespeare's devotion is also to the ideals of Urania, goddess of Astronomy, which is an art practiced largely at night. On that account, she could qualify as a candidate for Shakespeare's "dark lady."
13. "Distant a Thousand yeares" in I.M.S.'s poem of 1632 may express distance in units of time, which is analogous to the modern invention of a "light year." If so, it is an idea ahead of its time. In *Il Saggiatore* ("The Assayer") of 1622, Galileo wonders whether "light may have an instantaneous motion" (i.e., have an infinite speed), but soon thereafter, in *Due Nuoue Scienze* (which appeared eventually in 1638), he retracts his opinion and proposes an experiment to measure the speed (Drake, *Discoveries* 278n21).

Works Cited

Allen, Richard Hinckley. *Star Names: Their Lore and Meaning*. 1899. New York: Dover, 1963. Print.

Andersen, Peter Hvilshøj Vinilandicus. *Kunstvaerket*. Copenhagen: Books on Demand, 2009. Print.

Andrews, John F., ed. *Hamlet*. By William Shakespeare. London: Dent, 1993. Print.

Arbery, Glenn C. "The Displaced Nativity in *Cymbeline*." *Shakespeare's Last Plays: Essays in Literature and Politics*. Ed. Stephen W. Smith and Travis Curtright. Lanham, MD: Lexington Books, 2002. 157–178. Print.

Asimov, Isaac. *Guide to Shakespeare*. 2 vols. New York: Random House, 1993. Print.

Barber, Peter. "England II: Monarchs, Ministers, and Maps 1550–1625." *Monarchs, Ministers, and Maps: The Emergence of Cartography as a Tool of Government in Early Modern Europe*. Ed. David Buisseret. Chicago: University of Chicago Press, 1992. 57–98. Print.

Baxter, James Phinney. *The Greatest of All Literary Problems: The Authorship of the Shakespeare Works*. Boston: Houghton Mifflin, 1917. Print.

Beauregard, David N. "Shakespeare against the Skeptics: Nature and Grace in *The Winter's Tale*." *Shakespeare's Last Plays: Essays in Literature and Politics*. Ed. Stephen W. Smith and Travis Curtright. Lanham, MD: Lexington Books, 2002. 53–72. Print.

Bennett, Roger E. *The Complete Poems of John Donne*. Chicago: Packard, 1942. Print.

Bergin, Thomas G., and Jennifer Speake. *Encyclopedia of the Renaissance*. New York: Facts on File Publications, 1987. Print.

Berners, Juliana, and William Gryndall. *Hawking, Hunting, Fouling, and Fishing*. London: Islip, 1596. Print.

Berry, Arthur. *A Short History of Astronomy*. New York: Dover, 1961. Print.

Best, Jason S., Sara A. Maene, and Peter Usher. "Copernicus's Neglected Successor." *Mercury Magazine* 30.5 (2001): 38–40. Print.

Bevington, David, ed. *The Complete Works of Shakespeare*. New York: Harper Collins, 1992. Print.

Birch, Thomas. *The History of the Royal Society*. 4 vols. London: Millar, 1756–1757. Print.

Bloom, Harold, ed. *The Winter's Tale*. By William Shakespeare. New York: Chelsea House, 1987. Print.

Bogdanowicz, Wieslaw, Marie Allen, Wojciech Branicki, Maria Lembring, Marta Gajewska, and Tomasz Kupiec. "Genetic Identification of Putative Remains of the Famous Astronomer Nicolaus Copernicus." *Proceedings of the National Academy of Sciences USA* 106 (2009): 12279–12282. Print.

Breight, Curtis C. *Surveillance, Militarism and Drama in the Elizabethan Era*. New York: St. Martin's Press, 1996. Print.

Bricker, Charles. *Landmarks of Mapmaking*. New York: Crowell, 1976. Print.

Brown, John Russell, ed. *The Merchant of Venice*. By William Shakespeare. London: Arden Shakespeare, 2004. Print.

Bryson, Bill. *Shakespeare: The World as Stage*. New York: Atlas, 2007. Print.

Buisseret, David. *Henry IV*. London: Allen and Unwin, 1984. Print.

Bulfinch, Thomas. *Mythology*. New York: Avenel, 1978. Print.

Butler, Samuel. *Hudibras*. Ed. Alfred Milnes. London: Macmillan, 1881. Print.

Camden, William. *Annales Rerum Gestarum Angliae et Hiberniae Regnante Elizabetha, 1615*. Print.

Campbell, Oscar James, ed. *The Reader's Encyclopedia of Shakespeare*. New York: Crowell, 1966. Print.

Caspar, Max. *Kepler*. New York: Dover, 1993. Print.

Chambers, Robert. *The Book of Days*. 2 vols. London & Edinburgh: Chambers, 1832. Print.

Christianson, John R. *On Tycho's Island: Tycho Brahe and his Assistants*. Cambridge: Cambridge University Press, 2000. Print.

Cipolla, Carlo M. *Clocks and Culture: 1300–1700*. New York: Norton, 1978. Print.

Clark, Cumberland. *Astronomy in the Poets*. Bournemouth: Sydenham, 196?. Print.

Clarke, Charles, and Mary Cowden. *The Shakespeare Key: Unlocking the Treasures of his Style*. London: Sampson, Low, Marston, Searle, and Rivington, 1879. Print.

Clegg, Brian. *The First Scientist: A Life of Bacon*. London: Constable, 2003. Print.

Cockle, James. "Mathematical Bibliography." *Notes and Queries* Second Series, Vol. X (1860): 162. Print.

Coffin, Charles M. *John Donne and the New Philosophy*. New York: Columbia University Press, 1937. Print.

Copernicus, Nicolaus. *On the Revolutions of the Heavenly Spheres*. Trans. Charles Glenn Wallis. New York: Prometheus, 1995. Print.

Crombie, A. C. *The History of Science from Augustine to Galileo*. 2 vols. New York: Dover, 1995. Print.

Crompton, Samuel. "The Portrait of Tycho Brahe." *Nature* Vol. XVI (1877): 501–502. Print.

Crystal, David, and Ben Crystal. *Shakespeare's Words: A Glossary and Language Companion*. London: Penguin, 2002. Print.

Dampier, William C. *A History of Science and its Relation with Philosophy and Religion*. Cambridge: Cambridge University Press, 1961. Print.

Danielson, Dennis. *The First Copernican*. New York: Walker, 2006. Print.

Deacon, Richard. *John Dee*. London: Frederick Muller, 1968. Print.

Diagnostic and Statistical Manual of Mental Disorders. 4th ed. Washington DC: American Psychiatric Association, 1994. Print.

Dicks, David R. *Early Greek Astronomy to Aristotle*. Ithaca, NY: Cornell University Press, 1970. Print.

Digges, Leonard, and Thomas Digges. *An Arithmeticall Militare Treatise, named Stratioticus*. London: Bynneman, 1579. Print.

Dijksterhuis, E. J. *The Mechanization of the World Picture*. Princeton, NJ: Princeton University Press, 1986. Print.

Dolan, Frances E., ed. *The Winter's Tale*. By William Shakespeare. Penguin Books, 1999. Print.

Drake, Stillman. *Discoveries and Opinions of Galileo*. New York: Doubleday, 1957. Print.

———. "Galilei, Galileo." *Dictionary of Scientific Biography*. Ed. Charles C. Gillespie. New York: Scribner, 1971. 74–86. Print.

———. *Galileo at Work: His Scientific Biography*. New York: Dover, 1995. Print.

Dreyer, John L.E. *Tycho Brahe*. Gloucester: Peter Smith, 1977. Print.

Dunkling, Leslie. *Dictionary of First Names*. New York: Facts on File Publications, 1984. Print.

Edwards, Philip, ed. *Hamlet, Prince of Denmark*. By William Shakespeare. Cambridge: Cambridge University Press, 2003. Print.

Egert, Joseph. "Re: Gertrude done her in?" SHAKSPER: The Global Electronic Shakespeare Conference. 27 Mar. 2007. Web. 2 Oct. 2007.

Eissler, Kurt Robert. *Discourse on Hamlet and* HAMLET: *A Psychoanalytic Inquiry*. New York: International Universities Press, 1971. Print.

Eliot, Charles William, ed. *The Harvard Classics*. 51 vols. Volume 40. *English Poetry I: From Chaucer to Gray*. New York: Collier, 1909–1914. Print.

Ellis, Herbert A. *Shakespeare's Lusty Punning in Love's Labour's Lost*. The Hague: Mouton, 1973. Print.

Everitt, Alan. *Continuity and Civilization: The Evolution of Kentish Settlements*. Leicester: Leicester University Press, 1986. Print.

Freeman, Jessica. "Sorcery at court and manor: Margery Jourdemayne, the witch of Eye next Westminster." *Journal of Medieval History* Vol. 30 (2004): 343–357. Print.

Freire, Paulo. *Pedagogy of the Oppressed*. New York: Continuum, 1993. Print.

Gainer, Michael K. "Construction of a 16th-century telescope: An experiment in the history of astronomy." *Journal of the Royal Astronomical Society of Canada* 103.1 (2009): 18–21. Print.

Gainer, Michael, and Peter Usher. "Super-Early Reflectors?" *Sky and Telescope* 118.4 (2009): 10. Print.

Gilbert, William. *De Magnete*. Trans. P. Fleury Mottelay. New York: Dover, 1958. Print.

Goddard, Harold C. "Hamlet to Ophelia." *College English* Vol.16 (1955): 403–415. Print.

Gould, Peter and Rodney White. *Mental Maps*. Boston: Allen and Unwin, 1986. Print.

Granville-Barker, Harley, ed. *Cymbeline*. By William Shakespeare. Portsmouth, NH: Heinemann, 1995. Print.

Graves, Robert. *The Greek Myths*. 2 vols. London: Penguin Books, 1960. Print.

Graziani, René. "M. Marcadé and the Dance of Death. *Love's Labour's Lost*, v.ii.705–11." *The Review of English Studies* Vol. 37 (1986): 392–395. Print.

Green, David A., and F. Richard Stephenson. "The Historical Supernovae." *Supernovae and Gamma Ray Bursters*. Ed. Kurt W. Weiler. Berlin: Springer-Verlag, 2003. 7–19. Print.

Hall, Marie Boas. *The Scientific Renaissance 1450–1630*. New York: Dover, 1994. Print.

Halio, Jay L., ed. *The Merchant of Venice*. By William Shakespeare. Oxford: Oxford University Press, 1998. Print.

Hariot, Thomas. *A Briefe and True Report of the New Found Land of Virginia*. New York: Dover, 1972. Print.

Harting, James Edmund. *The Birds of Shakespeare*. Chicago: Argonaut Publishers, 1965. Print.

Heath, Thomas L. *Greek Astronomy*. New York: Dover, 1991. Print.

Heninger, Simeon Kahn, Jr. *Touches of Sweet Harmony: Pythagorean Cosmology and Renaissance Poets*. San Marino: Huntington Library, 1974. Print.

Hibbard, George R., ed. *Hamlet*. By William Shakespeare. Oxford: Oxford University Press, 1987. Print.

———, ed. *Love's Labour's Lost*. By William Shakespeare. Oxford: Oxford University Press, 1990. Print.

Hoff, Linda Kay. *Hamlet's Choice.* Hamlet: *A Reformation Allegory*. Lewiston, NY: Edwin Mellen Press, 1988. Print.

Holland, Peter, ed. *Cymbeline*. By William Shakespeare. New York: Penguin Books, 2000. Print.

Hotson, Leslie. *I, William Shakespeare Do Appoint Thomas Russell, Esquire*. New York: Oxford University Press, 1938. Print.

———. *Shakespeare by Hilliard*. Berkeley: University of California Press, 1977. Print.

Hughes, Stephanie H. Editor's Notes. *The Oxfordian* Vol. 8 (2005): 107–108. Print.

Huizinga, Johan. *Tien Studiën*. Haarlem: Tjeenk Willink, 1926. Print.

Hulme, Hilda M. *Explorations in Shakespeare's Language*. New York: Barnes and Noble, 1963. Print.

Hunter, G. K., ed. *All's Well That Ends Well*. By William Shakespeare. London: Arden Shakespeare, 2000. Print.

International Astronomical Union. Information Bulletin 101 (2008): 1–2. Print.

Jenkins, Harold, ed. *Hamlet*. By William Shakespeare. London: Arden Shakespeare, 1997. Print.

Johnson, Francis R. *Astronomical Thought in Renaissance England: A Study of English Scientific Writings from 1500 to 1645*. Baltimore: Johns Hopkins University Press, 1937. Print.

Johnson, Francis R., and Sanford V. Larkey. "Thomas Digges, the Copernican System, and the Idea of the Infinity of the Universe in 1576." *The Huntington Library Bulletin* Number 5 (1934): 69–117. Print.

Johnston, Stephen. "Making Mathematical Practice: gentlemen, practitioners and artisans in Elizabethan England." Diss. Cambridge University, 1994. Print.

Jones, Harold Spencer. *General Astronomy*. London: Arnold, 1951. Print.

Jones, Prudence, and Nigel Pennick. *A History of Pagan Europe*. London: Routledge, 1997. Print.

Jonson, Benjamin. *Poetaster*. London: Bradock, for Lownes, 1602. Print.

Kippis, Andrew, ed. *Biographia Britannica.* 5 vols. London: Strahan, 1793. Print.

Kitto, H. D. F. *Form and Meaning in Drama.* London: Methuen, 1959. Print.

Kliman, Bernice W. *Hamlet: Film, Television, and Audio Performance.* Rutherford, NJ: Fairleigh Dickinson University Press, 1988. Print.

Koestler, Arthur. *The Watershed.* New York: Macmillan, 1960. Print.

Koyré, Alexandre. *The Astronomical Revolution: Copernicus — Kepler — Borelli.* Trans. R.E.W. Maddison. Ithaca, NY: Cornell University Press, 1973. Print.

———. *From the Closed World to the Infinite Universe.* Baltimore: Johns Hopkins University Press, 1957. Print.

Lever, J. W., ed. *Measure for Measure.* By William Shakespeare. London: Thomson Learning, 2004. Print.

Levi, Peter. *The Life and Times of William Shakespeare.* New York: Holt, 1988. Print.

Levy, David H. "Shakespeare's Eclipses Return." *Sky and Telescope* 103.6 (2002): 75. Print.

———. *Starry Night: Astronomy and Poets Read the Sky.* Amherst, NY: Prometheus, 2000. Print.

Loades, D. M. *Two Tudor Conspiracies.* Cambridge: Cambridge University Press, 1965. Print.

Lynch, Mike. *Canada StarWatch: The Essential Guide to Our Night Sky.* St. Paul, MN: Voyageur Press, 2007. Print.

Lynn, W. T. "Leonard Digges." *Notes and Queries* Sixth Series, Vol. X (1884): 368, 515. Print.

———. "Thomas Digges." *Notes and Queries* Eighth Series, Vol. V (1894): 186. Print.

Maley, Willy. "Postcolonial Shakespeare: British Identity Formation and *Cymbeline*." *Shakespeare's Late Plays: New Readings*. Ed. Jennifer Richards and James Knowles. Edinburgh: Edinburgh University Press, 1999. 145–157. Print.

Marcus, Leah S. "*Cymbeline* and the Unease of Topicality." *Shakespeare: The Last Plays*. Ed. Kiernan Ryan. London: Longman, 1999. 134–168. Print.

Matus, Irvin Leigh. *Shakespeare, In Fact*. New York: Continuum, 1994. Print.

Meadows, Arthur J. *The High Firmament*. Leicester: Leicester University Press, 1969. Print.

Michel, Henri. *Scientific Instruments in Art and History*. Trans. R. E. W. Maddison and Francis R. Maddison. New York: Viking Press, 1967. Print.

Michell, John. *Who Wrote Shakespeare*? London: Thames and Hudson, 1996. Print.

Mitton, Jacqueline. *Dictionary of Astronomy*. London: Penguin, 1993. Print.

Mowat, Barbara, and Paul Werstine, ed. *The Merchant of Venice*. By William Shakespeare. New York: Washington Square Press, 1992. Print.

———. *Love's Labour's Lost*. By William Shakespeare. New York: Washington Square Press, 1996. Print.

Murry, John Middleton. "Shakespeare's Method: The Merchant of Venice." *Shakespeare, the Comedies: A Collection of Critical Essays*. Ed. Kenneth Muir. Englewood Cliffs, NJ: Prentice-Hall, 1965. 32–46. Print.

Nicholl, Charles. *The Reckoning: The Murder of Christopher Marlowe*. Chicago: University of Chicago Press, 1992. Print.

Noguères, Henri. *The Massacre of Saint Bartholomew*. New York: Macmillan, 1962. Print.

Nosworthy, J. M., ed. *Cymbeline*. By William Shakespeare. London: Methuen, 1955. Print.

Oertel, Abraham. *Theatrum Orbis Terrarum*. Antwerp: de Diest, 1570 Print.

Ogburn, Charlton. *The Mysterious William Shakespeare: The Myth and the Reality*. McLean, Virginia: EPM Publications, 1992. Print.

Olson, Donald W., Marilyn S. Olson, and Russell L. Doescher. "The Stars of Hamlet." *Sky and Telescope* 96.5 (1998): 67–73. Print.

Ovid. *Metamorphoses*. Trans. Charles Martin. New York: Norton, 2004. Print.

Palingenius, Marcellus. *The Zodiake of Life*. London: Newberie, 1576. Print.

Pannekoek, Anton. *A History of Astronomy (1961)*. New York: Dover, 1989. Print.

Payne-Gaposchkin, Cecilia. *Introduction to Astronomy*. Englewood Cliffs, NJ: Prentice-Hall, 1954. Print.

Pendergast, John S. *Love's Labour's Lost: A Guide to the Play*. Westport, CT: Greenwood, 2002. Print.

Perlstein, Arnie. "To be or not to be." SHAKSPER: The Global Electronic Shakespeare Conference. 14 February 2008. Web.

Philipott, Thomas. *Villare Cantianum*. London: William Godbid, 1659. Print.

Pumfrey, Stephen, and David Riley. "England's First Copernican: a new text by Thomas Digges on the 'New Star' of 1572." *British Journal for the History of Science*. 15 Jan. 2008. Web. 8 June 2009.

Rasor, Eugene L. *The Spanish Armada of 1588*. Westport, CT: Greenwood, 1993. 79–80. Print.

Reeves, Eileen. *Galileo's Glassworks*. Cambridge: Harvard University Press, 2008. Print.

Rienitz, Joachim. *Historisch-physikalische Entwicklungslinien optischer Instrumente*. Lengerich: Pabst, 1999. Print.

Roberts, John R. *John Donne: An Annotated Bibliography of Modern Criticism, 1979–1995*. Pittsburgh: Duquesne University Press, 2004. Print.

Robine, J. M. "A New Biodemographic Model to Explain the Trajectory of Mortality." *Experimental Gerontology* 36 (2001): 899–914. Print.

Ronan, Colin A. "The origins of the reflecting telescope." *Journal of the British Astronomical Association* Vol. 101 (1991): 335–342. Print.

Rowse, Alfred L., ed. *Cymbeline*. By William Shakespeare. Lanham, MD: University Press of America, 1987. Print.

———. *Shakespeare The Man*. New York: St. Martin's Press, 1988. Print.

Saxo Grammaticus. *The Nine Books of the Danish History*. Trans. Oliver Elton. 2 vols. London: Norroena Society, 1905. Print.

Schoenbaum, Samuel. *William Shakespeare: A Compact Documentary Life*. New York: New American Library, 1986. Print.

Seeger, Raymond J. *Galileo Galilei: His Life and his Works*. Oxford: Pergamon, 1966. Print.

Shakespeare, William. *All's Well That Ends Well*. Ed. G.K. Hunter. London: Arden Shakespeare, 2000. Print.

———. *Cymbeline*. Ed. Harley Granville-Barker. Portsmouth, NH: Heinemann, 1995. Print

———. *Cymbeline*. Ed. Peter Holland. Penguin Books, 2000. Print.

———. *Cymbeline*. Ed. J. M. Nosworthy. London: Methuen, 1955. Print.

———. *Cymbeline*. Ed. Alfred L Rowse. Lanham, MD: University Press of America, 1987. Print.

———. *Cymbeline*. Ed. Roger Warren. Oxford: Clarendon, 1998. Print.

———. *Hamlet*. Ed. John F. Andrews. London: Dent, 1993. Print.

———. *Hamlet, Prince of Denmark*. Ed. Philip Edwards. Cambridge: Cambridge University Press, 2003. Print.

———. *Hamlet*. Ed. George R. Hibbard. Oxford: Oxford University Press, 1987. Print.

———. *Hamlet*. Ed. Harold Jenkins. London: Arden Shakespeare, 1997. Print.

———. *Love's Labour's Lost*. Ed. George R. Hibbard. Oxford: Oxford University Press, 1990. Print.

———. *Love's Labour's Lost*. Ed. Barbara Mowat and Paul Werstine. New York: Washington Square Press, 1996. Print.

———. *Love's Labour's Lost*. Ed. H.R. Woudhuysen. London: Arden Shakespeare, 1998. Print.

———. *Measure for Measure*. Ed. J.W. Lever. London: Thomson Learning, 2004. Print.

———. *The Merchant of Venice*. Ed. John Russell Brown. London: Arden Shakespeare, 2004. Print.

———. *The Merchant of Venice*. Ed. Jay L. Halio. Oxford: Oxford University Press, 1998. Print.

———. *The Merchant of Venice*. Ed. Barbara Mowat and Paul Werstine. New York: Washington Square Press, 1992. Print.

———. *The Winter's Tale*. Ed. Harold Bloom. New York: Chelsea House, 1987. Print.

———. *The Winter's Tale*. Ed. Frances E. Dolan. Penguin Books, 1999. Print.

———. *The Winter's Tale*. Ed. Susan Snyder and Deborah T. Curren-Aquino. Cambridge: Cambridge University Press, 2007. Print.

Shapley, Harlow, and Heber Curtis. "The Scale of the Universe." *Bulletin of the National Research Council* Vol. 2, Part 3, Number 11 (1921): 171–217. Print.

Shirley, John W. *Thomas Harriot: A Biography*. New York: Clarendon Press, 1983. Print.

Shirley, Rodney W. *The Mapping of the World: Early printed world maps 1472–1700*. London: Holland Press, 1983. Print.

Showerman, Earl. "'Look Down and See what Death is Doing': Gods and Greeks in *The Winter's Tale*." *The Oxfordian* Vol. 10 (2007): 55–74. Print.

———. "Orestes and Hamlet: From Myth to Masterpiece, Part I." *The Oxfordian* Vol. 7 (2004): 89–114. Print.

Simonds, Peggy Muñoz. *Myth, Emblem, and Music in Shakespeare's* Cymbeline. Newark: University of Delaware Press, 1992. Print.

Simpson, Evelyn M. *A Study of the Prose Works of John Donne*. London: Oxford University Press, 1948. Print.

Singer, Charles. *A Short History of Science to the Nineteenth Century*. 1943. New York: Dover, 1997. Print.

Singer, Dorothea W. *Giordano Bruno: His Life and Thought*. New York: Schuman, 1968. Print.

Smith, William, and John Lockwood, eds. *Chambers Murray Latin-English Dictionary*. Edinburgh: Chambers, 1988. Print.

Snyder, Susan, and Deborah T. Curren-Aquino. *The Winter's Tale*. By William Shakespeare. Cambridge: Cambridge University Press, 2007. Print.

Sobel, Dava. *Longitude*. New York: Walker, 1995. Print.

Sohmer, Steve. *Shakespeare for the Wiser Sort*. Manchester: Manchester University Press, 2009. Print.

———. *Shakespeare's Mystery Play: The Opening of the Globe 1599*. Manchester: Manchester University Press, 1999. Print.

Sokol, Barnett J. "The Problem of Assessing Thomas Harriot's *A Briefe and True Report* of his Discoveries in North America." *Annals of Science* Vol. 51 (1994): 1–16. Print.

Steinberg, Saul. "View of the World from 9th Avenue." *New Yorker* cover, 29 March, 1976. Print.

Thoren, Victor E. *The Lord of Uraniborg*. Cambridge: Cambridge University Press, 1990. Print.

Thorne, Alison. "'To Write and Read / Be Henceforth Treacherous': *Cymbeline* and the Problem of Interpretation." *Shakespeare's Late Plays: New Readings*. Ed. Jennifer Richards and James Knowles. Edinburgh: Edinburgh University Press, 1999. 176–190. Print.

Trimble, Virginia, and Markus Aschwanden. "Astrophysics in 2004." *Publications of the Astronomical Society of the Pacific* Vol. 117 (2004): 311–394. Print.

Turner, Gerard L'E. "There was no Elizabethan Telescope." *Bulletin of the Scientific Instrument Society* Vol. 37 (1993): 3–5. Print.

Usher, Peter. *Hamlet's Universe*. San Diego: Aventine Press, 2007. Print.

———. "Jupiter and Cymbeline." *The Shakespeare Newsletter* 53.1 (2003): 7–12. Print.

———. "A New Reading of Shakespeare's Hamlet." *Bulletin of the American Astronomical Society* Vol. 28 (1996): 1305. Print.

———. "Shakespeare and Elizabethan Telescopy." *Journal of the Royal Astronomical Society of Canada* 103.1 (2009): 16–18. Print.

———. "Shakespeare's Support for the New Astronomy." *The Oxfordian* Vol. 5 (2002): 132–146. Print.

van Helden, Albert. "The Invention of the Telescope." *Transactions of the American Philosophical Society* Vol. 67 Part 4 (1977): 1–67. Print.

———. *Measuring the Universe*. Chicago: University of Chicago Press, 1985. Print.

———, trans. *Sidereus Nuncius*. By Galileo Galilei. Chicago: University of Chicago Press, 1989. Print.

Vendler, Helen. *The Art of Shakespeare's Sonnets*. Cambridge: Belknap Press, 1999. Print.

Voelkel, James R. *Johannes Kepler and the New Astronomy*. New York: Oxford University Press, 1999. Print.

Wagner, Jeffrey K. *Introduction to the Solar System*. Philadelphia: Saunders, 1991. Print.

Warren, Roger, ed. *Cymbeline*. By William Shakespeare. Oxford: Clarendon, 1998. Print.

Whalen, Richard F. "Cross-examining Leonard Digges on his Stratford Connections." *Shakespeare Oxford Newsletter* 37.1 (2001): 13–15. Print.

———. "The Stratford Bust: A Monumental Fraud." *The Oxfordian* Vol. 8 (2005): 7–24. Print.

Wilson, J. Dover. *What Happens in Hamlet*. Cambridge: Cambridge University Press, 2001. Print.

Withycombe, E. G. *The Oxford Dictionary of English Christian Names*. Oxford: Clarendon Press, 1977. Print.

Wood, Anthony à, with Philip Bliss. *Athenae Oxoniensis*. London: Rivington et al., 1813. Print.

Woudhuysen, H. R., ed. *Love's Labour's Lost*. By William Shakespeare. London: Arden Shakespeare, 1998. Print.

Illustration Credits

Figure 1.1. Peter Apian from *Cosmographia* (1539).

Figure 1.2. Author.

Figure 1.3. Author.

Figure 1.4. Nicholas Copernicus, from *De Revolutionibus* (1543).

Figure 1.5. Author.

Figure 1.6. Tycho Brahe, from *Liber Secundus* (1588).

Figure 1.7. Thomas Digges, from *A Perfit Description* (1576).

Figure 1.8. Thomas Digges, from *Prognositication Everlasting* (1576).

Figure 1.9. Michael Gainer, Pittsburgh, Pennsylvania.

Figure 1.10. Galileo Galilei, from *Sidereus Nuncius* (1610).

Figure 1.11. Author.

Figure 1.12. David Osborne, Wiltshire, England.

Figure 1.13. National Aeronautics and Space Administration and Jet Propulsion Laboratory of California Institute of Technology.

Figure 1.14. Author.

Figure 2.1. Michael Gainer.

Figure 3.1. Michael Gainer.

Figure 3.2. Author.

Figure 3.3. Jacob de Gheyn II (1586).

Figure 3.4. Jacob de Gheyn II (1586).

Figure 3.5. (a) Gordon Bradbury, Milton Keynes, England. (b) National Aeronautics and Space Administration, Jet Propulsion Laboratory of California Institute of Technology and the University of Arizona.

Figure 7.1. Reproduction by author.

Index

aberration, optical, 328, 334n8
action-at-a-distance, 265
Aeschylus (525–456 BC)
  Oresteia, 321
aether, 10, 51, 225
air (atmospheric), 96, 114, 119–120, 134, 199, 220, 225, 228, 275–276
Air (element), 10, 51, 119, 333n6
Albertus Magnus (1193–1280), 107
alchemy, 241
Alfonso X, King of Castile and Leon (1221–1284), 9, 356
allegory, xvi–xxiii, 1, 69, 79, 84, 108, 126, 135–136, 146, 153, 156, 161, 168, 178, 217, 226, 326, 335n2, 338n2, 345n16
Anaxagoras (*c.*500–*c.*428 BC), 3
Ancient Planets, xxi, 4–10, 19, 26, 41, 44, 52, 59, 205, 208, 214, 241, 205, 251, 265, 333n6
Anne of Denmark (1574–1619), 117
anomaly, 10, 12, 52, 84
anxiety, 84, 208, 262, 267
Aphrodite. *See* Venus
Apian, Peter (1495–1552), 356n2
  Cosmographia, 5
Apollo, xxiv, 66–67, 81, 172, 190, 259, 271, 273, 276–278, 290–291, 293, 295, 298–299, 304, 322, 325–326, 329, 343n11
  as *Theos*, 160
apology, xxiv, 166, 338n2
apple, 47, 50–52, 66
  of wisdom, 157
Aquinas, Thomas (1225–1274), 12, 107
Aquitaine, 48, 65
Archimedes (287–212 BC), 26, 194
Archytas (428–347 BC), 3
Aristarchus (*c.*310–250 BC), 16, 21
Aristotelian, xvi, 8, 11–12, 34, 40, 42, 44, 59–60, 72, 82, 84, 91, 111, 113, 117–118, 148, 194, 210, 225, 280, 338n2
Aristotle (384–322 BC), xxv, 4–11, 28, 33, 51, 74–75, 82, 113, 116, 132–133, 139, 148, 265, 357n6
  Metaphysics, 8, 132
  On the Heavens, 5
  On Interpretation, 148
  Organon, xxv
  Poetics, 139
  Politics, 116, 139
arithmetic, xxiii, 28, 114, 211, 345n17
Armada, Spanish, xxi, 43, 50, 287, 335n2
art, xxiv, 90, 110, 210–211, 300, 303, 326, 340n5, 358n12
Arts, 325
asterism, 204, 298
astrology, 224
astronomer, 30, 72, 90, 127, 143, 162, 169, 224, 237–238, 338n1, 342n9, 357n6
atomist, 3, 83, 345n17

Babylon, 5, 21
Bacon, Francis (1561–1626), xxv, 43

Bacon, Nicholas (1510–1579), 313
Bacon, Roger (*c.*1214–1294), 28–29, 42, 254, 296, 303, 309, 332n5
  Opus Majus, 28
Badovere, Jacques (1570/1580–*c.*1630), 192–193
Baer. *See* Bär
Bär, Reymers (*c.*1550–1600), 278–284, 287–290, 295, 354n4
  De Astronomicis Hypothesibus, 282–283, 289
  Fundamentum astronomicum, 279, 287–289
Barbary coast, 165
Barberini, Francesco (1597–1679), 40, 357n6
Barberini, Maffeo (1568–1644), 40
bardolatry, 318
Bellarmine, Robert (1542–1621), 39
Benson, John, 328
Bernard of Verdun (*fl.*13th century), 113
Bernardus Silvestris (*fl.*1147), 21, 71, 73, 87
  Cosmographia, 5, 73, 87, 375
Bessel, Friedrich Wilhelm (1784–1846), 334n8, 345n17
bias, xvii, 216, 276, 300, 338n1
Bible
  1 Corinthians, 291
  2 Corinthians, 296
  Ezekiel, 318
  Genesis, 54
  Hosea, 283
  Jeremiah, 108
birds, 55, 66, 97–98, 126, 189, 213, 222, 273, 298, 320, 323
  of Jove, 181, 184–185, 298
  Phoenix, 184–185
Boccaccio, Giovanni (1313–1375), 171–172, 187, 201
  Decameron, 171, 187, 200–201, 206
Bohemia, 20, 256–257, 259–260, 262, 266, 274–275, 278, 283, 291–293, 296, 298–299
  Prague, 20, 37, 257, 266, 279, 282, 304, 340n5
Book of Days, The, 299
Book of Nature, xx, 24
Bourne, William (1535–1585), 29–30, 254, 310
  Inuentions or Deuices, 29
  A treatise on the properties and qualities of glasses, 29, 310
Bradley, James (1693–1762), 334n8
Brahe, Tycho (1546–1601), xvi, 1, 17–20, 27, 71–72, 78, 92–93, 102, 109, 112, 117, 119–120, 122–127, 134, 138, 140, 145, 161–162, 254, 258, 279–291, 295, 304, 315, 332n4, 337n4, 338nn1–2, 340n5, 344n13, 351n18, 354nn3–4, 357n6, 358n11
  Astronomiae Instaurata Mechanica, 289, 291
  children, 284–285, 354n4
  De Stella Nova, 78, 286, 338n1
  death, 20, 102, 280, 283–284, 289, 315
  Epistolarum Astronomicarum, 289
  island. *See* Hven
  Liber Secundus, 18–19, 124, 279, 287–289
  observatory, 17, 279, 281, 337n4, 351n18
Brahmagupta (*c.*598–*c.*665), 21

Britain, 70, 171–173, 178, 185, 188, 193, 199, 216, 220, 225
Bruce, Edmund or Edward (*fl.*1604), 355n8
Bruno, Giordano (1548–1600), 112, 186–187, 345n17
Brutus. *See* Bruce
Burghley. *See* Cecil
Bürgi, Joost (1552–1632), 357n6
Butler, Samuel (1612–1680), 108
Hudibras, 108
Byrd, William (1543–1623), 357n6

Caesar, Augustus (63 BC–AD 14), 172, 174
Caesar, Julius (*c.*102–44 BC), 172, 174, 199
calendar, xxv, 100, 155, 345n17
Capella, Martianus (*fl.*410–439), 18
Cassini, Giovanni Domenico (1625–1712), 142–143, 253, 353n6
Catherine de' Medici (1519–1589), 44
Cavendish, William (1592–1676), 357n6
Cecil, William (1520–1598), 29, 186, 310–313
celestial pole, 77–78, 86, 100
celestial spheres, 27
Central Fire, 2– 4
Ceres, 300
Cesi, Federico (1585–1630), 39
change, 15, 38, 64, 78–80, 82, 84, 89, 96, 151, 190, 211–212, 216, 225, 228–230, 233, 241, 250, 259–260, 277, 301–302, 317, 345n17
Charles I of Britain (1600–1649), 178
Charles I of Spain (1500–1558), 47
Charles VII of France (1403–1461), 351n2
Charles IX of France (1550–1574), 44
chastity, 58, 88–90, 101, 105, 174, 188, 190, 200–201, 215, 217, 228, 276
Chaucer, Geoffrey (*c.*1340–1400), 204
Chettle, Henry (*c.*1560–*c.*1607), 324
Christian IV of Denmark (1577–1648), 20, 282, 285–286
chronology, xxi, 43, 51, 54, 58, 69, 89, 93, 100–101, 103–104, 128, 259–260, 340n5, 357n7
Churchill, Winston Leonard Spencer (1874–1965), 128
Cicero (106–43 BC), 225
Clavius, Christopher (1537–1612), 39
Clinton. *See* Fiennes
clocks, 102–103, 106, 300, 340n5
clown, 150–152–153, 155–160
Coke, Edward (1552–1634), 357n6
Coligny, Gaspard de (1519–1572), 44
Columbus, Christopher (1451?–1506), 45
comets, 19, 27, 80, 286, 288, 345n17
compass directions, 126, 258, 345n17
Conjunction, 14, 38, 87–88, 105, 149, 342n10, 344n13
constellations, 52, 74, 78, 100, 104, 204, 208, 257, 298, 322, 355n7
Copernicus, Nicholas (1473–1543), xix, 1, 11–18, 21, 24, 26, 32, 42, 56, 72, 76, 87, 112, 116, 139, 146–147, 151, 153–157, 223, 298, 310, 315, 332n4, 333n6, 334n8, 342n8, 344n14, 356n2
Commentariolis, 15, 154
De Revolutionibus, 12–13, 16, 56, 76–77, 87, 151, 153, 155, 356n2

Cosimo II de' Medici (1590–1620), 33, 36
Cosmological Principle, 332n4
cosmology, 1–2, 4, 11, 40, 42, 76, 82, 94, 99, 109, 258, 295, 307
Cosmos. *See* World
Counter-Earth, 2–4
counterfeit, 141, 213–214
craft, 143–144
Cremonini, Cesare (1550–1631), 39–40, 357n6
Crete, 53–54
Creti, Donato (1671–1749), 143
Crol, Hans (?–1591), 281, 344n13
cuckold, 66, 191, 213, 263, 265
Curtis, Heber (1872–1942), 331n3

Dante Alighieri (1265–1321), 73
Dark Lady, 358n12
Darwin, Charles (1809–1882), 314
Davis, John (*c.*1550–1605), 163
Death, 280, 282, 319
Dee, John (1527–1608), xv, 29–30, 57, 105, 124, 163, 310, 345n17, 357n5–n6
deer, xxi, 43, 48–53, 55, 262, 335n2, 357n9
Deity, xxii, 57, 117, 160, 169, 171, 176, 179–181, 184, 272, 298, 300, 329
Delos, 271, 274–275
Delphos, 260, 271, 273, 275
delver, 144, 187
Demeter, 300
Democritus (469–*c.*370 BC), 3, 10
Denmark, 1, 17, 70, 72, 79, 82, 88, 94, 96, 102, 109, 117, 122, 149, 168, 279, 282, 338n2, 345n17, 351n18
  Helsingor, 17, 73, 102
  Jutland, 119
design
  cosmic, 2–4, 30
  instrumental, 30–31, 200–201, 268, 292, 304
destiny, 354n3
dialectic, xxiv
Digge, Adomarus de (*fl.*1310), 150, 308
Digges, Dudley (1583–1639), 124, 313
Digges, James (d.1535), 307
Digges, Leonard (*c.*1521–1571?), xv, xxi, xxiii, 21, 26, 28, 30, 71, 79, 84, 92–93, 95, 98, 102, 127, 142–143, 152, 158–160, 221–224, 230, 237, 239, 247, 292–297, 299, 303, 307–309, 311–316, 319, 321, 323–326, 328–329, 332n5, 333n11, 354n3, 356n1
  Pantometria, 25–26, 93, 152, 295–296, 309, 312–314
  Prognostication Everlasting, 21, 27, 309, 312, 314
  Prognostication of Right Good Effect, 309, 312, 314
  Stratioticus, 28, 31–32, 312–315
  Tectonicon, 309
Digges, Leonard the Younger (1588–1635), 327, 356n1, 356n2
Digges, Thomas (*c.*1546–1595), xv, xxi, 1, 21, 23, 25–28, 30–33, 42, 57, 71–72, 74–75, 78, 81–82, 84, 92–93, 96, 99, 102, 112–113, 118, 124, 131, 139, 143, 150–151, 153, 157–159, 162–163, 169, 183, 191, 212, 217, 222–224, 226, 229, 237–238, 288, 292, 294, 296, 308–310, 312–315, 331n3, 338nn1–2, 342n10, 343n11, 344n14, 345n17, 354n3, 356n1, 357n5

Digges, Thomas (*continued*)
Alae seu Scalae Mathematicae, 78, 81, 292, 314, 338n1
Commentaries vpon the Reuolutions of Copernicus, 32, 310, 315
Letter sent by a gentleman of England, 338n1
Pantometria, 25–26, 93, 152, 295–296, 309, 312–314
Perfit Description, A, 21, 41, 75, 81–82, 151, 153, 309–310, 314, 332n4, 342n10, 344n14
Prognostication Everlasting, A, 21, 27, 309, 312, 314
Stratioticus, 28, 31–32, 312–315
Dionysus, 276
Dis, 300
disguise, 42, 59, 64, 124, 152, 157, 174, 176, 216–217, 239, 243, 246, 249–250, 253, 299, 301, 327
Donne, John (1572–1631), 34, 37, 333n6
Devotions upon Emergent Occasions, 333n6
First Anniversary, An Anatomy of the World, The, 333n6
Ignatius his Conclave, 336n6
Valediction: forbidding mourning, A, 336n6
Verse-Letter to the Countesse of Bedford, 336n6
Donne's Criterion, 334
doubt, 11, 30, 59, 75, 92, 108–113, 115, 162, 179, 224, 234, 262, 272, 282, 284, 332n5, 333n6, 356n2
dream, 99, 117–118, 169, 180, 183, 190, 198, 220, 256, 264, 277–278
Droeshout, Martin (b.1601?), 327–328
Dudley, Robert (*c.*1532–1588), 313
Dugdale, William (1605–1686), 317
dumb, 209–210, 233
dumb show, 135–136, 177

earth, 50, 73, 99, 116, 187, 216, 278
Earth (element), 10
Earth (planet), xix, xxv, 1–18, 20–23, 26–27, 34, 36, 38, 40, 50, 54, 59, 73, 75, 77, 80, 84, 87–88, 91, 99, 102, 104–105, 111–116, 119, 139, 170, 180, 190, 214, 228, 247, 264–266, 270, 279–280, 300, 304, 331nn1–2, 332n5, 333n6, 334n8, 340n5, 342n9, 344n12, 351n18, 352n4, 354n1
ecliptic, 209
Ecphantus (5th cent. BC), 3, 16
Edward II of England (1284–1327), 308
Edward III of England (1312–1377), 308
Edward IV of England (1442–1483), 308
Edward VI of England (1537–1553), 312
Elizabeth I of England (1533–1603), xv, xxiii, 30, 186, 309, 311–313, 345n17
ellipse, 20
Elsinore, 17, 72–73, 85, 95, 99–103, 106, 118–119, 121–122, 126–131, 162, 165, 167, 340n5, 351n18
empiricism, xv–xvii, xx, xxv, 4–5, 11–12, 20, 24, 33, 56, 61, 75, 108, 111–113, 127, 149, 183, 224, 230, 334n8, 338n2, 342n10
Empyrean, 20, 22, 24

England, xv, 1, 28, 41, 47, 50, 74, 78, 102, 106, 112, 138, 143–145, 148, 152, 162, 168, 172, 179, 186, 221, 232, 254, 298–299, 307, 310, 313, 333n6, 338n1, 342n10, 345n15, 345n17, 357n6
  Canterbury, 44, 307–308
  Eye (Ebury), 236
  Greenwich, 102–103, 340n5, 344n13
  Kent, 44, 307–308, 310, 312, 329, 332n5
  London, xv, 103, 172, 308–309
  Stratford, xxiii, 317, 321, 323, 325, 327
ephemeris, ephemerides, xxv, 9, 100, 102, 153, 155, 158–159, 338n1, 340n5, 345n15, 354n6
Epicurus (341–270 BC), 3
epistemology, xxi, 47, 81, 130, 134
Equation of Time, 340n5
equator, 36, 142, 340n5, 352n4, 354n1
ethics, 170, 282
Euclid (*fl.c.*300 BC), 117, 356n2
Euripides (484?–406 BC), 321–322
  Alcestis, 321
evil, 95–96, 224, 272, 338n2
Evil, 68, 134
eye, 5, 25, 28, 30, 47, 57–58, 75–76, 79, 98, 113, 118, 127, 130, 135, 139, 141–144, 194–195, 204, 208, 213, 218, 230, 236, 263, 331n3, 337n3, 342n10, 344n13, 345n17, 352n3
eyepiece, 30–31

fairy tale, 237, 273
falconry, 55, 97–98, 121, 126–127, 129, 158, 185, 237, 292, 298–299, 316–318, 320
Ferdinand V of Castile (1452–1516), 45
Field, John (*c.*1525–1587), 153, 158–159
  Ephemerides Trium Annorum, 153
  Ephemeris Anni 1557, 153, 158, 338n1, 345n15
Fiennes, Edward (1512–1585), 309, 312–313, 356n3
Fire, 10, 59, 112, 333n6
Firmament, 7, 17, 22, 93, 95, 118, 205
First Cause, 15, 184
fixed stars, 203
Fletcher, John (1579–1625), 357n7
Flora, 297
flowers, 297, 299–300
focal length, 26–27, 29, 253, 292
forgery, 112, 114
forgiveness, 169, 176, 178, 218–219, 230, 236, 250, 261, 284, 293, 326
Forman, Simon (1552–1611), 304, 356n2
Foucault, Jean (1819–1868), 334n8
France, xv, 44, 48–49, 65, 82, 88, 107, 148–149, 165, 168–169, 189, 193, 336, 356n3
  Aquitaine, 48, 65
  Calais, 336
  Normandy, 149
  Orleans, 188–191, 211–212, 227, 351n2
  Paris, 44, 51, 58, 65, 67, 82, 85, 107, 113, 164–166, 337n4, 338n2
Frederick II of Denmark (1534–1588), 17, 281–282, 285–286, 351n18, 354n4,
  children, 20, 282, 285–286, 354n4

Frederyke of Jennen, 172, 227
Frobenius, Georg Ludwig (1566–1645), 281

galaxy, 130, 204. *See also* Milky Way
Galilei, Galileo (1564–1642), xv–xvi, xx, xxii, 28, 32–42, 59, 64, 89, 130, 142, 171, 179–180, 183–184, 187, 189, 191–195, 197–198, 201, 203–210, 212–213, 217, 219, 226–230, 304, 333n6, 334n7, 338n2, 345n17, 351n1, 352n3, 353n1, 353n5, 355n8, 357n6, 358n13
  Dialogo, 40
  Due Nuove Scienze, 40, 358n13
  Epistolas de Maculis Solaribus, 39
  Il Saggiatore, 358n13
  Sidereus Nuncius, xxii, 33–34, 179–181, 183, 187, 189, 201, 203, 206, 212, 351n1
Galilei, Vincenzio (1520–1591), 32, 171
Galileo. *See* Galilei, Galileo
Galton, Francis (1822–1911), 314
geocentricism, xvi, 4–5, 8–10, 12–13, 15, 18–19, 41, 56, 72, 82–84, 109, 112, 115, 127, 129, 131, 135, 139–140, 149, 157, 161, 167, 264, 281, 286, 290, 333n6, 338n2, 345n16, 351n18
Germany, xv, 76, 102, 106, 178, 215, 221, 278, 283
  Augsburg, 179, 355n8
  Bavaria, 179
  Frankfurt, 301
  Holstein, 278
Germany (*continued*)
  Nuremberg, 154
  Rhaetia, 154
  Schleswig, 279
  Wittenberg, 16, 72, 76, 78, 84–85, 87, 94–95, 99, 101–103, 106, 119, 126, 145, 153–154, 159, 161–162, 281, 344n14
  Worms, 145
Gheyn, Jacob de, II (1565–1629), 123–125, 161–162
Gilbert, William (1540–1603), 23, 265, 345n17
  De Magnete, 23, 265
Gilbert, William Schwenck (1836–1911), 357n9
Giles of Rome (*c.*1247–1316), 13
Giulio Romano (*c.*1492– 1546), 303
glare, 18
glass, 29, 133, 187, 254, 326, 345n17
  concave, 30
  lens, 29–30, 133, 197, 254, 328
  mirror, 133, 326, 345n17
  perspective. *See* telescope
Globe Theatre, 97, 121
Golden Age, 222
Good, 68, 94, 134, 191, 230, 251, 254
Googe, Barnaby (1540–1594), 74
grace, 59, 141, 178, 184, 187, 190, 248, 261–263, 280, 293, 301, 303–304
grave, graveyard, 146–147, 150, 152, 236–237, 267, 313, 316–317, 325–327, 342n9
gravedigger, 112, 144, 150, 152, 158, 187
gravity, 157, 265, 354n1

Greece
  Aegean, 271, 274
  Athens, 109
  Delos, 271, 274–275
  Delphi, 271, 276
  Samos, 2, 16
  Stagira, 75
Greene, Robert (1558–1592), 112, 322–324
  Groats-worth of witte, bought with a million of Repentance, 322, 324
  Pandosto, 287, 291
Greenwich Mean Time, 102, 340n5, 344n13
Grenville, Richard (1542–1591), 345n17
Groot, Johan (1554–1640), 357n6
Grosseteste, Robert (*c.*1170–1253), 254
Gualdo, Paolo (1553–1621), 355n8
  Vita Joannis Vincentii Pinelli, 355n8
Gualterotti, Raffael (1548–1639), 310

Hades, 300
Hakluyt, Richard (1552–1616), 102, 345n17
  Principal Navigations, 102
Hales, James (*c.*1500–1554), 150
Halley, Edmund (1656–1742), 154, 345n17
harmony, 2, 5, 138, 247, 254, 326
Harriot, Thomas (1560–1621), xvi, 60, 71, 162–163, 166, 188, 229, 304, 338n2, 345n17
  Briefe and true report of the new found land of Virginia, A, 162, 345n17
Harvey, Gabriel (*c.*1545–1630), 74, 356n2
heliocentricism, xvi, 12–15, 18–19, 21, 34, 37, 40–42, 72, 76–77, 85, 94, 109, 114, 127–128, 130–131, 140, 145–146, 154–157, 161, 183, 194, 286, 290, 310, 321, 333n6, 334n8, 342n8, 354n6, 356n2
Helsingor, 17, 73, 102
Henderson, Thomas (1798–1844), 334n8
Henri of Lorraine (1550–1588), 44
Henry III of England (1207–1272), 308
Henry VII of England (1457–1509), 178, 308
Henry VIII of England (1491–1547), 308, 312, 356n3
Henry of Navarre (1553–1610), 44
heptad, 241, 250
Heraclides Ponticus (*c.*388–*c.*315 BC), 3, 16, 18
Heraclitus (*c.*540–480 BC), 10, 195
Hercules, 121, 332n5, 342n9
hermeneutics, xx
Hermes, 297, 343n11
Herschel, William (1738–1822), 30, 80
Hicetas (5th century BC), 3
Hildegard of Bingen (1098–1179), 21
Hilliard, Nicholas (1537–1619), 322, 343n11
Hipparchus (*c.*190–*c.*120 BC), 8, 337n4
Hobbes, Thomas (1588–1679), 357n6
Holinshed, Raphael (?–*c.*1580), 172
  Chronicles, 172

Holland
Amsterdam, 124
The Hague, 192
holy day, 101, 106, 128
Homer (before 700 BC), xxiv
homocentricism, 4, 10
Hooke, Robert (1635–1703), 142
Horace (65 BC–8 BC), 71, 168
horoscope, 282, 285–286
horticulture, 258
Huguenots, 44, 67
humanism, 33, 73, 153, 345n16
Humanities, 325
humors, 92
hunting, 45, 48–51, 54, 58, 74, 129, 175, 256, 262, 287, 290, 335n2
Huygens, Christiaan (1629–1695), 38–39
Systema Saturnium, 38
Hven, 17, 126, 162, 279–282, 290, 337n4
hyperbole, 117–118, 152, 227, 261
hypocrisy, 58, 108
hypothesis, xx, xxiii, 12–13, 15, 24, 51, 79, 81, 153, 157, 159, 203, 282–283, 288–289, 299, 310, 315

I.M., 326–327
To the memorie of M. W. Shake-speare, 326–327
imagination, 2–3, 10, 19–20, 24, 33, 49, 61, 65, 75, 77, 117–118, 125, 138, 185, 201, 215, 217, 242, 246, 262–263, 313, 317
immensum, 15, 18, 20, 334n8, 342n8
I.M.S. (friendly admirer), 328, 358n13
Index Librorum Prohibitorum, 40, 74
infinite, infinity, 3, 20–24, 28, 42, 76–77, 79, 109, 112, 114, 117–120, 131, 139, 147, 156–157, 159, 183, 280, 315, 328, 331n3, 338n2, 342n8, 342n10, 345n17, 355n8, 358n13
infinitum, 342n8
Inns of Court, 106, 152, 169, 277, 308, 324
Inquisition, 40, 45, 47
insects, 185, 194, 198, 200, 204
interpretation, xvi–xvii, xx–xxii, xxiv, 22, 35–36, 42, 45, 54, 69, 78, 84, 112, 120, 126, 147–148, 157, 176–179, 182, 204, 212, 221–223, 230, 263, 291, 311, 335n2, 342n7, 342n10, 343n11, 344n12, 345n17, 352n3
epistemology, xxi, 47, 81, 130, 134
hermeneutic-dialectic, xx, xxiv
hypothetico-deductive, xx
maximally reasonable, xx
irony, 18, 56–58, 64, 109–110, 114–116, 121, 132, 134, 139, 154, 162, 212, 265, 275, 282, 286, 291, 301, 336n2
Isabella I of Spain (1451–1504), 45
Italy, xv, 2, 12, 16, 35, 39–40, 171–172, 176, 179, 188, 191–193, 197, 215, 217, 237, 257, 355n8
Croton, 2
Florence, 32, 39
Padua, 33, 40, 179, 193, 243–244, 355n8
Pisa, 32–33, 357n6
Rome, 13, 15, 37, 39–40, 172–174, 176, 188, 191–192, 199–200, 207–208, 217, 222, 225–226

Italy (*continued*)
Trieste, 274
Venice, 34, 193, 231, 234–235, 239, 242–245, 247, 251, 253–254

James I of Britain (1566–1625), 171, 178–179, 222, 231
James VI of Scotland. *See* James I of Britain
Janus, 232
Joachim, Georg (1514–1576). *See* Rheticus
Joan of Arc (*c.*1412–1431), 351n2
Johnson, Samuel (1709–1784), 177
Jonson, Ben (1572–1637), 320–322, 327–328
Every Man out of his Humour, 320–321
Poetaster, 321–322
To the memory of my beloved, The Author, Mr. William Shakespeare, 321
To the Reader, 327
Jourdemayne, Margery (d.1441), 236–237
Jove. *See* Jupiter (deity)
Jupiter (deity), xxii, 139, 141, 171, 176, 178–184, 197, 205–211, 213, 216, 220, 222, 226, 272, 278, 283, 298
Jupiter (planet), xxii, 2, 5–6, 12, 15, 33–35, 52, 142–143, 171, 176, 183–184, 203, 205–210, 212–213, 229, 298, 334n6, 344nn12–13, 355n7
Great Red Spot, 142–143, 344n12
moons, xxii, 33–35, 171, 176, 179, 183–184, 203, 205–210, 212–213, 226, 228–229, 351n1
Keats, John (1795–1821), 326
Kepler, Johannes (1571–1630), 20, 34, 37, 76, 179, 265, 293, 295–296, 300–304, 338n2, 345n17, 351n1, 354n6, 355nn7–8
Astronomia nova, 20, 304–305
De Stella Nova, 355n7
Dioptrice, 304–305
Harmonices Mundi, 20
Mysterium Cosmographicum, 265, 354n6
Optica, 295–296, 301–303, 354nn6–7
Kronborg Castle, 17, 73, 117, 351n18

Lactantius (*c.*260–340), 16–17
ladder, 78, 292, 314
Plato's, 24, 118, 230, 292
Lavater, Ludwig (1527–1586), 99
Of ghostes and sprites walking by nyght, 99
lawyer, 169, 245, 277, 324, 357n9
lens. *See* glass
Leoviticus, Cyprianus (1524–1574), 337n4
Leucippus (*c.*500–428 BC), 3
Liddell, Duncan (1561–1617), 280–281, 283, 290–291
lift. *See* sky
Lincoln's Inn, 308, 324
lion, 221–223, 284, 293, 298, 323
Lippersey, Hans (1570–1619), 33, 192
Lodge, Thomas (*c.*1558–1625), 216
Rosalynde, 216
love, xxi, xxviii, 37, 39, 47–49, 56–58, 60–61, 67, 70, 74, 88–91, 107, 110–111, 114, 133, 135–137, 158, 173, 184, 189, 200, 210, 212, 218, 230, 232, 238, 240–241, 243,

love (*continued*)
246, 258, 263, 300, 302, 325, 330, 333n11, 355n7
Lucretius (*c.*99–*c.*55 BC), 3
Luther, Martin (1483–1546), 16, 145

Mabbe, James (1572–1642), 327
Macrobius (*fl.* AD 430), 358n11
mad, madness, xvi, xxi, 70, 94, 107–108, 110–112, 116, 126–127, 131, 134, 143–144, 152, 156–157, 166, 193–194, 196, 220
magic, 3, 28, 140, 151, 183, 196, 213, 227, 261, 303
magnetism, 23, 265–266, 355n8
Manwood, Roger (1525–1592), 32, 332n5
Manzoli, Pietro Angelo (*c.*1500–1543), 71, 73–76, 87, 117
Zodiacus Vitae, 73–74, 77, 87
Marlowe, Christopher (1564–1593), 44–45, 112, 226, 332n5
Epitaph to Sir Roger Manwood, 332n5
Massacre at Paris, The, 44
Mars (deity), 141
Mars (planet), 2, 5–6, 12, 14–15, 20, 51–52, 141, 143, 279, 304, 331n2, 333n6, 344n13, 355n7
Mary I of England (1516–1558), 47, 95, 308, 311–312
massacre, 44, 51–52, 337n4, 356n3
mathematics, xv–xvi, xxi, xxiii, 2, 5, 8–9, 11–12, 15–16, 20–21, 23–26, 28, 31, 33, 39, 60, 66, 78, 90, 102, 114, 117, 124, 153–155, 158, 162, 169, 211, 230, 259, 279–280, 282, 290, 292, 308, 313–314, 325–326, 345n17, 355n8
Maurice of Nassau (1567–1625), 192
melancholy, 47, 61–62, 106, 109, 134, 232–233, 241, 250
Melanchthon. *See* Schwarzerd
Mercator, Gerard (1512–1594), 102, 342n9, 357n6
Mercury (deity), 65–66, 297, 320, 322, 343n11
Hermes, 297, 343n11
Mercury (planet), 2, 5–6, 10, 12, 15, 18, 51–52, 59, 333n6
meridian, 14, 77–78, 102, 143, 257, 298, 340n5
metals
gold, xxii, 63, 119, 125, 201, 219, 222, 227, 231, 234, 239–241, 247, 249, 252–253, 345n17
lead, xxii, 231, 234, 241
mercury, 354n3
silver, xxii, 119, 231, 234, 240–241
metaphysics, xxi, 34, 43, 79, 108, 135, 190–191, 196, 226, 251, 297, 315
meteors, 91
microscope, 194, 204
Milesia, 332n4
Milky Way, 130, 203–204, 254
Minerva, 66
minutum, 20, 117
mirror. *See* glass
mole (epidermal), 174, 187, 206–207, 209, 211–213, 227–229
"of nature," 92, 338n2
Molyneux, Samuel (1689–1728), 334n8
monad, 3, 331n3

Moon (deity), 53–54
  Artemis, 53–54, 88, 90, 190
  Britomartis, 54
  Cynthia, 38, 53
  Diana (Dian), 53, 88, 190, 205, 207, 209–210
  Dictynna, 53–54
Moon (Earth's), xvi, xxi, 2, 5–6, 9–10, 12, 14, 19, 27, 33–34, 38, 51–55, 58–59, 63–64, 68–69, 73, 78, 80, 87–90, 100–101, 104–106, 110, 115, 133, 159, 189, 202–203, 205, 208–210, 213, 230, 241, 246, 259, 333n6, 337n3, 344n13, 345n17, 354n1
  Syzygy, 87–88, 105
Moses, 54, 182
Mousetrap, The, 134, 136–137, 141, 144
Müller, Johann (1436–1476), 345n16
Munosius, Hieronymus (*fl.*1572), 100, 103, 105, 340n4
Murder of Gonzago, The, 130, 132, 134, 136–137
Muses, 325–326
music, 2, 66, 138, 247–248, 303
  of the spheres, 5, 205, 248

narcissism, 140
Nashe, Thomas (1567–1601), 334n1
necromancy, xv, 236–237, 356n2
Neoplatonism, 94, 205, 322
Neptune (deity), 80, 94
Netherlands, 192
New Astronomy, xix, xxv, 1–42, 56–57, 61, 73, 76–77, 81, 84, 87–88, 94–95, 100, 103–104, 106, 108, 112–114, 118–119, 121, 127, 134, 145, 153, 155, 157–158, 168–169, 187–189, 191–192, 194, 214, 219, 223, 226, 314, 333n6, 344n14
"New Mythology", 325
New Organon, xxv, 17, 114
New Philosophy, xxv, 11, 17, 114, 144, 157, 169, 268, 274–276, 290, 302, 343n11
New Physics, xxv, 265
New Star
  of 1572, 17, 78, 91, 93, 100–101, 103–106, 124, 151, 290, 337n4, 338n1, 340n4, 357n5
  of 1604, 34, 355n7
Newton, Isaac (1642–1727), 20, 155, 157, 308, 334n8
  Principia Mathematica, 155
Nicholas of Cusa (1401–1464), 23, 117
North Celestial Pole, 77–78, 100
North Pole Star, 257
Norway, 82, 109, 146
Norwood, Richard (1590–1675), 357n6
Novara, Domenico Maria da (1454–1504), 12, 344n14
Nut, 117

Occam's razor, 13
Old Astronomy, 15, 34, 78, 112, 118, 152, 187–191, 206, 227, 261
Old Philosophy, 17
Old Physics, 10, 51, 59, 112, 119–120, 333n6
omen, 17, 79–80, 88, 91, 150, 338n2
One, 94, 170
Opposition, 14, 84, 87–88, 105, 208, 304
Oresund Sound, 17, 351n18

Osiander, Andreas (1498–1552), 16, 154
Ovid (43 BC–AD 18), 293, 317

Palingenius. *See* Manzoli
paradigm, 99, 118, 132
paradox, 60, 148, 205, 228
parallax angle, 6–7, 15, 17, 280, 331n1, 332n4, 334n8, 342n8
Paul V, Pope (1552–1621), 39–40
peace, 147, 152, 176–178, 180, 192, 220, 222, 226
pedantry, xvi–xvii, xxi, 46–47, 49–50, 56, 64, 67, 72, 112, 114, 148, 187, 195, 240, 314, 319, 335n2
pentad, 212, 229
perception, 8, 27, 30, 75, 118, 170, 190, 194, 205, 215, 264, 331n3
  appearance, xix–xx, 8, 13, 22, 24, 64, 67, 75–76, 83, 91, 96–97, 112–113, 133–134, 137, 165, 167, 169, 195, 214, 233, 241, 253, 256, 303, 319, 322, 353n5
  reality, xx, 2, 4, 8, 16, 24, 33, 40, 50, 75, 83, 89, 96, 112, 134, 165, 167, 183, 195, 214, 233, 253, 256, 277, 303, 319, 353n5
Percy, Henry (1564–1621), 60, 345n17
perfection, 2, 5, 8–9, 11, 60, 62, 110, 187, 190–191, 194, 198, 202, 206, 209, 215, 227, 293, 314, 333n6, 340n5, 352n4
Persephone, 300
perspective glass. *See* telescope
Petrarca, Franceso (1304–1374), 71, 73
Peurbach, Georg (1423–1461), 345n16
Philip II of Spain (1527–1598), 47, 308
Philolaus (*c.*480–*c.*385 BC), 2, 16
Pius V, Pope (1504–1572), 311
Plato (428–347 BC), 4, 10, 12, 16, 24–25, 56, 118, 170, 230, 292, 326
play-within-a-play, 96
Plutarch (*c.* AD 46–120), xxii–xxiii, 3, 255–256, 276
  Life of Cleomenes, 255
  Life of Lucullus, 255
  Lives of the Noble Grecians and Romans, xxii, 255
Pluto (deity), 300
poison, 137–138, 150, 165, 167, 175, 267, 277, 354n3
Poland, 12, 76, 109, 146–147, 156, 168, 338n2
  Ermland, 12
  Frauenburg, 76, 154, 344n14
  Frombork, 147
  Krakow, 12
Polos. *See* North Pole Star
Pope, 192
Pope, Alexander (1688–1744), xxiv
pox, 63, 90
Prime Mover, 8
Prince of Orange. *See* Maurice of Nassau
prison, 40, 95, 116–118, 173, 176, 180, 271
Privy Council, 308–309
Prosperina. *See* Persephone
Ptolemy, Claudius (AD 90–168), 8–9, 12–13, 15, 17–19, 27, 40, 71–72, 90, 99, 107, 113–114, 117, 139–140, 151, 161, 265, 281, 315, 352n3, 356n2
  Almagest, 8, 140, 352n3, 356n2
Pullen, Robert (d.*c.*1147), 71, 107

Purgatory, 95, 99, 106
Pygmalion, 300
Pythagoras (*c.*582–*c.*507 BC), 2–4, 84, 170, 334n8
Pythagorean, xx, 2–4, 8, 10, 16, 20–21, 24, 32, 119, 138, 168, 170, 247, 266, 280, 338n2

Raleigh, Walter (*c.*1552–1618), 71, 345n17
Ramus, Petrus (1515–1572), 44, 67
Rare Triumphs of Love and Fortune, The, 172
Regiomontanus. *See* Müller
Regnans in Excelsis, 311
Reinhold, Erasmus (1511–1553), 153–160, 345n15
resolution, 307–329
  optical, 26, 36, 98, 127, 130, 228, 253, 292, 307, 309, 332n4, 337n3, 344n13
  plot, 168, 219, 251, 302
  puzzle, 60, 182, 307–329
retrograde, 84, 321, 334n1
  planetary motion, 6, 10, 13, 15, 84–85, 149, 321, 331n2
Reymers. *See* Bär
Reynolds, John (1549–1607), 71, 107
Rheticus, 16, 76, 112, 139–140, 153–160, 344n14
  Ephemeris, 155, 158
  Narratio Primus, 140, 154
Richard I of England (1157–1199), 222, 308
riddle, xxi, 43, 53–55, 182, 186, 220, 225, 335n2
rodents, 62, 98–99, 141, 152, 185
Rothmann, Christopher (*fl.*1580), 290–291
Royal Society, 28, 311
Rudolph II (1552–1612), 257, 283, 300

Sannazaro, Jacopo (1458–1530), 340n3
  Arcadia, 340n3
Saturn (deity), 45, 51, 62–63, 65, 68, 214–215, 222, 278, 297, 300
Saturn (planet), 2, 5–6, 9, 12, 15, 18, 35–39, 47, 51–52, 62–63, 68, 214–215, 227, 229, 233–234, 241–242, 249–254, 300, 315, 333n6, 352n4, 353n1, 353nn5–6, 355n7
Saturnia, 215, 257
Savile, Thomas (*c.*1560–1593), 124
Saxo Grammaticus (*fl.*1188–1201), 70–71, 87, 146
  Historia Danica, 70, 146
schoolmen, 11, 50, 56–57, 107, 112, 170, 311
Schroeter, Johann Hieronymus (1745–1816), 337n3
Schuler, Wolfgang (d.1575), 102–104, 106
Schwarzerd, Philipp (1497–1560), 16, 154
Seleucus the Babylonian (*fl.*150 BC), 21
Seneca (4 BC–AD 65), 21, 322
Shakespeare, William, xvi–xvii, xix–xxv, 1–2, 17, 24, 35, 41–47, 49, 51–54, 56, 59–60, 62, 64–65, 68–72, 76–78, 81, 84, 86–87, 89–90, 92–93, 95, 97–102, 105–108, 111, 113–114, 118, 120–121, 124–128, 130, 132, 134, 139–143, 151–152, 156–157, 159,

Shakespeare (*continued*)
162, 164–165, 168–172, 177–179, 182–188, 190–193, 197, 200–201, 205–206, 210, 212, 215–219, 221–232, 235–237, 242, 246, 248–249, 251–252, 254–257, 262, 265, 268, 270, 275, 280, 283–285, 287–289, 291, 293–295, 297, 300, 303–305, 307, 311, 315–319, 321–324, 326–329, 334n1, 335n2, 337n3, 337n4, 338n2, 342n10, 344n13, 345nn15–17, 351n2, 351n18, 352n3, 354n1, 354n6, 355n7, 356n2, 357nn7–9, 358n12
All's Well that Ends Well, 355n7
Antony and Cleopatra, 88
As You Like It, 61, 182
Comedy of Errors, The, 335n2
Cymbeline, xxii, 35, 171–230, 249, 292, 298, 304, 315–316, 321, 333n6, 351n2, 356n2
Hamlet, xx–xxiv, 1, 17, 23, 42, 45, 68–69–170, 180, 183, 186
Henry IV part 2, 142
Henry VI part 1, 304, 351n2
Henry VI part 2, 194, 236, 324
Henry VI part 3, 323
Henry VIII, 357n7
Julius Caesar, 335n2
King Lear, 228
Love's Labour's Lost, xvii, xxi, 42–43–68, 90, 292, 297, 315, 342n10, 345n17
Lucrece, 200, 219, 320
Measure for Measure, 324
Merchant of Venice, The, xxii, 118–119, 231–254, 292, 316
Merry Wives of Windsor, The, 50, 198

Shakespeare (*continued*)
Midsummer Night's Dream, A, 50
Much Ado about Nothing, 316
Othello, 62, 266
Pericles, 316
Sonnets, 124, 320, 325, 328
Titus Andronicus, 62
Troilus and Cressida, 41, 221
Twelfth Night, 102
Two Gentlemen of Verona, 335n2
Two Noble Kinsmen, The, 357n7
Venus and Adonis, 200
Winter's Tale, The, xxii–xxiii, 68, 198, 219, 228, 255–305, 315–316, 321, 326, 354n6
Shake-speare, 320, 323, 326–327
Shakspere, William (1564–1616), xii, xxiii, 316–321, 323, 325
Shapley, Harlow (1885–1972), 331n3
shepherds, 100, 259, 283, 287–297, 299–302
shooting stars. *See* meteors
Sicilia, Sicily, 11, 256–258, 260, 262, 267–269, 274–276, 278, 296, 298–299, 301–303
Sizzi, Francesco (1585–1618), 39
Smith, Thomas (1513–1577), 356n2
Socrates (469–399 BC), 2, 4–5
Sophocles (*c.*496–*c.*406 BC), 273, 322
Antigone, 273
Spain, 11, 44–45, 47, 100–102, 192–193, 308, 335n2
Cadiz, 232
Navarre, 44
Valencia, 100–103
Spinola, Ambrogio (1569–1630), 192
spy, 70, 91, 107, 110–111, 115, 128, 133–134, 292

stars, xxii, 3–7, 9–10, 12, 15, 17–18, 20, 22–23, 28, 34, 38, 41, 46, 51, 58–59, 69, 75–78, 80, 86, 91, 94–95, 100, 102–103, 105, 109–115, 117–118, 127, 130, 139–140, 157, 183–184, 188–189, 197, 203–205, 209, 212–213, 219, 224, 233, 247, 254, 257, 259–261, 264, 268, 279–280, 298, 309–311, 322, 330, 331nn1–3, 332n4, 332n6, 333n6, 334n8, 337n4, 338nn1–2, 340n5, 342n8, 342n10, 352nn3–4, 355nn7–8, 357n9
Stellatus. *See* Manzoli
Stevin, Simon (1548–1620), 357n6
Stuart, Charles, 178. *See also* Charles I of Britain
Stuart, Elizabeth (1596–1662), 178
Stuart, Henry Frederick (1594–1612), 178
Sun-god, 85, 141. *See also* Apollo
  Helios, 88, 141
  Hyperion, 85, 141
  Phoebus, 81
Sun (star), xix, xxi, xxv, 1–3, 5–7, 9–10, 12–15, 18–20, 27, 35, 38–39, 41, 51–52, 54, 58–62, 69, 72, 78–81, 83–85, 87–89, 100–102, 104–105, 110–112, 115–116, 133, 140–141, 189, 202, 209, 214, 241, 248, 265, 279, 301, 331nn1–3, 332n4, 333n6, 340n3, 340n5, 345n17, 352n4, 354n1, 355n8
sunlight, 14, 59, 62, 214, 253, 337n3
sunspots, 39, 80–81, 345n17
  omens, 80
supernatural, xxi, 24, 94–96, 104, 120, 167, 183–184, 214, 233, 239
supernatural beings, 104
  angel, 24, 82, 120, 149, 190, 205, 207, 247, 296, 318
  cherub, 145, 204, 247–248, 317–318
  demon, 98, 227, 270
  devil, 61, 130, 132, 215, 227, 236
  fairy, 236–237, 273
  fiend, 236, 238
  ghost, xxii, 35, 76, 96, 98, 119, 135, 143, 168, 171, 176, 180–181, 183, 197, 239, 294, 343n11, 351n1
  goblin, 269–270
  phantom, 80, 96, 181, 237
  specter, xxii, 77, 80, 94–96, 98, 106, 176, 180, 207, 210, 356n2
  spirit, 51, 79, 83, 86, 92–96, 98–99, 105, 130, 140, 143, 152, 183–184, 191, 212–213, 227, 236, 239, 254, 269, 272–273, 278, 292, 335n2
  spook, 239
  sprite, 99, 269
  zombie, 237, 239, 247–248
supernatural space, 24, 183, 239
Sweden, 17
Syzygy, 87–88, 105

Tables. *See also* ephemeris
  Alfonsine, 9, 286
  Prutenic (Prussian), 155, 157–158, 286, 356n2
telescope, xvi, xxi, 28–29, 36, 39, 98, 113, 143, 151, 203, 228, 239, 253, 331n3, 337n3, 342n10, 345n17, 352n4. *See also* glass
  Bacon, 28
  Bourne-Digges, 29–30

telescope (*continued*)
Digges, 31, 63, 228
Elizabethan, 29
Fracastorius, 29
Galileo, 33–36, 39, 42, 171, 184, 194, 201, 204, 207, 338n2, 345n17
Gualterotti, 310
Herschel, 30
Kepler, 304
Lippersey, 33, 192
perspective glass, xxi, xxiii, 25–26, 30, 32, 127, 133–134, 184, 197, 239, 253, 268, 292, 295, 309, 311, 315, 326, 332n5, 338n2, 342n10, 345n17, 358n9
spyglass, 35–36, 39, 42, 171, 184, 192, 194, 201, 204, 207, 310, 338n2, 345n17
two-element, 28
Terence (*c.*185 or *c.*195 BC–*c.*159 BC), 320, 322
Thomist, 168
Time (chorus), 293
Time (deity), 45, 65, 67, 214, 256–257, 297, 300
time series, 230
To Mr. William Shake-spear, 320
torture, 40, 117, 175, 186, 301
treason, xxiii, 102, 248, 269–270, 308, 313
trunk, 197–198, 200–201, 207–208, 215, 268, 292
Turkey, 2
Troy, 172
twilight, 18, 85, 100–101, 104–105, 248, 355n7
Tycho. *See* Brahe

Universal Time. *See* Greenwich Mean Time
Universe, xix, xxv, 3–4, 7–12, 15, 18–21, 23–24, 28, 34, 56, 72, 76–77, 109, 112, 114, 118–119, 121, 131, 139, 157, 165, 183, 194, 265, 286, 315, 331n3, 332n4, 345n17, 351n18, 355n8, 358n11. *See also* World
Urania, xxiv, 103, 325–326, 329, 358n12
Urban VIII, Pope. *See* Barberini, Maffeo
Ursus. *See* Bär

variability. *See* change
Venus (deity), 37–38, 59–61, 89, 200–201, 300
Aphrodite, 59, 300
Cytheria, 200–201
Venus (planet), 5–6, 10, 12, 14–15, 18, 37–38, 51–52, 59–62, 64, 68, 78, 89–90, 101, 104, 159, 333n6, 334n7, 337n3
Vere, Edward de (1550–1604), 356n2
Vergil (70–19 BC), 225

Wales, 172–176, 228
Milford Haven, 174–175, 178
Walsingham, Francis (1532–1590), 311
Wanderers, wandering stars. *See* Ancient Planets
Webster, John (1580?–1625?), 320
Wedderburn, John (1583–1651), 194
Wilford, Sarah Bridget (d.*c.*1550), 308, 356n2
William of Occam (*c.*1284–1349), 13
wings, 78, 128, 185, 226, 292, 314, 317
witch, 196, 236

Witelo, Erazmus (*c.*1230–*c.*1275), 296, 303
Wits Recreation, 320
Wittich, Paul (*c.*1546–1587), 280–281, 290
world, xxii, 3, 17, 24–25, 34, 64, 66–67, 92, 94, 116, 119–120, 129, 134, 146, 181, 183, 189–190, 192, 197, 213–214, 216–217, 230, 233–234, 237, 254, 259–260, 290–291, 296, 318, 322, 331n3, 337n4, 340n5, 342nn9–10, 345n17, 351n18, 357n9
- political, 97, 121
- supernatural, 96, 140, 183, 214, 237–238

World, xxv, 3–4, 11, 18, 20, 22, 24, 27, 34, 40, 42, 74-76, 97, 99, 107, 112, 118, 121, 131, 137, 139, 149, 169–170, 183, 214, 230, 279–280, 287, 290, 304, 315, 331n3, 333n6, 342n9. *See also* Universe
World models
- Aristotelian, 5, 11, 34, 117–118, 338n2
- Bounded. *See* geocentric
- Capellan, 18
- Copernican, xvi, 1, 11–18, 20, 26–27, 69, 99, 117, 161, 168, 214, 280–281, 290, 315, 333n6, 334n8, 342n8
- Diggesian, 1, 21–27, 69, 72–73, 76, 99, 118, 161, 168–169, 214, 315, 331n3, 332n4, 334n8, 338n2, 342n8
- Egyptian, 18, 117
- geocentric, xvi, 4–10, 12–13, 17–20, 41, 56, 72, 82–83, 109, 112, 115, 127, 129, 131, 135,

World models (*continued*)
- 139–140, 149, 157, 161, 167, 264, 281, 286, 290, 333n6, 338n2, 345n16, 351n18
- heliocentric. *See* Copernican, Diggesian
- homogeneous, 28
- infinite. *See* Diggesian
- isotropic, 28
- modern, 41–42, 331n3, 333n6
- Philolaic, 4
- Platonic, 12, 24, 56, 118
- primitive, 264
- Ptolemaic, 8–9, 14–15, 18, 27, 69, 72–73, 99, 107, 114, 117, 139–140, 161, 214, 281, 290, 315
- Pythagorean, 4, 280, 334n8
- standard, 84, 332n4
- two-sphere, 4
- Tychonic, 1, 17–20, 27, 69, 72–73, 117, 119, 122, 127, 140, 145, 161–162, 214, 279–280, 283–284, 287, 280–281, 290, 304, 315, 358n11
- Ursusian, 279–280, 283–284, 287, 290

worldview, xix, xxv, 24, 35, 41, 43, 47, 170, 230, 265, 272, 307
Worldview, xxv, 3, 39, 42, 51, 74, 79, 82, 111, 119, 126, 129, 190, 305, 314, 356n2
Wotton, Henry (1568–1639), 34
Wyatt's Rebellion, xxiii, 95, 102, 297, 309, 343n11
Wyatt, Thomas Jr. (d.1554), 308
Wyatt, Thomas Sr. (*c.*1503–1542), 308

zodiac, 37, 73–74, 214, 224, 298

www.ingramcontent.com/pod-product-compliance
Lightning Source LLC
Chambersburg PA
CBHW051255200926
51612CB00001B/50
9781604977332